Nutri-Horticultu

Nutri-Horticulture

Editor
Professor K.V. Peter

2012
DAYA PUBLISHING HOUSE®
New Delhi - 110 002

© 2012 K.V. PETER (b. 1948–   )
ISBN 9789351241256

Published by   : **Daya Publishing House®**
                        A Division of
                        Astral International Pvt. Ltd.
                        – ISO 9001:2008 Certified Company –
                        4760-61/23, Ansari Road, Darya Ganj,
                        New Delhi - 110 002
                        Phone: 23245578, 23244987
                        Fax: (011) 23260116
                        E-mail: dayabooks@vsnl.com
                        website: www.dayabooks.com

Laser Typesetting   : **Classic Computer Services**
                        Delhi - 110 035

Printed at   : **Chawla Offset Printers**
                        Delhi - 110 052

PRINTED IN INDIA

Devotion

The edited book NUTRI-HORTICULTURE is devoted to my wife Mrs. Vimala Peter who entrusted me to the safe and protective hands of Prof. P.I. Peter Chairman, NoniBiotech, Chennai to continue my aptitude for science writing and editing. In the process she suffered loneliness and solitude.

Nutri-Horticulture

Editor
Professor K.V. Peter

2012
DAYA PUBLISHING HOUSE®
New Delhi - 110 002

Published by   : **Daya Publishing House®**
                            A Division of
                            Astral International Pvt. Ltd.
                            – ISO 9001:2008 Certified Company –
                            4760-61/23, Ansari Road, Darya Ganj,
                            New Delhi - 110 002
                            Phone: 23245578, 23244987
                            Fax: (011) 23260116
                            E-mail: dayabooks@vsnl.com
                            website: www.dayabooks.com

Laser Typesetting   : **Classic Computer Services**
                            Delhi - 110 035

Printed at   : **Chawla Offset Printers**
                            Delhi - 110 052

PRINTED IN INDIA

Devotion

The edited book NUTRI-HORTICULTURE is devoted to my wife Mrs. Vimala Peter who entrusted me to the safe and protective hands of Prof. P.I. Peter Chairman, NoniBiotech, Chennai to continue my aptitude for science writing and editing. In the process she suffered loneliness and solitude.

Acknowledgement

I acknowledge with gratitude Prof. P.I. Peter Chairman NoniBiotech, Chennai for support and facilities extended. The academic atmosphere he has created around me is highly motivating and stimulating. I have the highest sense of appreciation for all the 31 scientists from India, Australia, Indonesia and Thailand who toiled hard to produce excellent chapters within stipulated time. Dr. Kirti Singh, my teacher and now Chairperson World Noni Research Foundation is always supportive in my academic pursuits. I express my gratitude to the publisher Mr. Anil Mittal of Daya Publishing House, New Delhi for patience, quality printing and above all gentlemanliness.

Foreword

The transformation of horticulture from a way of rural life, art, practice and skill oriented endeavor to a front line science based hort-business is a milestone in the history of human development. Nutritional security through horticulture is the mantra. In earlier days it was a part of science of botany, later life sciences and now a symphony of genetics, soil science, soil physics, microbiology, virology, bacteriology, entomology, biotechnology, nanotechnology, molecular biology, micro and macroeconomics. Many areas of applied sciences are made use of to synthesize plant types to yield high under stressed conditions-biotic and abiotic. The growth of the Division of Horticulture into four scientific divisions-Vegetable Science, Fruit Science, Floriculture and Landscape Planning, Plantation crops and Medicinal Plants-is reflected in the high rate of growth in the above crops during the 11th plan period. Many Agricultural and Horticultural Universities have an additional Division of Post Harvest Technology and Processing. The financial allocation for the horticulture sector rose from Rs.24.7 crores in IVth plan to Rs 16,000 crores in XIth plan in addition to sizeable allocations to tea, coffee, rubber, coir, spice and coconut. The sector launched central sector schemes like National Horticultural Mission, Technology Mission for North Eastern and Himalayan States, Bamboo Mission, Construction expansion and modernization of cold storages, Micro irrigation, Use of Plastics, National Agriculture Insurance Scheme (NAIS) etc. The Horticultural Research and Development infrastructure in India today is one of the best in the world with ten Central Research Institutes, 13 Project Directorates, 13 All India Coordinated Research Projects in addition to separate Universities for Horticulture in Himachal Pradesh, Andhra Pradesh, Karnataka, Tamil Nadu and several Colleges of Horticulture. National Horticulture Board, National Horticulture Research and Development Foundation, Agriculture Produce Export Development Agency, State Farming Corporation of India, Institute for Organic Farming, Ghaziabad, Central Institute of Horticulture, Medziphema, Nagaland; International Horticulture Innovation and Training Centre, Jaipur are established to provide R and D support to Indian Horticulture. Many Scientific Societies were established to provide forum for organizing discussion and exchange of ideas and develop fraternity among Horticulture Scientists, farmers and extension workers. The premier ones are South Indian Horticulture Association, Horticulture Society of India, Indian Society of Vegetable Science, Orchid Society of India, Indian Society of Spices, Indian Society of Plantation Crops etc. Many journals of high impact factor now carry referred research articles. Indian Journal of Horticulture, Vegetable Science and Indian Journal of Plantation Crops are journal much read and referred. India also witnessed emergence of a large number of books, series and monographs authored and edited by Indian Scientists of repute.

The present book NUTRI-HORTICULTURE carries 17 chapters with the first chapter on 'Improving nutrition security and health for all-the important role of horticulture' and authored by 31 scientists and teachers of high standing. The emphasis on a science based horticulture calls for books on horticulture which focus on the science behind the processes and practices of horticulture. Dr. K.V. Peter, a Former Professor of Horticulture; Director, Indian Institute of Spices Research, Calicut; Director of Research Kerala Agricultural University; Vice-Chancellor KAU and now Director World Noni Research Foundation, Chennai was successful in bringing together 31 eminent scientists even from FAO, Rome to contribute chapters to this edited book.

I congratulate the publisher Daya Publishing House, New Delhi for publishing the much valued book.

Prof. K.R. Dhiman
Vice-Chancellor
Dr. Y.S. Parmar University of Horticulture & Forestry
Solan, H.P.

Preface

Horticultural crops-vegetables, fruits, tubers, mushrooms-are the nature's gift for nutritional security. Plantation crops and spices provide the much needed foreign exchange to the country and livelihood security by enhanced purchasing power to all the stakeholders. Ornamentals and aromatic crops lend much needed life to living to the inhabitants of the country. India sustains a large number of horticultural crops of considerable economic value. The costliest saffron in Kashmir, the cheapest curry leaf in the tarai forests to ghats of south India, the divine Noni plant in Andaman and Nicobar Islands; the high priced cardamom and pepper of South India and an array of ornamentals make India a distinct paradise on earth. Despite the numerous advantages India has in terms of biodiversity, climate, soil, water and above all the second largest domestic market, the annual growth rate in production is lesser than 4 per cent. Productivity of horticultural crops are the lowest except for a few crops like rubber, potato and cabbage. Many of the fruit trees are of seedling origin and less productive. Seed replacement ratio in vegetables are very low and spread of hybrids and high yielding varieties very thin. Water, fertilizer, pests and diseases are limiting factors to higher productivity. Post harvest losses range 20 to 40 per cent in fruits and vegetables. Marketing of perishable fruits, vegetables and cut flowers is still a gray area. Many initiatives have been launched by the Government of India, a few are National Horticultural Mission(NHM), National Horticulture Board(NHB), Rashtriya Krishi Vikas Yojana (RKVY), National Mission on Micro Irrigation(NMMI), National Bamboo Mission(NBM) and Technology Mission for North Eastern and Himalayan States. The National Information Commission is mandated with reaching information to all in the country including the vast unreached. Use of ICT in the transfer of knowledge in horticulture would empower farmers, students, scientists, trade, processors and consumers as well to acquire nutritional security. The present book NUTRI-HORTICULTURE edited by Prof. K.V. Peter focuses on Nutritional security, science behind horticultural practices, water requirement of crops, rainwater harvesting, biological control of pests and diseases and related subject areas. Thirty one eminent scientists have contributed to the 17 chapters. I recommend the book to all the stakeholders in Horticulture.

Dr. Brahma Singh
Former Director, Life Sciences
Defence Research and Development Laboratory
Tezpur

Contents

List of Contributors

Albert, Janice
Nutrition Officer, Nutrition and Consumer Protection Division, Food and Agriculture Organization of the United Nations, Rome
E-mail: janice.albert@fao.org

Amaresh, Y.S.
Department of Plant Pathology, College of Agriculture, UAS, Raichur, Karnataka, India

Ashok, K.R.
Department of Agricultural Economics, Tamil Nadu Agricultural University, Coimbatore – 641 003, T.N., India
E-mail: ashok10tnau@yahoo.com

Asrey, Ram
Division of Postharvest Technology, Indian Agricultural Research Institute, New Delhi – 110 012, India
E-mail: ramu_211@yahoo.com

Barman, Kalyan
Division of Postharvest Technology, Indian Agricultural Research Institute, New Delhi – 110 012, India

Chavan, R.L.
Department of Genetics and Plant Breeding, College of Agriculture, University of Agricultural Sciences, Raichur – 584 101, Karnataka, India
E-mail: rajuchavanuasr@gmail.com

Jayaraj, Renish
Department of Pomology and Floriculture, College of Horticulture, P.O. KAU, Kerala – 680 656, India

Joseph, E.J.
Centre for Water Resource Development and Management, Kozhikode, Kerala, India
E-mail: jej@cwrdm.org

Mani, K.
Department of Agricultural Economics, Tamil Nadu Agricultural University, Coimbatore – 641 003, T.N., India

Manjunatha, S.V.
Department of Plant Pathology, College of Agriculture, UAS, Raichur, Karnataka, India
E-mail: manjunaik2000@yahoo.co.in

Naik, M.K.
Department of Plant Pathology, College of Agriculture, UAS, Raichur, Karnataka, India

Nair, R.V.
Central Plantation Crops Research Institute, Kasaragod – 671 124, Kerala, India
E-mail: rvncpcri@gmail.com

Parameswaran, N.K.
Department of Pomology and Floriculture, College of Horticulture, P.O. KAU, Kerala – 680 656, India
E-mail: parameswarannk@yahoo.co.in

Parvatha Reddy, P.
Former Director, Indian Institute of Horticultural Research, Bangalore – 560 089, Karnataka, India
E-mail: reddy_parvatha@yahoo.com

Patil, Suresh
Department of Plant Pathology, College of Agriculture, UAS, Raichur, Karnataka, India

Pebam, Nongallei
Institute of Bioresources and Sustainable Development (IBSD), Takyelipat Institutional Area, Imphal – 795 001, Manipur, India
E-mail: rk_ryan2000@yahoo.com

Rajmohan, Jijii
Professor, Department of Entomology, College of Agriculture and Research Institute, P.O. Vellayani, Trivandrum, India
E-mail: jijirajmohan2004@yahoo.co.in

Ramakrishnan, Vibin
Institute of Bioinformatics and Applied Biotechnology, Biotech Park, Electronic City Phase-I, Bangalore, Karnataka, India
E-mail: vibin@ibab.ac.in

Roshan, R.K.
Krishi Vigyan Kendra, Churachandpur, Personmun Village, Churachandpur Dt. Manipur – 795 128, India

Samsudeen, K.
Central Plantation Crops Research Institute, Kasaragod – 671 124, Kerala, India

Sebastian, Aimy
Institute of Bioinformatics and Applied Biotechnology, Biotech Park, Electronic City Phase-I, Bangalore, Karnataka, India

Shareefa, M.
Central Plantation Crops Research Institute, Kasaragod – 671 124, Kerala, India

Sreeletha, S.
Rubber Research Institute of India, Kottayam, Kerala, India

Taji, Acram
Queensland University of Technology, Brisbane, Australia
E-mail: acram.taji@qut.edu.au

Tapingkae, Tanya
Faculty of Agricultural Technology, Chiang Mai Rajabhat University, Thailand

Tembhune, B.V.
Department of Genetics and Plant Breeding, College of Agriculture, University of Agricultural Sciences, Raichur – 584 101, Karnataka, India
E-mail: bvtembhurne@gmail.com

Thomas, Molly
Rubber Research Institute of India, Kottayam, Kerala, India

Thomas, Pious
Division of Biotechnology, Indian Institute of Horticultural Research, Hessarghatta Lake, Bangalore – 560 089, Karnataka, India
E-mail: piousts@iihr.ernet.in; piousts@yahoo.co.in

Usha, Nair N.
Rubber Research Institute of India, Kottayam, Kerala, India
E-mail: usha@rubberboard.org.in

Vanaja, T.
College of Agriculture, Kerala Agricultural University, P.O. Padannakad, Kasaragod, Kerala, India
E-mail: vtaliyil@yahoo.com

Visalakshi, K.P.
Professor, Agricultural Engineering, Kerala Agricultural University, P.O. KAU Thrisur, Kerala – 680 656, India
E-mail: visalam2009@yahoo.co.in

Zulkarnain, Zul
Agriculture Faculty, University of Jambi, Indonesia

Introduction

Science based Horticulture encompasses and targets increased productivity and enhanced income to farmers at optimum and efficient use of space, time, water, energy and information, communication technology. Fruits, vegetables, ornamentals, plantation crops, spices, herbs, tuber crops, medicinal and aromatic plants, mushrooms and processing into value added products come under the preview of general horticulture. Majority of crops are non-calorific and provide the much needed protective and curative foods. The distinction between food security and nutrition security takes into cognizance the availability of fruits, vegetables and spices to diet of common man. In 2010, FAO estimated that a total of 925 million people were undernourished. These numbers indicate that millions of people who do not have access to sufficient quantities of food to meet their energy needs, let alone meet their nutrient requirements (Janice Albert-Chapter 1). The World Health Organization (WHO) categorized the major health risks and found that high blood pressure, high blood glucose, overweight and obesity are the leading causes of mortality irrespective of income levels. Underweight and micronutrient deficiencies cause millions of child death each year in poor communities. Infectious diseases are another major cause of death and they are often associated with low nutritional status consequent to negligible intake of fruits and vegetables. Horticulture can play a vital role in preventing the above diseases and promoting health and wellness. Community level projects in Africa and Asia demonstrated that promotion of nutrition/kitchen garden led to availability and consumption of vegetables and fruits which improved the nutritional status of children.

All the Horticultural crops-for that matter all the members of plant kingdom-are associated with micro and macro organisms both beneficiary and parasitic. Being an integral part of the biosphere, there is no independent and exclusive living for horticultural crops. Endophytes-prokaryotic bacteria, eukaryotic fungi and yeasts-have been isolated from a diverse group of plants and different plant organs, more frequently from roots. There is an emerging interest in endophytes in view of their potential significance as agents of plant growth promotion, stress alleviation, phytoremediation and as sources of bio-molecules and novel genes (Pious Thomas-Chapter 2). Of the nearly 3 lakh plant species which exist on earth, each individual plant is host to one or more endophytes. The capability of colonizing internal host tissues has made endophytes valuable for horticulture as a tool to improve crop performance.

Productivity of majority of horticultural crops is low in developing countries like India except seasonal crops like cabbage and perennial tree crops like rubber. After the discovery of heterosis- hybrid vigour-, attempts are made to exploit the phenomenon in vegetables, fruits, ornamentals, spices, tubers and plantation crops. Hand emasculation and pollination being costly, male sterility was employed to develop female lines. Genetic, cytoplasmic and genic cytoplasmic male sterility were reported. Cytoplasmic male sterility being maternal in inheritance, its utility is enormous keeping the female characters intact in the progenies. Cytoplasmic male sterility was reported in tomato, chilli, brinjal, onion, chive, radish, carrot, Chinese cabbage, cabbage, sugar beet, potato, cucumber, plantago, alfalfa and petunia (Tembhune-Chapter 3). The cytoplasmic male sterility is affected by temperature as the responsible genes are temperature sensitive.

Another matter of concern for low productivity and crop loss is incidence of pests and diseases. In India alone annual crop loss due to pests, diseases and weeds are estimated to Rs. 600, 000 million in 2005. (Parvatha Reddy-Chapter 4). Integrated pest management is an important principle on which sustainable crop protection can be based. IPM is defined as "a pest management system that in the context of the associated environment and the population dynamics of the pest species utilizes all suitable techniques and methods, in a compatible manner as possible and maintains the pest populations at levels below causing economic injury". Biointensive IPM incorporates ecological and economic factors into agricultural system design and decision making and addresses public concerns about environmental quality and food safety. It is defined as a systems approach to pest management based on an understanding of pest ecology. It begins with steps to accurately diagnose the nature and source of pest problems and then relies on a range of preventive tactics and biological controls to keep pest populations within acceptable limits. Reduced risk pesticides are used if other tactics have not been adequately effective, as a last resort and with care to minimize risks".

Quality of seeds/planting materials governs upto 20 per cent of crop productivity. One reason for poor total production is low seed/planting material replacement ratio. One reason may be high cost of seed coupled with preference for local seeds for consumption. Micropropagation through tissue culture has come as an economically sustainable practice and TC plants are available in several fruits, vegetables, ornamentals, spices, tubers, aromatic and medicinal plants and plantation crops. Homogeneinity, earliness, resistance to biotic and abiotic stresses are a few desirable characters. *In vitro* breeding has been now standardized to develop recombinants and variants with marker and distinguishing traits (Zul Zulkarnain *et al.*-Chapter 5). Clonal micro propagation, somatic embryogenesis, somaclonal variation, *in vitro* micro grafting, haploid plant production, embryo rescue technology, *in vitro* flowering, *in vitro* pollination and fertilization, protoplast technology, plant genetic engineering and transformation etc. are areas of applied science of much relevance to Horticulture.

In perennial plants like fruit trees, root stock breeding has come to stay. Use of seedlings as root stocks has several disadvantages as the seedlings are heterozygous, heterogeneous and many times incompatible at late stage. Polyembryonic seedlings consisting of zygotic and nucellar can be easily identified by the faster growth of zygotic seedlings and they can be culled out at first sight. First reported in citrus by Leeuwenhoek (1719) and then confirmed by Strasburger (1878), polyembryony was described as formation of more than one embryo in a seed. Nucellar seedlings are true to types, they form genetically uniform rootstocks, more vigourous seedlings, virus free seedlings and budwood and above all have a tap root developing into a better root system(Roshan and Nongallei Pebam-Chapter 6).

Computational biology is a recent application of bioinformatics in scientific research. Applicable in all branches of science, its utility in plant science was realized lately; medical, space and atomic-nano sciences being the early users. Biological databases available online can be mainly classified into nucleotide sequence data bases, protein sequence databases, protein structural databases, protein-protein interaction databases, metabolic pathway databases, microarray databases and specialized databases which are developed for some very specific applications (Aimy Sebastian and Vibin Ramakrishnan-Chapter 7). Genome Sequencing and annotation-sequencing techniques; genomic annotation; Proteomics applications-protein structure prediction; protein function prediction; Evolutionary studies in plants; Microarray data analysis; Mass spectrometric Analysis; Computational Primer Designs and a glossary of computational terms would be reader friendly to biological scientists now introduced to Computational Biology.

Horticultural crop improvement is a continuing scientific pursuit. Floral biology reveals the nature of pollination-self, cross, often cross-. Methods of crop improvement through conventional breeding depend largely on the system of pollination. Self pollination is defined as intra floral, intraplant –interfloral, interplant pollen transfer when plants are apomictic and homozygous. Hermaphroditism, cleistogamy, absence of anaemophily and entomophily make self pollination inevitable. Introduction, selection-pure line, mass-;hybridization and selection-pedigree, bulk, single seed descent, back cross, multiple cross, dihaploidy-are the common methods of crop improvement in self pollinated crops.(Vanaja-Chapter 8).

Natural rubber is one of the industrial crops introduced to India from the tropical rain forests of Central and South America and now the second largest producer and the highest productivity recorder. The Rubber Research Institute of India, Kottayam has done pioneering research on introduction and evaluation of clones; management of fungal diseases like abnormal leaf fall and physiology and biochemistry of rubber. There are not many literature available to readers in plantation crops pertaining to biochemistry and physiology of latex production in rubber (Usha Nair *et al.*-Chapter 9). Latex is extracted by tapping the trunk of tree before sun rise. It is coagulated into semi solid mass by adding acid; crushed to remove water; smoked and dried. The chemistry of latex is interesting to read.

Water is one of the most important inputs in crops production. Each cell and tissue has free water and bound water. Free water is the ideal solvent of water soluable compounds and maintains turgour pressure so vital for the posture. Free water movement through osmosis has been dealt extensively elsewhere. Water constitutes 80 to 90 per cent of most plant cells and tissues in which there is active metabolism. Soil serves as the storage reservoir for water to be used by plants. Soil is a three phase system comprising of the solid phase made of mineral and organic matter, the liquid phase called the soil moisture and the gaseous phase called the soil air. Water is retained by a soil particle in the form of a thin film around it and is contained in the numerous small pores of the soil matrix with forces such as surface tension capillary, cohesion and adhesion. Hygroscopic, Capillary and Gravitational are the three forms in which water is held in the soil. Information on water requirement of horticultural crops are limited. The Chapter 10 by Joseph gives basic information on soil-plant-water relationship and methods of predicting water requirement of crops.

Rainwater is the main source of water for human domestic use, irrigation, hydro-energy generation, eco-biological diversity maintenance and sustenance. Water is the most precious and unique natural resource in our planet. A few statistics mention loss to the extent of 90 per cent flowing down to sea and ocean;4 per cent to fauna use and only 6 per cent to irrigation in hilly to semi-hilly terrain. In scanty rain fall areas like Africa; Middle East Countries; deserts of India rainwater is the only source of drinkable water. Desalinised water is costly and unavailable for irrigation. All these indicate the need for scientific methods of rainwater harvesting (Visalakshi-Chapter 11). The chapter describes rainwater harvesting for direct use and for augmenting the groundwater storage.

Perennial trees are propagated vegetatively for uniformity in stand, fruiting and more tree density in high density orcharding. Dicots rather than monocots are more amenable for grafting. On planting grafted portion is held above ground to prevent rooting of scion. Biology of Grafting in dicot plants is elaborated with details in diagrams (Parameswaran and Renish Jayaraj-Chapter 12). Factors influencing the success of graft union are plant species and type of grafts; environmental conditions following grafting; growth activity of the root stock and virus contamination, insects and diseases. Graft incompatibility results from adverse physiological responses between grafting partners; virus transmission and anatomical abnormalities of the vascular tissues in the callus bridge.

Among biotic stresses affecting horticultural crops, fungi and bacteria cause significant damage and loss. Attempts to screen crop germplasm to sort out resistant genotypes are carried out in many laboratories. Among bacterial diseases wilt by *Ralstonia solanacearum* is devastating in tomato, brinjal and tomato. Among fungal diseases *Phytophthora infestans* causes leaf fall in many crops. The interdisciplinary research with plant breeders joining plant pathologists has yielded useful results in many crops. The screening techniques for disease resistance against fungal and bacterial diseases of vegetables are elaborated by Naik *et al.* (Chapter 13). The diseases covered are dieback and anthracnose, Phytophthora fruit rot and leaf blight, fusarium wilt in chilli and bell pepper; buck eye rot, late blight, early blight and bacterial wilt in tomato; foliar and fruit diseases, fusarium wilt in cucurbits; black rot and stalk rot in cruciferous vegetables and alternaria blight in cabbage.

There is a school of thought that the present estimate of post harvest loss of 30-40 per cent if brought down to an achievable 10-15 per cent, the horticultural crops availability can be increased to 20-25 per cent, not a mean achievement in any scale. Pre-harvest treatment, stage of harvest, genotype of cultivars, environmental parameters like temperature, humidity and rainfall, methods of packing, pre-packing preparations and above all controlled atmospheric treatments determine the extent of loss till the produce reaches the consumer in edible form. Shelf life is a European terminology emanated from the wooden cabinets in British homes to keep fruits and vegetables. The period range in which the cosmetic appeal of the produce keeps satisfying to the consumer was called shelf life. Post-harvest factors affecting shelf life are increased metabolic activity resulting from high humidity, high temperature and pre-contamination with fungi, bacteria and insect eggs/larvae. Hot water treatments are recommended to control different pathogens in fruits. Vapour treatment, fumigation, chemical treatments-washing with ozonized water, chlorine solution, calcium application and use of growth regulators-can minimize post harvest loss (Ram Asrey and Kalyan Barman-Chapter 14).

It is observed that scientists, teachers and students are not kept informed about the history of both basic and applied sciences and resultant innovations. A knowledge on history will throw light on the men, methods and materials, background information and the way results were interpreted and later applied for human welfare. Two chapters (Nair *et al.*-Chapter 15 and Jijii Rajmohan-Chapter 16) deal with history of coconut breeding and history of Nematology Research. Coconut provided thirst quenching drink to early mankind. It is called tree of life for its diverse uses of every part of the palm with more than 100 different products including for food and drink, fodder for livestock, fibre, cosmetics and timber. The root has medicinal properties, leaf has many uses from thatching to fuel to composting etc. Organized coconut breeding started in India in 1916 at the erstwhile Central Coconut Research Station, Kasargod (presently CPCRI) and Research Stations now under Kerala Agricultural University. In Sri Lanka, research work on coconut began with the inception of the Coconut Research Institute in 1928 (Chapter 15-Nair *et al.*). Breeding new varieties and hybrids is a long drawn and slow process because of its floral biology, heterozygous nature, low rate of sexual propagation with one seedling per nut, lack of selection procedures for isolation of superior hybrid seedlings, lack of reproducible asexual methods for rapid multiplication, prolonged interval between generation and long juvenile phase before flowering. Methods of breeding followed are selection-motherpalm selection, seedling selection, identification of prepotent palm-and heterosis breeding. Tall x Dwarf hybrids and Dwarf x Tall hybrids were developed. Dwarf x Tall hybrids are found efficient in point of view of hybrid production. Breeding for special characteristics like disease resistance, drought resistance and nut water quality have made much headway.

Chapter 16 (Jijii Rajmohan) elaborates history of nematode research. Nematodes are triploblastic, bilaterally symmetrical, unsegmented, pseudoelomate and vermiform animals. They exist almost every where in nature. The different groups of nematodes are fungal feeders, bacterial feeders, predators, animal parasites, algal feeders, omnivores and plant parasites. Landmarks in the History of Nematology Research are reviewed from literature since 1873 when Butschli described the morphology of free-living nematodes. In India Barber, then the economic botanist at Coimbatore, studied on root knot infesting tea in South India in 1901. He identified root knot nematode in black pepper in 1906. Extent of crop loss due to Nematode attack is estimated 5 per cent in oil seeds, 8 per cent in cereals and pulses, 10 per cent in fruit crops and 12 per cent in vegetables. Symptoms of attach vary from leaf discolouration, dead or devitalized buds, seed galls, twining of leaves and stem and lesions on leaves and stem. Above ground symptoms are stunting, discolouration of foliage, decline or dieback and wilting.

The 17th Chapter by Ashok and Mani pertains to marketing management of crops. It deals with marketing environment, marketing mix, marketing segmentation, marketing information system and marketing potential. Forecasting in marketing is also dealt with.

The edited book NUTRI-HORTICULTURE carries 17 chapters authored by 31 scientists from 15 Universities/Research Institutes.

K.V. Peter

2012, Nutri-Horticulture
Editor: Professor K.V. Peter
Published by: DAYA PUBLISHING HOUSE, NEW DELHI

Pages 1–6

Chapter 1

Improving Nutrition Security and Health for All: The Important Role of Horticulture

Janice Albert

Nutrition Officer, Nutrition and Consumer Protection Division,
Food and Agriculture Organization of the United Nations
E-mail: janice.albert@fao.org

Horticulture products – fruits, vegetables, roots, tubers, nuts and legumes–can make important contributions to achieving nutrition security and healthy diets. Nutrition security means that all people have access to a diversity of foods to meet their needs for essential nutrients, as well as sufficient energy to grow and develop, carry out activities and maintain the body. The quality, as well as the quantity of foods consumed, are vitally important.

Global Food and Nutrition Security Situation

In spite of advances in science and technology and economic and social progress in many countries, the number of people who suffer from food insecurity worldwide is still unacceptably high. In 2010, FAO estimated that a total of 925 million people were undernourished (FAO 2010).

These numbers indicate that there are millions of people who do not have access to sufficient quantities of food to meet their dietary energy needs, let alone meet their nutrients requirements. Most of our dietary energy comes from foods that are high in carbohydrates, like grains, roots and tubers. These foods alone cannot provide for all nutritional needs. Other plants and animals are required to provide the protein, fats, vitamins, minerals essential for human health. When households are unable to obtain the variety of foods which they require, nutritional deficiencies occur which can seriously affect health, particularly of small children and women of reproductive age (15-45 years). Even when the amount of dietary energy is sufficient, poor quality diets can lead to poor health.

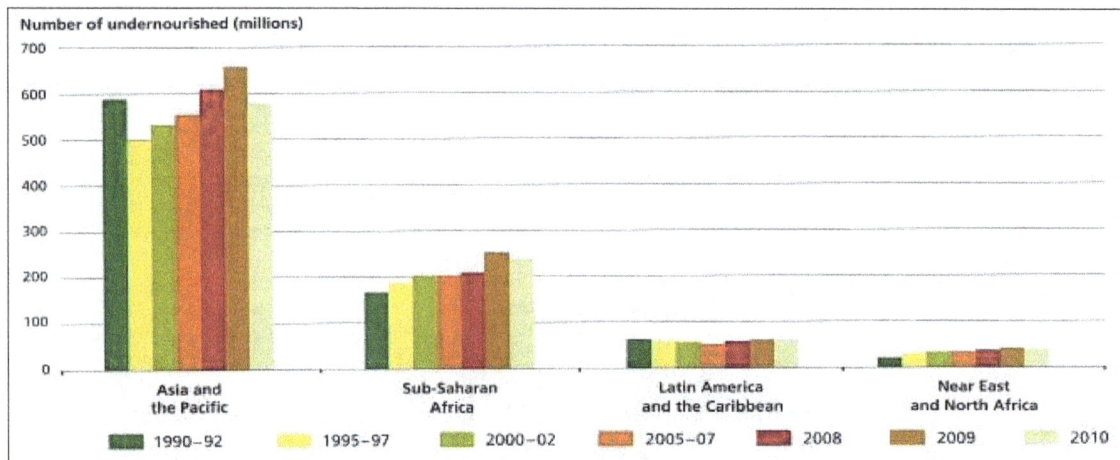

Figure 1.1: Regional Trends in the Number of Undernourished from 1990-92 to 2010

Major Health Problems Worldwide

The World Health Organization categorized the major health risks throughout the world and found that high blood pressure, high blood glucose, overweight and obesity are leading causes of mortality affecting all income groups regardless of a country's level of economic development (WHO 2009a). Underweight and micronutrient deficiencies cause millions of child deaths each year in poor communities. Infectious diseases are another major cause of death and they are often associated with lower nutritional status. For several of these widespread causes of premature death, horticulture can play a role in preventing the diseases and promoting health. These contributions are described below.

Vitamin A Deficiency

Three nutritional deficiencies, vitamin A, iron and iodine are widespread and serious threats to human health. Horticulture products provide a range of vitamins which contribute to human health (*e.g.* consuming vitamin C- rich fruit helps the absorption of iron). Horticulture is a key strategy in preventing one vitamin deficiency that is a major public health concern in lower income countries, vitamin A deficiency (VAD).

VAD is the leading cause of preventable childhood blindness and it leads to night blindness in adults, particularly pregnant women. The weakened resistance to infection caused by this deficiency makes children more vulnerable to infectious diseases and death. Table 1.1 shows the prevalence of this deficiency worldwide.

Consumption of vegetables, fruits and tubers rich in carotenoid precursors (mainly beta-carotene) like green leaves, carrots, ripe mangoes and orange-yellow vegetables and fruits can help to prevent this deficiency. Community level projects in Africa and Asia have demonstrated that promotion of gardening and consumption of foods rich in carotenoid precursors can improve the nutritional status of children (Thompson and Amoroso, 2011).

Table 1.1: Vitamin A Deficiency* Prevalence (Aerum retinol under 0.70 µmol/l) in 156 Countries with Incomes Under $15000/year

WHO Regions	Pre-school Age Children		Pregnant Women	
	Prevalence (Per cent)	# Affected (millions)	Prevalence (Per cent)	# Affected (millions)
Africa	44.4	56.4	13.5	4.18
Americas	15.6	8.68	2.0	0.23
South-East Asia	49.9	91.5	17.3	6.69
Europe	19.7	5.81	11.6	0.72
Eastern Mediterranean	20.4	13.2	16.1	2.42
Western Pacific	12.9	14.3	21.5	4.90
Global	**33.3**	**190**	**15.3**	**19.1**

Source: World Health Organization 2009b.

Diet-related Chronic Diseases

Chronic diseases like cardiovascular diseases, type 2 diabetes and cancer generally affect the middle aged and elderly segments of the population. Globally, the number of persons aged 60 or above is expected nearly to triple, increasing from 673 million in 2005 to 2 billion by 2050 (United Nations, 2007). The vast majority of these people will be in developing countries. To ensure that these people enjoy health and that their medical needs do not become a devastating drain on economies, it is critically important to make efforts today to prevent chronic diseases. The rising levels of overweight and obesity among children in some countries imply that these diseases are occurring at earlier ages as well.

Preventive Action is Needed

Contributions of horticulture products to prevention of deficiencies are well-understood by scientists, the obstacles to resolving these problems are usually related to the availability, affordability and consumer awareness. With regard to the prevention or treatment of chronic illnesses like heart disease, type 2 diabetes and cancer, the scientific evidence is still emerging and caution is needed in making sweeping statements about the potential of foods to reduce these diseases.

Focusing specifically on fruits and vegetables, health experts are convinced of the multiple benefits of consuming fruits and vegetables and the urgency of the need to take preventive actions and they are encouraging increased consumption of these foods. Public health agencies have stressed that at least 400 grams of fruits and vegetables should be eaten per day (WHO 2003). Some recommended that the amount should be raised to 600 grams per day (World Cancer Research Fund 2009).

Research on Fruits and Vegetables

Fruits and vegetables are very complex and this paper focuses on only a few components where substantial research has been carried out recently. The particular components are beta carotene (precursor for vitamin A), lycopene, alpha-tocopherol (Vitamin E), folate and flavonoids. Current research suggests that each of these compounds may have great public health significance in terms of preventing disease, disability and premature mortality. This research includes epidemiological studies, animal experiments and clinical trials with humans.

Cardiovascular Diseases

Although cardiovascular diseases (CVD) usually affect middle age and elderly adults, the diets consumed early in life can influence the risk of stroke and coronary heart disease. Children's consumption of a heart healthy diet can reduce the prevalence of obesity, high cholesterol and high blood pressure in adults. When fruits and vegetables replace foods high in animal fat, salt and sugar this can lower health risks.

Consumption of these foods can help people to manage their weight and salt and sugar content; however, this is not the main reason that these foods are believed to be protective. In adolescents and adults, high consumption of fruits and vegetables is associated with lower risk of cardiovascular diseases. Flavonoids, a subclass of polyphenols, are found in parsley, spinach, peppers, cauliflower, brussel sprouts, onions, apples, berries and grapes, among other foods (Holt *et al.*, 2009). They are present in the skins and peels of fruits and vegetables and give these foods their deep colours. Some flavonoids have antioxidant properties.

Vitamins C (ascorbic acid), E (alpha-tocopherol) and folate have favourable effects on vascular system and risk factors for heart disease (George *et al.*, 2009). Several studies have found that lycopene, a carotenoid which is found in tomato, has a preventive effect on cardiovascular disease and risk of myocardial infraction. Foods containing beta-carotene and vitamin A derivatives also reduce the risk of CVD.

Type 2 Diabetes

It is estimated that there are 180 million cases of diabetes mellitus (DM) worldwide (over 90 per cent are type 2 DM); without strong preventive actions, this number may double by 2030 (Barnard *et al.*, 2009 and Wylie-Rosett and Vinicor, 2007). This increase is largely the result of changing diets and more sedentary lifestyles. Observational and clinical studies have shown that diets, comprised of an appropriate diversity of plant foods (whole grains, legumes, vegetables and fruits) can improve the management of type 2 diabetes. Some of the benefits of plant based diets include: reduced obesity, reduced fat intake and increased intake of dietary fibre (Barnard *et al.*, 2009).

Cancers

The International Agency for Research on Cancer, the World Cancer Research Fund and American Institute for Cancer Research believe that it is probable that high fruit and vegetable intakes may reduce a number of cancers (namely, oesophageal, colo-rectal, oral, pharynx, stomach, larynx, lung, ovary and kidney cancers). However, the evidence over the past two decades is conflicting and not yet considered convincing. More research is needed to identify the specific types of fruits and vegetables and their particular components, which have an impact on disease.

Based on current knowledge, it appears that dark green leafy vegetables, yellow and red vegetables, cruciferous vegetables, as well as red and purple fruits and vegetables may be protective against cancer. It is plausible that the vitamins C, E and folate have an effect on cancer. The phytochemicals such as carotenoids, phenolics, isoflavonoids, isothiocyannates (ITC) and indoles also have anticarcinogenic effects (Kim and Park, 2009).

Cruciferous Vegetables

Cruciferous vegetables like broccoli, cauliflower, cabbage, kale and mustard are common worldwide and they are rich in carotenoids, vitamin C, folate and soluble fibre. There is evidence from animal studies that these foods contain phytochemicals having anticarcinogenic effects (Kim and

Park 2009). The latest research suggests that these vegetables, especially Brassicas, may lower the risk of gastric and lung cancers.

Confounding Factors

It is also important to note that genetics, age, gender and environmental exposures affect the influence of these foods on disease. Food preparation and storage can also affect the health benefits of vegetables and fruits.

Concluding Remarks

The world faces two sets of challenges, hunger and nutrient deficiencies, which usually afflict the very young and vulnerable groups, and diet-related chronic diseases which affect adults in all social strata. Horticulture must play an important role in both arenas.

Severity of the problems of micronutrient deficiencies is a compelling reason to call for greater production and consumption horticulture products, especially vegetables and fruits. It is well-documented that these foods can contribute to prevention of deficiencies which can be debilitating and can lead to premature deaths. Although we do not have all the evidences we would like regarding the health benefits of consuming vegetables and fruits in relation to chronic diseases, it is clear that these foods are associated with healthier populations. Research on the impact of these foods on chronic disease is ongoing and with time, we will have a better understanding of the mechanisms which link specific components of foods to health.

FAO encourages countries to promote the production and consumption of these horticulture products. Nutritionists and health promoters are actively engaged in raising awareness about the benefits of consuming fruits and vegetables through food-based dietary guidelines and nutrition education campaigns. We have reason to believe that interventions to promote gardening and educate young people will lead to increased awareness and demands for these foods in the future (Laurie and Faber 2008 and Parmer *et al.,* 2009). FAO is promoting school-based garden learning to educate young people about the benefits of vegetables and fruits with the aim of instilling lifelong healthy eating patterns, as well as stimulating appreciation of nature and agriculture. As consumer awareness rises, the horticulture sector will be challenged to meet the growth in demand. We count on the horticulture sector to develop strategies to make these foods more available, accessible and affordable to the world's people.

References

Barnard, N.D., Katcher, H.I., Jenkins, D.J.A., Cohen, J. and Turner-McGrievy, G., ., 2009. "Vegetarian and vegan diets in type 2 diabetes management". *Nutrition Reviews,* 67(5): 255–263.

Campbell, A.A., Thorne-Lyman, A., Sun, K., de Pee, S., Kramer, K., Moench-Pfanner, R., Sari, M., Akhter, N., Bloem, M.W. and Semba, R.D., 2009. "Indonesian women of childbearing age are at greater risk of clinical vitamin A deficiency in families that spend more on rice and less on fruits/vegetables and animal-based foods". *Nutrition Research,* 29: 75–81.

Food and Agriculture Organization, 2010. *The State of Food Insecurity in the World Addressing Food Insecurity in Protracted Crises.* FAO, Rome.

George, T.W., Niwat, C., Waroonphan, S., Gordon, M.H. and Lovegrove, J.A., 2009. "Postgraduate symposium effects of chronic and acute consumption of fruit- and vegetable-puree-based drinks on vasodilation, risk factors for CVD and the response as a result of the *e*NOS G298T polymorphism" In: *Conference on 'Multidisciplnary Approaches to Nutritional Problems" Proceedings of the Nutrition Society*, 68: 148–161.

Holt, E.M., Steffen, L.M., Moran, A., Basu, S., Steinberger, Ross, J.J., Hong, C.-P. and Sinaiko, A., 2009. "Fruit and vegetable consumption and its relation to markers of inflammation and oxidative stress in adolescents". *Journal of the American Dietetic Association*, 109(3): 414–421.

Kim, M.K. and Yoon Park, J.H., 2009. "Cruciferous vegetable intake and the risk of human cancer: epidemiological evidence" In: *Conference on 'Multidisciplinary Approaches to Nutritional Problems' Symposium on 'Nutrition and Health' Proceedings of the Nutrition Society*, 68: 103–110.

Laurie, S.M. and Faber, M., 2008. "Integrated community-based growth monitoring and vegetable gardens focusing on crops rich in β-carotene: Project evaluation in a rural community in the Eastern Cape, South Africa". *Journal of the Science of Food and Agriculture*, 88: 2093–2101.

Parmer, S.M., Salisbury-Glennon, J., Shannon, D. and Struempler, B., 2009. "School gardens: An experiential learning approach for a nutrition education program to increase fruit and vegetable knowledge, preference and consumption among second-grade students." *Journal of Nutrition Education and Behavior*, 41(3): 212–217.

Thompson, B. and Amoroso, L., (Editors), 2011. *Combating Micronutrient Deficiencies: Food-based Approaches*. Food and Agriculture Organization of the United Nations and CAB International. Rome.

United Nations, Department of Economic and Social Affairs, Population Division, 2007. *World Population Prospects: The 2006 Revision, Highlights*, Working Paper No. ESA/P/WP.202.

World Cancer Research Fund/American Institute for Cancer Research, 2007. *Food, Nutrition, Physical Activity, and the Prevention of Cancer: A Global Perspective*. AICR, Washington DC.

World Cancer Research Fund/American Institute for Cancer Research, 2009. *Policy and Action for Cancer Prevention, Food, Nutrition and Physical Activity: A Global Perspective*. AICR, Washington DC.

World Health Organization, 2003. *Diet, Nutrition and the Prevention of Chronic Diseases Report of a Joint WHO/FAO Expert Consultation*. WHO Technical Report Series 916, Geneva.

WHO, 2009a. *Global Health Risk. Mortality and Burden of Disease Attributable to Selected Major Risks*. World Health Organization, Geneva.

WHO, 2009b.. *Global Prevalence of Vitamin A Deficiency in Populations at Risk 1995–2005*. WHO Global Database on Vitamin A Deficiency, Geneva.

2012, Nutri-Horticulture
Editor: **Professor K.V. Peter**
Published by: **DAYA PUBLISHING HOUSE, NEW DELHI**

Pages 7–24

Chapter 2

Endophytic Microorganisms in Agriculture and Horticulture

Pious Thomas

Division of Biotechnology, Indian Institute of Horticultural Research,
Hessarghatta Lake, Bangalore – 560 089, INDIA
E-mail: pioust@iihr.ernet.in; piousts@yahoo.co.in

By definition, endophytes are microorganisms which colonize the plants internally without any apparent adverse effects on the host plant (Hallmann *et al.*, 1997; Bacon *et al.*, 2002). They include prokaryotic bacteria and eukaryotic fungi and yeasts. Endophytes are isolated from healthy plants after tissue surface sterilization and they are reportedly present in much low numbers in comparison to the pathogens (Hallmann, 2001). Evidence is now emerging that plants harbor a series of endophytes in considerable numbers in different organs, and they include both culturable and normally non-culturable organisms (Thomas *et al.*, 2008a, Thomas and Soly, 2009). Endophytes have been isolated from a diverse group of plants and different plant organs, more frequently from roots (Hallmann, 2001; Bacon *et al.*, 2002). Endophytic fungi and bacteria may confer benefits to the plant, and the benefits may be reciprocal, resulting in an enhanced symbiotic system for the plants. There is an emerging interest in endophytes in view of their potential significance as agents of plant growth promotion, biocontrol, stress alleviation, phytoremediation and as sources of bio-molecules and novel genes (Hallmann *et al.*, 1997; Sturz *et al.*, 2000; Hallmann 2001; Bacon *et al.*, 2002; Ryan *et al.*, 2008).

Endophytes are detected in plants growing in tropical, temperate and boreal forests with the hosts ranging from herbaceous plants in various habitats including extreme arctic, alpine and xeric environments to temperate and tropical forests. Vogl (1898) first recorded the presence of endophytes that revealed a mycelium residing in the seeds of grass *Lolium temulentum*. In the early years, studies on endophytes focused on filamentous fungi. Research on endophytic bacteria began in the 1870s with Pasteur and others. When compared to the bacterial endophytes, the research of endophytic fungi has a long history and their diversity among plants was considerably large (Hallmann, 2001). Since 1940, there are numerous reports on indigenous endophytic bacteria in various plant tissues, including seeds and ovules, tubers, roots, stems, leaves and even fruits. Evidence of plant-associated microorganisms found in the fossilized tissues of stems and leaves has revealed that endophyte - host associations might have evolved from the time that higher plants first appeared on the Earth (Strobel, 2003).

Isolation of Endophytes

Several methods are employed to isolate and culture endophytic bacteria such as planting surface sterilized tissue on bacteriological media (Bacon *et al.*, 2002), tissue homogenization and plating (Bell *et al.*, 1995; Reiter *et al.*, 2002) or physical methods like vacuum (Bell *et al.*, 1995), pressure extraction (Hallmann *et al.*, 1997) and centrifugation (Bacon *et al.*, 2002). Tissue culture system is proved to be a valuable tool for the isolation of endophytic bacteria, which included some uncommon endophytes (Thomas *et al.*, 2007a, 2008b). The simplest technique for isolating bacterial endophytes is their culturing on artificial media. The final population can then be estimated by counting the colonies developing on the media after considering the dilution factor. Direct isolations of cultivable bacterial endophytes from seed or plant tissue can be accomplished on media like nutrient agar, glucose-yeast extract, tryptic soy agar, King's B medium and MacConkey agar (Bacon *et al.*, 2002). Bacterial isolation based on artificial media only considers viable microorganisms able to utilize nutrients of the selected medium. The isolation medium is often supplemented with antifungal agents such as Nystatin and cycloheximide to avoid rapidly growing fungal contaminants.

Endophytic fungi usually require complex, non-defined media. Fungal endophyte assay is usually carried out by including antibiotics in the assay plates to suppress the abundant bacterial growth which otherwise may interfere with fungal development. The commonly used culture media for fungal isolation include malt extract agar, Corn meal agar, V8 agar, potato dextrose agar, potato carrot agar, Czapek's agar and peptone dextrose agar (Bacon *et al.*, 2002).

Theoretically, any isolation procedure should recover the complete internal population of organisms and only the internal population. Several techniques are employed for the isolation of endophytic bacteria with each one having advantages and disadvantages. The most frequently utilized method to isolate and quantify endophytes involves isolation from surface-sterilized host plant tissue. Surface sterilization is usually accomplished by treatment with a disinfectant like sodium hypochlorite, ethanol, hydrogen peroxide, mercuric chloride or a combination of two or more of these (2-5 min) followed by several washes in sterile water or buffer solutions. Surfactants such as Tween 80, Tween 20 and other liquid detergents like Teepol are used as wetting agents to enhance surface sterilization of the host plants. The effectiveness of surface sterilization is ascertained by plating the wash solutions and through tissue imprinting on nutrient media and observing for any colony growth.

The tissue is homogenized in sterile water, buffer solutions, or liquid media with a mortar and pestle, or using mechanical device like a fruit blender. The triturate is then plated on a nutrient medium of choice for the isolation of fungi or bacteria. Distinct colony morphotypes are selected after a few days (1-7) to several weeks (1-4) of incubation at 25-30 or 37°C. The plates during this period of incubation should be sealed properly to avoid medium desiccation and to protect from other contaminants.

Another approach is vacuum or pressure extraction of plant sap, which mainly consists of fluid from the conducting elements and adjacent intercellular spaces. Centrifugation is yet another approach for recovery of endophytic bacteria. This method is commonly used to collect intercellular fluid of plant tissue, but would also extract endophytic bacteria from intercellular and vascular fluid. This technique also requires surface disinfection of the plant material, since the plant tissue will come into contact with the extracted fluid during the centrifugation procedure (Hallmann *et al.*, 1997).

The potential of tissue culture as a tool for isolating endophytic bacteria is being increasingly recognized now (Thomas *et al.,* 2007a; 2008a, b). Tissue cultures are initiated after the surface sterilization of plant tissue which step is very similar to the surface disinfection practiced during the isolation of endophytes. The cultures displaying contamination *in vitro* could serve as the source of some novel organisms (Panicker *et al.,* 2007; Thomas *et al.,* 2007a; 2008b). Tissue cultures also offer the potential to bring some of the originally non-culturable organisms to cultivation thus offering a larger spectrum of coverage compared to conventional isolation procedures (Thomas *et al.,* 2008a, b).

Tissue Colonization by Endophytic Microorganisms

Endophytic bacteria are known to colonize the intercellular region in diverse plants with roots suggested to be the major point of entry and niche. They are also known to enter the aerial parts of plants through natural openings like stomata and wounds (Hallmann, 2001; Bacon *et al.,* 2002; Rosenblueth and Martinez-Romero, 2004). Endophytes have been isolated from various organs including roots, stem, leaf, buds, flowers and tender fruits of diverse plants. The extent of endophytic colonization observed with root tissue is estimated at log 5 CFU g^{-1} fresh wt, which is considerably low compared to the colonization by pathogenic bacteria, estimated at log 8 - log 10 CFU g^{-1} tissue (Hallmann, 2001). It is postulated that all plants harbor endophytic bacteria and the non-recovery of endophytes from any plant could be due to the inappropriateness of the procedures followed or media employed (Hallmann, 2001).

Endophytic fungi are more studied compared to bacteria. Endophytic fungi live asymptomatically, and sometimes systemically, within plant tissues. Fungal endophytes are ubiquitous and extremely diverse in host plants. Every plant examined to date harbors at least one species of endophytic fungus. Many plants, especially woody plants, may contain literally hundreds or thousands of species (Faeth and Fagan, 2002). By definition, an endophytic fungus lives in mycelial form in biological association with the living plant, at least for some time.

Organisms Associated with Different Plants

(a) Bacteria

Studies based on cultivation and cultivation-independent approaches have brought out a diverse array of endophytic bacteria associated with different plant species including proteobacteria, firmicutes, actinobacteria, acidobacteria, Cytophaga/Flexibacter/Bacteriodes (CFB) phylum, Deinococcus-Thermus groups and archaea (Chelius and Triplett, 2001; Hallmann, 2001; Hallmann *et al.,* 1997; Bacon and Hinton, 2006; Mano and Morisaki, 2008; Reiter and Sessitsch, 2006; Thomas *et al.,* 2007a, 2008b; Thomas and Soly, 2009). Organisms like *Bacillus, Enterobacter, Klebsiella, Pseudomonas, Burkholderia, Pantoea, Agrobacterium, Methylobacterium* spp. constitute the endophytes commonly isolated from diverse plants like rice, wheat, maize, cotton, clover, potato, sugarcane, tomato, cucumber which have been the focus of several past investigations.

Thomas and Soly (2009) studied endophytic bacteria present in deep-seated shoot-tips of banana yielding diverse organisms. They included *Bacillus, Brevibacillus, Paenibacillus, Virgibacillus, Staphylococcus, Cellulomonas, Micrococcus, Corynebacterium, Kocuria, Paracoccus, Pseudomonas* and *Acinetobacter* spp. Each shoot tip showed one to three different organisms and no specific organism appeared common to different sucker tips. Investigations with tissue-cultured bananas have brought out *Enterobacter, Klebsiella, Ochrobactrum, Pantoea, Staphylococcus, Bacillus, Brevundimonas, Methylobacterium, Alcaligenes, Ralstonia, Pseudomonas, Corynebacterium, Microbacterium, Staphylococcus, Oceanobacillus, Brachybacterium, Brevibacterium, Kocuria, Tetrasphaera* and *Staphylococcus* spp. as bacterial endophytes, some of which were hazardous to *in vitro* cultures while others remained in covert form without any apparent deleterious effects (Thomas *et al.*, 2008b).

Table 2.1: Endophytic Bacteria Isolated from some Common Crop Plants

Sl.No.	Plant Species/Organ	Bacterial Taxa	Source
1.	Rice (*Oryza sativa*); Seed, root and stems	*Pantoea* sp., *Klebsiella oxytoca*, *Sphingomonas* sp., *Agrobacterium* sp., *Azorhizobium* sp., *Azospirillum lipoferum*, *A. brasilense*, *Bacillus* sp., *Bradyrhizobium* sp., *Burkholderia graminis*, *Herbaspirillum rubrisub-albicans*, *H. seropedicae*, *Ideonella dechloratans*, *Enterobacter cancreogenus*, *Pseudomonas* sp., *Rhizobium leguminosarum*	Mano and Morisaki (2008)
2.	Wheat (*Triticum aestivum*); root and stem	*Bacillus polymyxa*, *Burkholderia cepacia*	Hallmann *et al.*, 2001; Bacon and Hinton (2006)
3.	Maize (*Zea mays* L.) root and stem	*Arthrobacter* sp., *Aureobacterium* sp., *Bacillus mojavensis*, *B. thuringiensis*, *Bacillus* spp. *Burkholderia cepacia*, *Burkholderia* sp., *Corynebacterium* sp., *Enterobacter* sp., *Klebsiella terrigena*, *K. pneumoniae*, *Pseudomonas* sp., *Staphylococcus* sp.	Hallmann (2001); Chelius and Triplett (2001)
4.	Cotton (*Gossypium hirsutum* L.) root and stem	*Acinetobacter baumannii*, *Agrobacterium radiobacter*, *Bacillus* sp., *Bacillus endophyticus*, *Burkholderia cepacia*, *B. gladioli*, *B. pickettii*, *Cellulomonas* sp., *Chryseobacterium* sp., *Comamonas testosteroni*, *Curtobacterium* sp., *Enterobacter cloacae*, *Enterobacter* sp., *Escherichia coli*, *Hydrogenophaga* sp., *Klebsiella* sp. *Kluyvera* sp., *Methlobacterium* sp., *Pseudomonas* sp., *Stenotrophomonas* sp., *Ochrobacterum anthropi*, *Pantoea* sp., *Phyllobacterium* sp., *Pseudomonas saccharophila*, *P. stutzeri*, *P. chloroaphis*, *Ralstonia japonicum*, *Rhizobium japonicum*, *Rhizobium* sp., *Serratia* sp., *Sphingomonas paucimobilis*, *Staphylococcus* sp., *Xanthomonas campestris*, *Yersinia frederiksenii*	McInroy and Kloepper (1995a, b); Hallmann *et al.* (1997); Hallmann (2001)
5.	Soybean (*Glycine max*), stem, leaves, root	*Pseudomonas citronellolis*, *P. oryzihabitans*, *P. staminea*, *Erwinia* sp., *Klebsiella pneumoniae*, *K. oxytoca*, *Agrobacterium*, *Caulobacter* sp., *Enterobacter agglomerans*, *Pantoea* sp.	
6.	Sugarcane (*Saccharum officinarum*), root and stem	*Acetobacter* sp., *Herbaspirillum* sp., *Gluconacetobacter* sp., *Diazotrophicus* sp.	Bacon and Hinton (2006)*
7.	Potato (*Solanum tuberosum*), tuber	*Acidovorax* sp., *Acinetobacter* sp., *Actinomyces*, *Agrobacterium* sp., *Alcaligenes* sp., *Arthrobacter ureafaciens*, *Bacillus alclophialus*, *B. pasteurii*, *B. sphaericus*, *Capnocytophaga* sp., *Comamonas* sp., *Corynebacterium* sp., *Curtobacterium citrenum*, *C. leteum*, *Deleya* sp., *Enterobacter* sp., *Erwinia* sp., *Flavobcaterium* sp., *Kingella kingae*, *Klebsiella* sp., *Leuconostoc* sp., *Methylobacterium* sp., *Micrococcus* sp., *Pantoea* sp., *Pasteurella* sp., *Photobacterium* sp., *Pseudomonas tolaasii*, *Psychrobacter* sp.	Bacon and Hinton (2006)*

Contd...

Table 2.1–Contd...

Sl.No.	Plant Species/Organ	Bacterial Taxa	Source
8.	Banana (*Musa* sp.); Shoot tips, tissue cultures	*Bacillus* sp., *Brevibacillus* sp, *Paenibacillus* sp, *Virgibacillus* sp., *Staphylococcus* sp., *Cellulomonas* sp., *Micrococcus* sp., *Corynebacterium* sp., *Kocuria* sp., *Paracoccus* sp., *Pseudomonas* sp., *Acinetobacter* sp., *Enterobacter* sp., *Klebsiella* sp., *Ochrobactrum* sp., *Pantoea* sp., *Brevundimonas* sp., *Methylobacterium* sp., *Alcaligenes* sp., *Ralstonia* sp., *Pseudomonas* sp., *Corynebacterium* sp., *Microbacterium* sp., *Oceanobacillus* sp., *Brachybacterium* sp., *Brevibacterium* sp., *Kocuria* sp., *Tetrasphaera* sp.	Thomas *et al.* (2008); Thomas and Soly (2009)
9.	Papaya (*Carica papaya* L.)	*Pantoea ananatis, Enterobacter cloacae, Brevundimonas aurantiaca, Sphingomonas* sp., *Methylobacterium rhodesianum* and *Agrobacterium tumefaciens, Microbacterium esteraromaticum* and *Bacillus benzoevorans*	Thomas *et al.* (2007a)
10.	Grape (*Vitis* spp.)	*Bacillus fastidiosus, B. insolitus, Clavinacter* sp., *Comamonas* sp., *Curtobacterium* sp., *Enterobacter* sp., *Klebsiella ozaenae, K. pneumoniae, K. terrigena, Moraxella bovis, Pantoea* sp., *Pseudomonas cichorii, Rahnella agquatilis, Rhodococcus luteus, Staphylococcus* sp., *Xanthomonas* sp.	Bell *et al.* (1995)
11.	Tangerine (*Citrus reticulata*) and sweet orange (*C. sinensis*); stem	*Curtobacterium flaccumfaciens, Enterobacter cloacae, Methylobacterium* sp., *M. zatmanii, Nocardia* sp., *Pantoea agglomerans, Xanthomonas campestris, Bacillus pumilus*	Bacon and Hinton (2006)*
12.	Cucumber (*Cucumis sativus* L.), root and fruit	*Agrobacterium* sp., *Bacillus* sp., *Burkholderia* sp., *Citrobacter* sp., *Clavibacter* sp., *Erwinia carotovora* sp., *Proteus mirabilis, Serratia* sp., *Xanthomonas* sp.	McInroy and Kloepper (1995b); Bacon and Hinton (2006)*
13.	Tomato (*Lycopersicon esculentum*) stem and fruit	*Pseudomonas syringae, Escherichia coli*	Bacon and Hinton (2006)*
14.	Beans, peas (*Phaseolus vulgaris,Pisum sativum*)	*B. polymyxa, B.* spp.	Bacon and Hinton (2006)*

* Secondary source.

Fourteen distinct endophytic bacteria which expressed visibly on tissue culture medium at culture initiation were isolated from papaya 'Surya' cultures. These were identified based on 16S rRNA gene sequence analysis as Gram-negative *Pantoea ananatis, Enterobacter cloacae, Brevundimonas aurantiaca, Sphingomonas* sp., *Methylobacterium rhodesianum* and *Agrobacterium tumefaciens* and two Gram-positive organisms, *Microbacterium esteraromaticum* and *Bacillus benzoevorans* (Thomas *et al.*, 2007a) (Figure 2.3). *P. ananatis* was the most frequently isolated organism (70 per cent cultures) followed by *B. benzoevorans* (13 per cent). From chrysanthemum 'Arka Swarna', three colony types were isolated all of which proved to be the morphotypes or strains of the same organism, *Curtobacterium citruem* (Panicker *et al.*, 2007).

(*b*) Endophytic Fungi

Endophytic fungi are a group of fungi which live asymptomatically inside plant tissues. These fungi were first noticed in the 1940s, but only at the turn of the 21st century was the ubiquity of these fungi fully recognized. They live like *fungi imperfecti*, much of the time, producing mostly conidial spores or simply cloning themselves. Many endophytic fungi become rotting agents upon the death of their host plant. Sometimes they may produce toxins in response to pests which feed on their host's tissues. These toxins benefit the host plant. For example, there are endophytes which make grass poisonous to grazing animals. Other endophytes, in oak leaves, ward off gall midges. In the sense that these 'endophytes' can protect their host, they are symbionts. Endophytes belong to several fungal taxonomic groups, but most are ascomycetes (White, 2009).

Common endophytic fungi isolated and characterized from different crop plants include *Fusarium moniliforme, Cladosporium sphaerospermum, Cryptosporiopsis* sp., *Phomopsis* sp., *Septoria alni, Ophiovalsa suffusam, Phyllosticta, Mycosphaerella punctiformis, Melanconis alni, Rhabdocline parkeri, Cryptocline* sp., *Leptrostroma* sp. (Bacon *et al.*, 2002).

Detection/Localization of Endophytes within Plant Tissue

Endophytic bacteria in plants may be visualized by using the tetrazolium stains (Figure 2.4). Surface sterilized plant tissues are incubated overnight in the triphenyl tetrazolium chloride (TPTZ) stain 1.5 g l^{-1} containing 0.625 g l^{-1} malic acid in 0.5M potassium phosphate buffer. The bacterial cells stained dark red to purple can be visualized in free hand or microtome sections under 100x objective (Bacon *et al.*, 2002). Acridine orange is another stain that can be used for the direct counting of endophytes *in situ*, and when compared with epifluorescence microscopy it is also a vital stain. Counterstaining the TPTZ-stained tissue with the aniline blue stain for a minute, removing the aniline blue, and then making a wet mount can enhance both bacterial and fungal endophytes (Bacon *et al.*, 2001). Labeling of the endophytic bacteria with fluorescent probes like green fluorescent fusion protein and monitoring the extent of tissue colonization through fluorescence or confocal microscopy facilitate the tracking of specific target organisms (Compant *et al.*, 2008).

Electron microscopy is a powerful tool to recognize and localize endophytic bacteria in plant tissue. Conventional scanning electron microscopy, cryo-SEM and transmission electron microscopy have been successfully used to detect endophytic bacteria in various plants. Within cross-sections, endophytic bacteria can be distinguished from plant organelles and plant-associated structures by a skilled electron microscopist. This technique has been widely used to detect endophytic bacteria in several plant species. Immunological techniques have been used for a long time to detect plant-associated microorganisms. Antibodies can be raised against proteinous components of bacteria or whole bacterial cells by injecting them into rabbits or mice. For visualization, the antibodies themselves or secondary antibodies raised against primary antibodies are coupled with fluorochrome such as fluorescein isothiocyanate (FITC) or colloidal gold. Endophytic bacteria can also be detected and identified within plant tissue by fluorescent *in situ* hybridization (FISH), a technique that detects specific bacterial DNA or RNA sequences in plant tissue (Amann *et al.*, 1995). Nucleic acid probes are labeled by fluorescent tags such as FITC (flourescein iso-thio cyanate) and detected using a fluorescent microscope or confocal microscope.

For the detection of fungal endophytes, surface sterilized plant material or seeds may be examined with a light microscope for internal hyphae. Fungal endophytes are stained with an aniline blue-lactic acid stain. Another stain used to see endophytic fungi within plant tissue is rose bengal. FISH can also be used for the detection and localization of endophytic fungi.

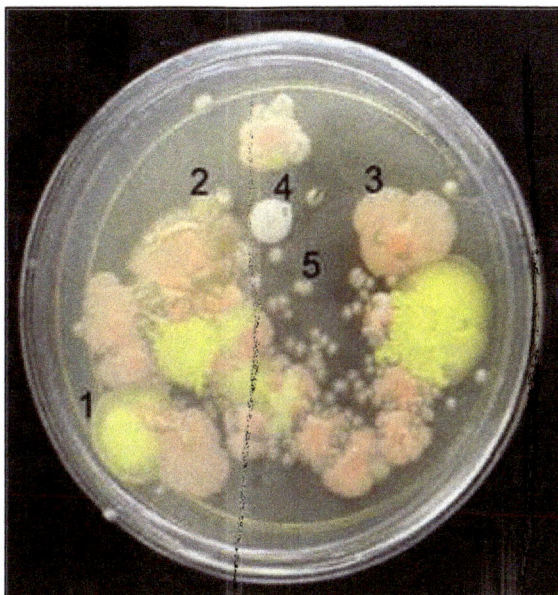

Figure 2.1: Diverse Endophytic Bacteria Derived after the Plating of Tissue Homogenate from Banana

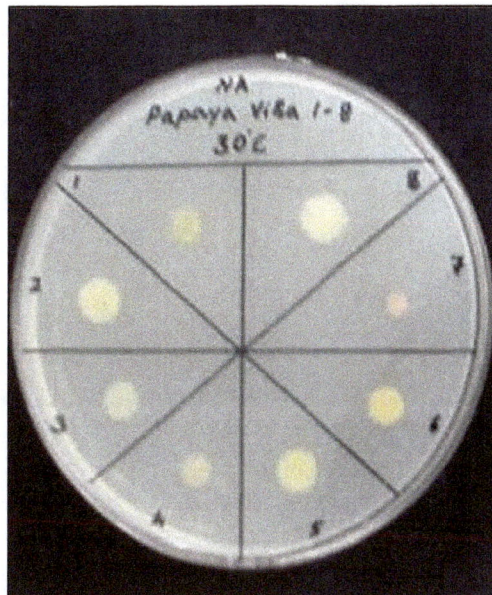

Figure 2.2: Different Endophytic Bacteria Isolated from Shoot Tips of Papaya (1-8 indicate *Microbacterium esteraromaticum, Pantoea ananatis, Enterobacter cloacae, Brevundimonas aurantiaca, Bacillus benzoevorans, Sphingomonas* sp., *Methylobacterium rhodesianum* and *Agrobacterium tumefaciens* in that order).

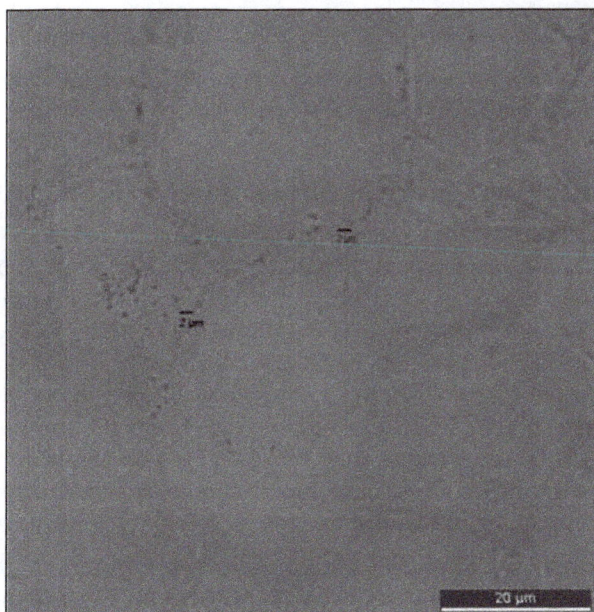

Figure 2.3: Internal Tissue Localization of Endophytic Bacteria in Banana Following TPTZ Staining

Figure 2.4: Promotion of Root Growth and Seedling Vigor in Papaya by Endophytic Bacteria
Microbacterium esteraromaticum (10), *Pantoea ananatis* (2) and *Sphingomonas* sp (6)

Identification of Endophytic Microorganisms

One of the primary aspects after the isolation of endophytic organisms is establishing the identity of the organisms. Once the identity is established, it becomes easy to get information about the functions attributable to the organism if it is an already characterized organism. Endophytic fungi are often identified using a combination of morphological and molecular methods. The morphological features of some fungi are usually medium dependent, and some cultural conditions can affect substantially vegetative and sexual incompatibility. Thus, the morphological character of endophytes should be coupled with the available molecular evidence to enable significant differentiation between closely related species. Morphological examinations are performed by assessing the culture growth, spore production, and characteristics of the spores. To obtain a better understanding of microbial diversity, other approaches, which complement the traditional microbiological procedures, are needed.

The application of molecular biological techniques to detect and identify microorganisms by certain molecular markers is now more and more frequently used to explore the microbial diversity and to analyze the structure of microbial communities. The tools that have been developed for identifying microbes and analyzing their activity can be divided into those based on nucleic acids and other macromolecules and approaches directed at analyzing the activity of complete cells. The nucleic acid-based 18S–28S rDNA ITS region sequences can be used to place organisms into a phylogenetic framework and can be linked to databases providing large deposits of sequences. Endophytic fungi which neither grow nor sporulate in culture can only be detected and identified by other means such as a comparison of ribosomal DNA (rDNA) gene sequences. The sequence-based approach also facilitates the analysis of genetic relatedness and to construct phylogenetic relationship trees.

Identification of bacteria may be facilitated with a combination of cultural characteristics, microscopic data and biochemical characteristics. With the number of bacterial species isolated expanding rapidly, methods based on cultural and biochemical characteristics is becoming cumbersome and they prove to be less dependable. Methods based on differences in the fatty acid composition and the use of Biolog plates involving a series of biochemical tests assisted with computerized data analysis are in vogue now (Leifert and Woodward, 1998). The most reliable method available today is the molecular based approach, which relies on the 16S rRNA gene sequence data. 16S rRNA-gene sequence based approach is a powerful and reliable method for identifying any newly isolated bacteria.

Figure 2.5: Visibly Clean Watermelon Culture (*a*) Revealing Covertly Associated Bacteria in the Medium during the Indexing of Culture Medium Using Bacteriological Media (*b*)

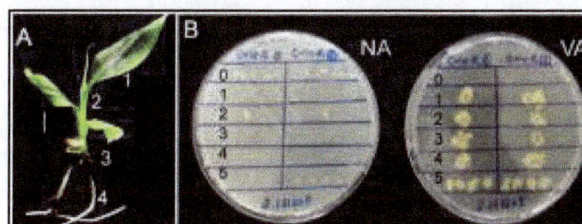

Figure 2.6: Medium-indexed Watermelon Cultures (Top) Showing Bacteria Internally Inside the Plant during Tissue Indexing (Below)

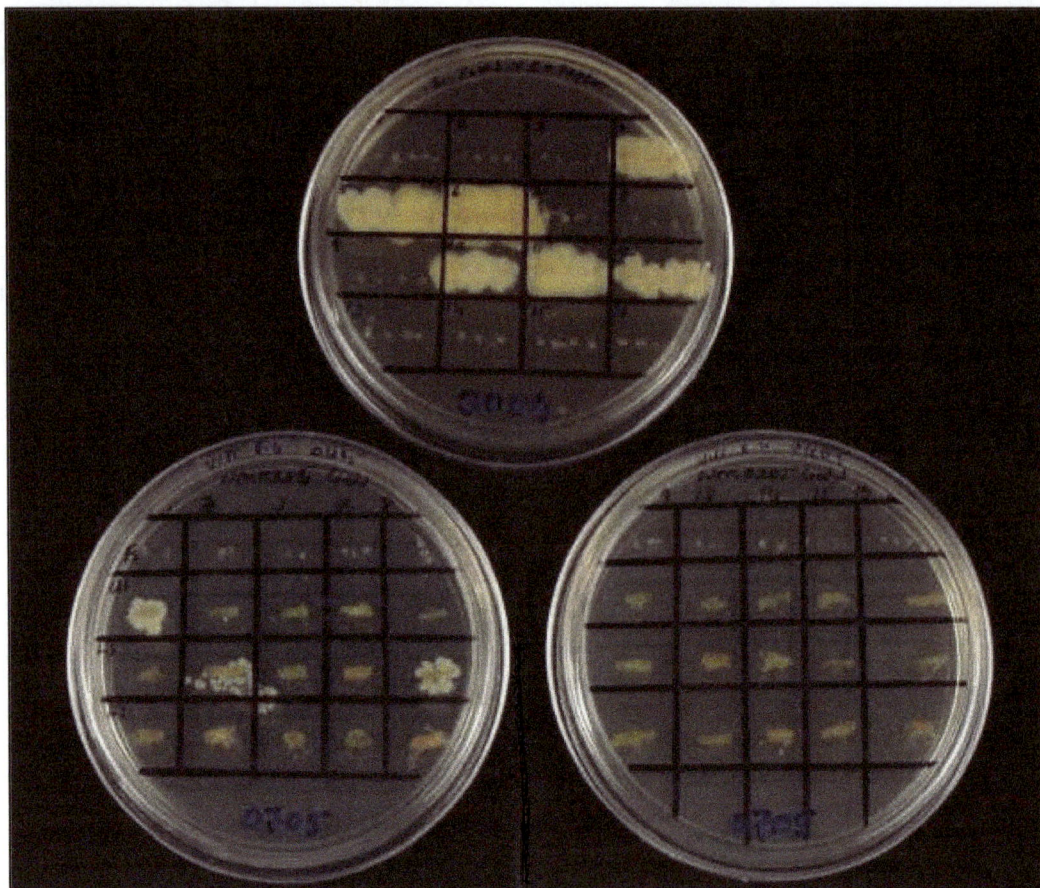

Figure 2.7: Indexing Procedure for the Detection of Covert Endophytic Bacteria in Banana Cultures: (A) An *in vitro* plant showing the parts used in indexing (1, leaf; 2, pseudostem; 3, corm; 4, root); (B) banana stocks indexed for covert endophytic bacteria after nine *in vitro* passages nutrient agar (NA) or Viss-agar (VA) showing covertly associated bacteria in one medium

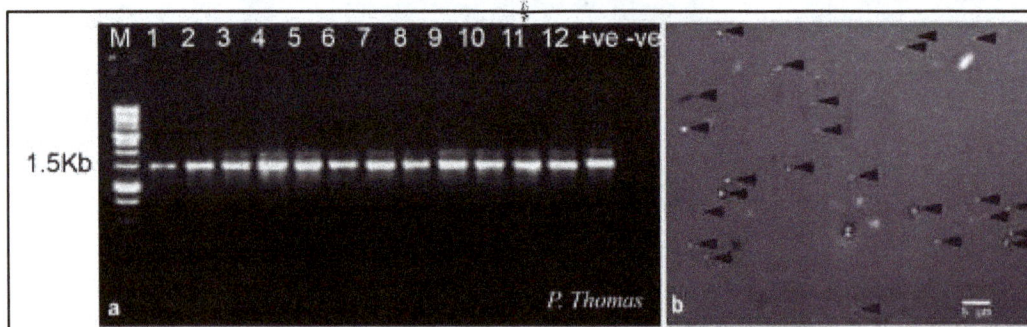

Figure 2.8: Molecular Screening of Index-negative Papaya Cultures Using Bacterial 16S rNA Gene Primers Yielding a Band of Approx. 1.5 Kb indicating the presence of non-culturable bacteria (*a*) and motile bacterial cells detected in the tissue sap under phase contrast microcopy (100x objective) as indicated by arrow heads (*b*)

Molecular based bacterial identification involves the following steps: extraction of genomic DNA from pure bacterial culture, amplification of 16S rDNA using universal bacterial primers, sequencing the purified PCR product and search of public gene bank databases for the homologous sequences and arriving at the identity of the organism based on the extent of sequence homology. It is essential that the culture is a pure one which is achieved through a series of single colony selection involving dilution plating or spotting and two-round of single colony selection after streaking on bacteriological media. Overnight culture or 2-3 days old culture derived from a single colony purified culture is used for DNA extraction. There are standard methods for DNA extraction. The protocol as per Ausubel *et al.* (2005) is described briefly below:

Reagents

TE buffer (Tris: 100mM; EDTA: 1mM; pH: 8.0); 10 per cent SDS; 20 mg/ml proteinase K; 5M Sodium chloride (Dissolve 297.2g l^{-1}; Autoclave the solution); CTAB/NaCl solution (Dissolve 4.1g NaCl in 80 ml water and add 10 g CTAB (Cetrimide) slowly while heating with stirring; heated to 65°C to dissolve the chemicals completely. The final volume is adjusted to 100 ml and then autoclave); 24:1 Chloroform: Isoamyl alcohol (store at 4°C); 25:24:1 Phenol: Chloroform: Isoamyl alcohol (store at 4°C); Isopropanol, 70 per cent Ethanol and RNAase (heat-treated).

Protocol

Bacterial culture is grown in nutrient broth or nutrient agar and the growth from 4-5 colonies is transferred to a microfuge tube with 1.5 ml TE or sterile saline, spin down for 2 min at 14000 rpm and collect the pellet; resuspend the pellet in 567µl of TE buffer by repeated pipetting; add 30 µl of 10 per cent SDS and 3 µl of proteinase K (20 mg ml^{-1}), mix well and incubate at 37°C for 1 hour. After adding 100 µl of NaCl and 80 µl of CTAB/NaCl solution, mix by inverting the tube and incubate at 65°C for 10 min. Extract with equal volume of chloroform: isoamyl alcohol mixture and spin at 12000 rpm for 5 min. Collect the supernatant in a fresh 1.5 ml tube and add RNAase (heat-treated) to a final concentration of 10 ìg/ml and incubate at 37°C for 20-30 min. Add equal volume of phenol: chloroform: isoamyl alcohol (500-600 ul), mix (5 min) and spin down for 5 min (12000 rpm) and transfer the supernatant to a fresh tube. DNA is precipitated by adding 0.6× volume of isopropanol and incubating at ⁻20°C for 20 min. DNA is pelleted by spinning at 12000 rpm for 5 min. Wash the pellet with 1 ml of 70 per cent ethanol and air-dry in laminar airflow (with the lid open). Dissolve the pellet in 100 µl TE buffer and then check the yield and quality of DNA.

PCR amplification of 16S rDNA is carried out in 40 µl reaction using 200 ng total genomic DNA, 10 pmol each of bacterial primers 27F (5′-AGAGTTTGATCCTGGCTCAG-3′) and 1492R/Y (5′-GGYTACCTTGTTACGACTT-3′; Y = C/T) and 50 µM dNTPs. The PCR amplicon is purified using a PCR purification kit, and then single-end sequenced, preferably using 27F primer. Similarity of partial 16S rDNA nucleotide sequences (> 500 bp) with known sequences in the NCBI Genbank database (http://www.ncbi.nlm.nih.gov/) is determined using BLASTn program. The organism is assigned to a species if the sequences are =99 per cent similar to a valid species sequence deposited with NCBI Genbank at the time of analysis or to genus if the species identity is not conclusive but the similarity is =97 per cent (Thomas *et al.*, 2008b). (In earlier publications, cut off levels of 97 per cent for species status and 75 per cent for genus identity used to be prescribed).

The Endophytic Role

Endophytes usually get nutrition and protection from the host plant. In return, they confer profoundly enhanced fitness to the host plants by producing certain functional metabolites. Some endophytes improve the ecological adaptability of hosts by enhancing their tolerance to environmental stresses and resistance to phytopathogens and/or herbivores including some insects feeding on the host plant. Endophyte-infected grasses usually possess an increased tolerance to drought and aluminium toxicity. Furthermore, some endophytes are able to provide the host plant with protection against some nematodes, mammal and insect herbivores as well as bacterial and fungal pathogens. Certain endophytes are capable of enhancing the hosts' allelopathic effects on other species.

It is generally accepted that endophytic microbial communities play an important beneficial role in the physiology of host plants. Plants infected with endophytes are often healthier than endophyte-free ones. This effect may be partly due to the endophytes' production of phytohormones such as indole-3-acetic acid (IAA), cytokinins and other plant growth-promoting substances like vitamins and/or partly owing to the fact that endophytes can enhance the hosts' absorption of nutritional elements such as nitrogen and phosphorus and that they regulate nutritional qualities such as the carbon - nitrogen ratio. Protective effects on endophyte-infected host plants greatly enhance their resistance to unfavorable challenges. The evidence suggests that plants infected with endophytic fungi often have a distinct advantage against biotic and abiotic stress over their endophyte-free counterparts. Beneficial features have been offered in infected plants, including drought acclimatisation, improved resistance to insect pests and herbivores, increased competitiveness, enhanced tolerance to stressful factors such as heavy metal presence, low pH, high salinity and microbial infections. Endophyte-infected plants also gain protection from herbivores and pathogens due to the bioactive secondary metabolites which endophytes generate in plant tissue. An increasing number of antimicrobial metabolites biosynthesized by endophytic microorganisms such as alkaloidal mycotoxins and antibiotics have been detected and isolated.

Exploitation of Endophytes in Plant Growth Promotion

Endophytic plant growth promoting bacteria (PGPB) are documented in several crops. Growth promotion may be facilitated by a variety of microbial processes such as nutrient solubilization, fixation of atmospheric nitrogen, production of plant growth hormones and warding off hazardous organisms (Hallmann 2001; Compant *et al.*, 2005). It would warrant extensive investigations to elucidate the mechanism of action. The ability to colonize roots and to promote root growth are two important considerations which determine the efficacy of an endophyte in crop growth, yield enhancement and for disease control and there is an emphasis on selection of plant-beneficial bacteria that are rhizosphere competent or that colonize the root system effectively (Compant *et al.*, 2005). *Pantoea, Sphingomonas* and *Microbacterium* isolated from papaya showed significant root and shoot growth promotion effect in papaya seedlings (Figure 2.5). The feasibility of seed inoculation with endophytes is an added advantage for bacterization as seed treatment is simple, inexpensive, rapid, reliable, and practically and economically viable in introducing bacteria in crop production or biological control.

Exploitation of Endophytes in Alleviation of Biotic Stress

There are several instances of biocontrol of pathogens and pests by the endophytic bacteria and fungi, notably those involving the interactions between fungi and grazing grasses from temperate countries (Azevedo *et al.*, 2000). The mechanisms by which endophytic fungi control insect attacks include toxin production as well as the influence of these compounds on plant and livestock. Entomopathogenic endophytic fungi also serve for insect control. Protection of elm trees against the beetle *Physocnemum brevilineum.*by endophytic fungus *Phomopsis oblonga* and protection of the perennial ryegrass *Lolium perenne* against the sod webworm are examples of the importance of microorganisms in controlling insects-pests in agriculture.

Endophytic bacteria as biocontrol agents may work in two ways, (*i*) colonizing internal tissue and occupying the ecological niche required by the pathogen, and (*ii*), colonizing the tissue and producing metabolites which are suppressive to the pathogen or stimulate the plant defense mechanism (Hallmann, 2001; Compant *et al.*, 2005). Preemptive colonization by *Agrobacterium radiobacter* could offer protection against *A. tumefaciens*, and *Pseudomonas fluorescens* against *P. syringae*. Sentilkumar *et al.* (2009) identified endophytic *Paenibacillus* sp. and *Bacillus* sp. as potential candidates for charcoal rot biocontrol as well as plant growth promotion in soybean. Endophytc fungi control the vascular wilt pathogen, *Fusarium oxysporum* in cotton and pea, and *Verticellium albo-atrum* and *Rhizoctonia solani* in potato (Hallmann, 2001).

Control of plant parasitic nematodes is often more complex and difficult compared to fungal and bacterial pathogens. Biocontrol of *Meloidogyne incognita* on cotton by *Pseudomonas fluorescens* and *Brevundimonas vesicularis* has been successful (Hallmann, 2001).

Natural Products from Endophytes

Endophytes are relatively unstudied and potential sources of novel natural products for exploitation in medicine, agriculture and industry. Secondary metabolites, defined as low-molecular-weight compounds not required for growth in pure culture, are produced as an adaptation for specific functions in nature. Endophytes are currently considered to be a wellspring of novel secondary metabolites offering the potential for medical, agricultural, and/or industrial exploitation (Strobel, 2003; Strobel and Daisy, 2003).

Some species of endophytic fungi have been identified as sources of anticancer, antidiabetic, insecticidal and immunosuppressive compounds. Often, endophytes act as a source of antibiotics. Natural products from endophytic microbes inhibit or kill a wide variety of harmful disease-causing agents including phytopathogens, as well as bacteria, fungi, viruses, and protozoans which affect humans and animals. A unique antifungal peptide antimycotic, termed cryptocandin, was isolated and characterized from *Cryptosporiopsis quercina*, a fungus commonly associated with hardwood species in Europe. Wide-spectrum antibiotics are produced by *Streptomyces* sp. strain NRRL 30562, an endophyte *Kennedia nigriscans*.

Paclitaxel and some of its derivatives represent the first major group of anticancer agents, produced by endophytes. Paclitaxel, a highly functionalized diterpenoid, is found in each of the world's yew (*Taxus*) species. This compound is the world's first billion-dollar anticancer drug. The diterpenoid taxol has been approved by FDA as one of the most potent anticancer drugs, but the supply of this drug has been limited for the destructive collection of yew tree, the main source of taxol. Undoubtedly, one of the most surprising findings of endophyte studies is the isolation of taxol-producing endophyte *Taxomyces andreanae*. The finding of *Taxomyces andreanae* provides another alternative approach for taxol production by fermentation technology (Strobel, 2003).

Plant Tissue Cultures and Endophytes

One of the fundamental requisites for plant tissue cultures is asepsis, which literally means freedom from all microorganisms. Microbial contamination is identified as the most serious limiting factor in tissue cultures (Leifert and Woodward, 1998; Leifert and Cassells, 2001). The general impression in tissue culture circles is that microbial contamination occurs due to inefficient surface sterilization at the time of culture initiation or due to faulty sterile practices during culture handling.

The organisms which survive the disinfection treatment are normally believed to express on the tissue culture medium, and only such cultures are generally considered as contaminated.

Endophytes are now increasingly recognized as another source of microbial contamination in plant tissue cultures (Thomas *et al.*, 2008a, 2008b). Their isolation from surface sterilized tissue implies that the disinfection practice followed at the initiation of plant tissue cultures would not eliminate such organisms. Although endophtyes are generally considered to be harmless to the host, under the modified conditions in tissue cultures, they may override the stocks and thus become hazardous. The term 'vitropath' was suggested by Herman (2004) to differentiate microorganisms harmful to *in vitro* cultures from those pathogenic to intact plants.

It is often assumed that the organisms that escaped the initial decontamination treatment would express on the culture medium, thus allowing the identification and exclusion of such stocks. Recent studies have indicated that several bacteria including endophytes could survive in the cultures in covert form in the medium or as tissue-colonizers. The studies have also brought out the frequent introduction of endophytes in tissue cultures and their omnipresence in totally unsuspecting form.

Covertly Surviving Bacteria in the Cultures and Need for Detection Methods

Incorporation of nutrients in the medium or streaking the base of plantlets on bacteriological medium at culture transfer are often practiced to facilitate the detection of organisms that do not normally express on tissue culture medium. Investigating the cause of decline in the performance of *in vitro* stocks of watermelon, Thomas (2004a) attributed the unexplained degeneration of tissue cultures to the association of a series of bacteria which survived in the cultures without any apparent indications of their presence in the medium. The covert survival was elucidated by indexing the cultures, *i.e.*, by transferring a bit of the culture growing medium to one or more enriched bacteriological media and observing the plates for any bacterial growth after a couple of days of incubation at 30-37°C (Figure 2.6). Subsequent testing of the cultures of different crops including grape, banana, papaya, capsicum, eggplant, chrysanthemum, etc. proved the rampant covert or quiescent association of diverse bacteria with such stocks.

Following the above finding, a three-step screening procedure was developed for the detection of such unsuspecting but covert bacteria-harboring cultures (Thomas, 2004b). This procedure involved diligent visual examination of cultures for any inconspicuous growth (step-1), indexing the medium of visually clean cultures using bacteriological media (step-2) and subsequent tissue-indexing (step-3) using split segments from different plant parts. Use of two bacteriological indexing media (BIM) namely nutrient agar (BIM_1) and 523 medium of Viss *et al.* (BIM_2; see Thomas, 2004b) differing in nutrient constituents, pH (7.0 and 6.4 respectively) and gel strength (20 and 10 gl^{-1} agar), post-indexing incubation at two different temperatures (37°C and 25-30°C respectively) and sterility testing of tools prior to use were the other considerations during indexing. (NA and Trypticasein soya agar- TSA are now recommended as BIM1 and BIM2 for better detection of bacteria during indexing; Thomas *et al.*, 2008a). Step-2 indexing of grape, watermelon, papaya, capsicum, eggplant and gerbera cultures revealed bacteria in 0-100 per cent cultures in different batches. Varying proportions of cultures that passed step-2 indexing turned positive during step-3 indexing (Figure 2.7). This screening procedure practiced for two–four cycles allowed reliable scrutiny of plant tissue cultures for freedom from cultivable bacteria at culture initiation or while sanitizing contaminated cultures (Figure 2.8).

Based on the above study, four types of bacterial association were found prevalent in tissue cultures. This included obviously visible growth that could easily be picked up, inconspicuous growth which might be detected if one resorts to careful visual examination against background light/dark background, covert bacteria in the medium which might be brought out through medium-indexing and endophytic bacteria which are detected through indexing of tissue (Thomas, 2004b). Covertly associated bacteria of the last three types pose threat to tissue cultures that they may be handled by the workers as if they are normal healthy stocks. Such stocks serve as the source of inoculum causing gradual spread to more cultures through dissection plates and tools. Some of the organisms tend to display active growth if they enter the cultures of other crops with changes in incubation temperature, pH of the medium or gelling agent used (phytagel *vs* agar), or with the ageing of stocks.

Non-cultivable Endophytes

Estimates from environmental microbiology suggest that a vast spectrum of bacteria are currently not amenable for cultivation and recent studies with endophytes endorse the same (Reiter *et al.*, 2002; Reiter and Sessitsch, 2006). While cultivation-independent studies give an estimate of the extent of community diversity, it warrants cultivation independent molecular techniques to study such non-cultivable organisms. The approach followed in environmental microbiology to study such non-cultivable organisms relies on PCR amplification of a gene of choice like 16S rRNA gene from the soil DNA, cloning and identification of distinct clones through ribo-typing and the sequencing of distinct clones (Figure 2.9). Direct application of cultivation independent approach using 16S rRNA gene mediated approach has often been met with interference from plant chloroplast and mitochondrial sequences (Thomas and Soly, 2009), warranting modified approaches to study the endophytic microbial diversity. Further, it is warranted that the organisms are brought to cultivation to facilitate their exploitation in agriculture or fermentation-based applications and in this respect use of host tissue extract has proved beneficial to activate some of the non-cultivable organisms to cultivation. It was also observed that several originally non-cultivable organisms were brought to cultivation during normal tissue culture operations.

The capability of colonizing internal host tissues has made endophytes valuable for agriculture as a tool to improve crop performance. An understanding of the endophytic organisms associated with different crops assumes importance in agriculture, human health and microbial ecology taking into account the potential significance of such organisms in plant growth promotion, protection against biotic and abiotic stresses, as contaminants and growth modifiers in plant tissue cultures, sources of novel biomolecules and agents in bioremediation and determinants of soil and environmental health.

Of the nearly 300,000 plant species which exist on earth, each individual plant is host to one or more endophytes. Only a few of these plants have ever been completely studied relative to their endophytic biology. Consequently, the opportunity to find new and interesting endophytic microorganisms among myriads of plants in different settings and ecosystems is great. It is important to explore other plant species for the associated and uncommon organisms, particularly long duration crops, which have the advantage of perennial colonization once inoculated.

Acknowledgements

I thank Ms. Aparna Shekar, SRF, for her help during the compilation of this chapter.

References

Amann, R.I., Ludwig, W. and Schleifer, K.H., 1995. Phylogenic identification and *in situ* detection of individual microbial cells without cultivation. *Microbiology and Molecualr Biology Reviews*, 59: 143–169.

Ausubel, F.M., Brent, R., Kingston, R.E., Moore, D.D., Seidman, J.G., Smith, J.A. and Struhl, K., 2005. *Current Protocols in Molecular Biology* Vol. 1. John Wiley and Sons, Inc., New York, pp. 2.4.1–2.4.2.

Azevedo, J.L., Maccheroni, W.Jr., Pereira, J.O. and Araújo, W.L., 2000. Endophytic microorganisms: A review on insect control and recent advances on tropical plants. *Electronic Journal of Biotechnology* 3: 40–65 (http://bioline.utsc.utoronto.ca/archive/00000242/).

Bacon, C.W. and Hinton, D.M., 2006. Bacterial endophytes: The endophytic niche, its occupants and its utility. In: *Plant Associated Bacteria*, (Ed.) S.S. Gnanamanickam Springer, New Delhi, pp. 155–194.

Bacon, C.W., Glenn, A.E. and Hinton, D.M., 2002. Isolation, in planta detection and culture of endophytic bacteria and fungi. In: *Manual of Environmental Microbiology, 2nd edn.*, (Eds.) C.J. Hurst, R.L. Crawford, M.J. McInerney, G.R. Knudsen and L.D. Stetzenbach. ASM Press, Washington, D.C., pp. 543–653.

Bell,C.R., Dickie, G.A., Harvey, W.L.G. and Chan, J.W.Y.F., 1995. Endophytic bacteria in grapevine. *Canadian Journal of Microbiology*, 41: 46–63.

Chelius, M.K. and Triplett, E.W., 2001. The diversity of archaea and bacteria in association with the roots of *Zea mays* L. *Microbial Ecology*, 41: 252–263.

Compant, S., Duffy, B., Nowak, J., Clément, C. and Ait Barka, E., 2005. Use of plant growth-promoting bacteria for biocontrol of plant diseases: Principles, mechanisms of action, and future prospects. *Applied and Environmental Microbiology*, 71: 4951–4959.

Compant, S., Kaplan, H., Sessitsch, A., Nowak, J., Ait Barka, E. and Clément, C., 2008. Endophytic colonization of *Vitis vinifera* L. by *Burkholderia phytofirmans* strain PsJN: from the rhizosphere to inflorescence tissues. *FEMS Microbiology Ecology*, 63: 84–93.

Faeth, S.H. and Fagan, W. 2002. Fungal endophytes: Common host plant symbionts but uncommon mutualists. *Integrative and Comparative Biology*, 42: 360–368.

Hallmann, J., 2001. Plant interactions with endophytic bacteria. In: *Biotic Interactions in Plant-Pathogen Associations*, (Eds.) M.J. Jeger and N.J. Spence. CABI Publishing, Wallingford, Oxon, pp 87–119.

Hallmann, J., Quadt-Hallmann, A., Mahaffee, W.F. and Kloepper, J.W., 1997. Bacterial endophytes in agricultural crops. *Canadian Journal of Microbiology*, 43: 895–914.

Herman, E.B., 2004. *Recent Advances in Plant Tissue Culture. VIII. Microbial Contaminants in Plant Tissue Cultures: Solutions and Opportunities 1996–2003*. Agritech Consultants, Inc., Shrub Oak, USA.

Leifert, C. and Cassells, A.C., 2001. Microbial hazards in plant tissue and cell cultures. *In vitro Cellular and Developmental Biology Plant*, 37: 133–138.

Leifert, C. and Woodward, S., 1998. Laboratory contamination management: the requirement for microbiological quality assurance. *Plant Cell Tissue and Organ Culture*, 52: 83–88.

Mano, H. and Morisaki, H., 2008. Endophytic bacteria in the rice plant. *Microbes and Environment*, 23: 109–117.

McInroy, J.A. and Kloepper, J.W., 1995. Survey of indigenous bacterial endophytes from cotton and sweet corn. *Plant and Soil*, 173: 337–342.

McInroy, J.A. and Kloepper, J.W., 1995. Population dynamics of endophytic bacteria in field-grown sweet corn and cotton. *Canadian Journal of Microbiology*, 41: 895–901.

Panicker, B., Thomas, P., Janakiram, T., Venugopalan, R. and Sathyanarayana, B.N., 2007. Influence of cytokinin levels on in vitro propagation of shy suckering chrysanthemum 'Arka Swarna' and activation of endophytic bacteria. *In vitro Cellular Developmental Biology Plant*, 43: 614–622.

Reiter, B. and Sessitsch, A., 2006. Bacterial endophytes of the wildflower *Crocus albiflorus* analyzed by characterization of isolates and by a cultivation-independent approach. *Canadian Journal of Microbiology*, 52: 140–149.

Reiter, B., Pfeifer, U., Schwab, H. and Sessitsch, A., 2002. Response of endophytic bacterial communities in potato plants to infection with *Erwinia carotovora* subsp. *atroseptica*. *Applied and Environmental Microbiology*, 68: 2261–2268.

Rosenblueth, M. and Martinez-Romero, E., 2006. Bacterial endophytes and their interactions with hosts. *Molecular Plant-Microbe Interactions*, 19: 827–837.

Ryan, R., Germaine, K., Franks, A., Ryan, D.J. and Dowling, D.N., 2008. Bacterial endophytes: Recent development and applications. *FEMS Microbiolgy Letters*, 278: 1–9.

Senthilkumar, M., Swarnalakshmi, K., Govindasamy, V., Lee, Y.K. and Annapurna, K., 2009. Biocontrol potential of soybean bacterial endophytes against charcoal rot fungus, *Rhizoctonia bataticola*. *Current Microbiology*, 58: 288–293.

Strobel, G.A., 2003. Endophytes as sources of bioactive products. *Microbes and Infection*, 5: 535–544.

Strobel, G.A. and Daisy, B., 2003. Bioprospecting for microbial endophytes and their natural products. *Microbiology and Molecular Biology Reviews*, 67: 491–502.

Sturz, A.V., Christie, B.R. and Nowak, J., 2000. Bacterial endophytes: Potential role in developing sustainable systems of crop production. *Critical Reviews in Plant Science*, 19: 1–30.

Thomas, P., 2004a. *In vitro* decline in plant cultures: detection of a legion of covert bacteria as the cause for degeneration of long-term micropropagated triploid watermelon cultures. *Plant Cell Tissue and Organ Culture*, 77: 173–179.

Thomas, P., 2004b. A three-step screening procedure for detection of covert and endophytic bacteria in plant tissue cultures. *Current Science*, 87: 67–72.

Thomas, P. and Soly, T.A., 2009. Endophytic bacteria associated with growing shoot tips of banana (*Musa* sp.) cv. Grand Naine and the affinity of endophytes to the host. *Microbial Ecology* DOI: 10.1007/s00248–009–9559–z.

Thomas, P., Kumari, S., Swarna, G.K. and Gowda, T.K.S., 2007a. Papaya shoot tip associated endophytic bacteria isolated from *in vitro* cultures and host-endophyte interaction *in vitro* and *in vivo*. *Canadian Journal Microbiology*, 53: 380–390.

Thomas, P., Kumari, S., Swarna, G.K., Prakash, D.P. and Dinesh, M.R., 2007b. Ubiquitous presence of fastidious endophytic bacteria in field shoots and index-negative apparently clean shoot-tip cultures of papaya. *Plant Cell Reports*, 26: 1491–1499.

Thomas, P., Swarna, G.K. and Patil, P., 2008a. Ubiquitous presence of normally non-cultivable endophytic bacteria in field shoot-tips of banana and their gradual activation to quiescent cultivable form in tissue cultures. *Plant Cell Tissue Organ Culture,* 93: 39–54.

Thomas, P., Swarna, G.K., Roy, P.K. and Prakash, P., 2008b. Identification of cultivable and originally non-culturable endophytic bacteria isolated from shoot tip cultures of banana cv. Grand Naine. *Plant Cell Tissue and Organ Culture,* 93: 55–63.

Vogl, A.E., 1898. Mehl und die anderen mehlprodukte der cerealien und leguminosen. *Nahrungsm. Unters. Hyg. Warenk,* 12: 25–29.

White, D.A., 2009. *Taxonomy of Life* (http://www.ontarioprofessionals.com/weird3.htm).

2012, Nutri-Horticulture
Editor: Professor K.V. Peter
Published by: DAYA PUBLISHING HOUSE, NEW DELHI

Pages 25–45

Chapter 3

Genetics of Cytoplasmic Male Sterility as Affected by Temperature

B.V. Tembhune[1] and R.L. Chavan[2]

Department of Genetics and Plant Breeding, College of Agriculture,
University of Agricultural Sciences, Raichur – 584 101, INDIA
E-mail: [1]bvtembhurne@gmail.com; [2]rajuchavanuasr@gmail.com

Cytoplasmic male sterility (CMS) refers to the inability of plants to produce functional pollen, one of the most important traits in crop breeding, has been used for commercial seed production by F1 hybrid cultivars in vegetable crops. CMS is maternally inherited male sterility which is common in various plant species. In some examples of CMS, floral organs identity is unperturbed, but the anther tissue degenerate by process of programmed cell death (PCD) or necrotic cell death. In addition abiotic stress dominantly affects male reproductive development, in particular high temperature stress cause male sterility in many plant species. This type of injury relates to premature progression of early development programs in anthers and includes proliferation arrest, degradation of anther wall and progressive to meiosis in pollen mother cell (PMCs) all of which requires comprehensive alterations in transcription. Ultrastructural and morphometric analysis clearly showed that mitochondria in both tapetum and PMC are seriously affected in CMS plants, which is reflected in changes of their number, size and structure. The results of RFLP studies and of hybridizations with specific mitochondrial probes revealed, that the organization of several genes such as *atpA, atp6, atp9, coxI, coxII, coxIII* and *cob* is different in sterile plants in comparison with their fertile counterparts.

Male reproductive development might be more sensitive to environmental stresses than female reproductive development and vegetative growth (Sakata and Higashitani, 2008). High temperatures also reduce yields in many crops, but the genetics and physiology of heat tolerance in reproductive tissues have received comparatively little attention (Hall 1992). Temperature increases of 4°C are projected over the next few decades, potentially reducing the reproductive fitness of plants in both natural and agro ecosystems, especially in tropical and sub tropical regions. The mineral nutrient, light and temperature condition influenced the sex expression in hermaphrodite, monoecious and dioecious plants (Heslop 1957).

Cytoplasmic male sterility (CMS) has been widely exploited for hybrid seed production of a number of agronomical and horticultural crops (Havey, 2007). Male sterility systems have been created by different mechanism, most of these affect tapetum and pollen development (Yui *et al.*, 2003; Zheng *et al.*, 2003). Unfortunately, additional severe phenotypic alterations that were due to interference with general metabolism and development had precluded its use in agriculture (Napoli *et al.*, 1999; Goetz *et al.*, 2001).

Hybrid seed production of hermaphrodite or monoecious plants requires the maintenance of separate lines of male and female parents. Emasculation must be performed on the seed parents before pollination with the desired pollen (George 1985). Whether by hand or using chemicals, emasculation is expensive, time consuming, and labour intensive, thereby contributing significantly to the high cost of hybrid seeds (Sawhney 2004). However, most of these lines are genic male sterile mutants (GMS), and hence face the problem of maintenance. Normally, a GMS line is maintained by backcrossing with a heterozygous maintainer line, but the progeny produced are 50 per cent fertile and 50 per cent sterile. In addition to the need for a third line, the maintainer line, an extra problem is created of roguing the fertile plants.

Ideally, the male sterile parent should be facultative so that it can be induced to self pollinate when desired, thereby avoiding the maintenance of the male sterile trait in the heterozygous condition. The ability to manipulate the restoration of fertility in CMS lines by environmental control is a desirable approach for the maintenance of these lines. Selfing as well as crossing of the male sterile individual may be achieved through temperature manipulation because temperature alteration may induce a degree of variation in male sterility, ranging from complete to partial (Gulyas *et al.*, 2006). Breeder must be used restorer as well as maintainer lines for various environments to secure a completely sterile female parent but fully fertile hybrids. Hence genetic study of cytoplasmic male sterility as affected by temperature is prerequisite to understand the effect of temperature on CMS and their maintainer and restorer.

Tomato

Tomato (*Lycopersicon esculentum* Mill.) is an example of a species whose productivity in warm summer areas is likely to be adversely affected by even slight temperature increases. Induction of male sterility in the seed parent has been a useful way to circumvent the problems of high emasculation costs in F1 tomato hybrid seed production (Lasa and Bosemark 1994). A number of male sterile tomato lines have been developed and are currently being used with some level of success in hybrid seed production (Dhaliwal *et al.*, 2004). Fruit set in tomatoes is reduced when maximal daily temperatures average above 30°C and minimal daily temperatures average above 21°C (Moore and Thomas 1952). Variable male sterility was encountered in a spontaneous mutant of *Lycopersicon esculentum* Mill. 'San Marzano.' The trait, conditioned by a single recessive gene *vms*, was proved by trisomic and linkage tests to lie on chromosome 8, probably in the longer arm between the markers *bu* and *dl* (Charles and John 1967). Differences in temperature account for the varied response of *vms* to the environment, minimal temperatures of 30°C in the field and of 32°C in the greenhouse being required to evoke the sterile phenotype.

A temperature dependent, photoperiod independent male sterile mutant, T-4, which is sterile in spring (with some residual fertility) and shows partial restoration of fertility in autumn, has been isolated by Masuda and Ojiewo (2006). Two line hybrid seed production system the risk of contamination is minimal and the selfed seed progeny are easily distinguishable at the seedling stage by narrow leaf markers.

Number of explanations has been offered for the poor reproductive performance of tomatoes at high temperatures. These include reduced or abnormal pollen production, abnormal development of the female reproductive tissues, reduced supply of photosynthates, and poor production of growth regulators in sink tissues which are summarized below.

Effects on Reproductive Structure and Function

Reductions in pollen production at high temperatures have been noted in many crops, including tomatoes. In a comparison of high temperature-resistant and susceptible tomato cultivars, Dane *et al.* (1991) found that prolonged periods of high temperature caused drastic reductions in pollen fertility in most genotypes. Peet and Bartholomew (1996) found that total pollen production and the percentage of normal grains were reduced in plants grown at high night temperatures. In tomatoes, *in vivo* pollen germination is optimal near 20°C (Charles and Harris, 1972), but the decrease in pollen germination from 20-27°C is much less dramatic than the decrease in fruit set. At temperatures above 30°C, flower formation, pollen grain and ovule formation, style elongation, pollen germination, fertilization, and seed formation are all adversely affected and it was not possible to identify a single process as causing poor fruit set (Dane *et al.*, 1991). Microspores were more affected than macrospores by high temperature in tomato (Levy *et al.*, 1978).

Effect of Temperature and Hormonal Imbalance on Male Sterility

Low temperatures were reported to restore male fertility in the stamenless-2 (sl-2) tomato mutant (Singh and Sawhney, 1998). Gomez *et al.* (1999) similarly reported that the stamenless tomato mutant sl had its fertility restored in more than 15 per cent of flowers that developed under low temperature conditions. Thus, the ideal male-sterile seed parents show complete sterility during hybrid seed production and an appropriate way of maintenance when hybrid-seed is not required. Additionally, female fertility must be normal with no defects in physiological or morphological defects that would compromise cross compatibility or fruit appearance. The BC1F3 progeny of the male sterile tomato T-4 mutant had been observed to be fertile in autumn and sterile in spring (Masuda *et al.*, 2000). Vegetative and floral parts (except pistil) of male sterile stamenless sl-2 contain more abscisic acid (ABA) than the normal wild type. The maximum difference in ABA content between sl-2 and normal tissues was in stamens, and the increase in ABA level in sl-2 stamens coincided with first signs of abnormalities in the anthers. Low temperatures restored male fertility in sl-2 and there was a concomitant drop in ABA level in sl-2 leaves and stamens. Hence it is suggested that male sterility in sl-2 is a manifestation of hormonal imbalance involving high ABA, and that low temperature regulation of male sterility is mediated through reduction in ABA content (Singh and Sawhney, 1998).

Influence of Night Temperature on Fertility Restoration

Low temperature has been reported to inhibit pollen germination and pollen tube growth both *in vitro* and *in vivo* (Maestro and Alvarez, 1988). As temperature rose gradually in spring, which is usually the season for hybrid-seed production and fruit-set, the numbers of seeded fruits and seeds per fruit were very minimal. Parthenocarpic fruit-set was almost completely absent. The low winter temperatures that proceeded this season could have hampered the development and viability of the T-4 mutant pollen. On the other hand, the extreme day temperatures towards the end of spring increased flower abortion but hampered parthenocarpic fruit development. Fertility restoration was always better enhanced in autumn than in spring even when the night temperature during pollination was similar in both seasons. There were higher fruit sets and frequencies of seeded fruits when growth chamber night temperatures were set at both 12°C and at 24°C in autumn than when the set-up was repeated in spring. Restoration of pollen fertility in the T-4 mutant is night-temperature-sensitive and that, for fertility restoration, the temperature condition during plant growth was as important as that during and after pollination. The optimum night temperature for fertility restoration lies somewhere above 12°C and below 18°C. A high night temperature (24°C) resulted in a high fruit set but most fruits were parthenocarpic. The environmental modulation of fertility provides a further advantage to produce tomato hybrid seed using the T-4 mutant by a two line system. (Masaharu *et al.*, 2007). This may be associated with the low winter night temperatures between November and March. The male sterile female line can be propagated by growing it under low night temperature conditions (above 12°C and below 18°C) that restore fertility.

Chilli (*Capsicum annuum* L.)

Cytoplasmic male sterility the only source to produce hybrid pepper cultivars in chilli (*Capsicum annuum* L.) was first found by Peterson (1958) in USDA accession PI 164835 which segregated spontaneously. The trait was found to be controlled by major recessive *rf* gene (Gulyas *et al.*, 2006) and it is restored by one major dominant nuclear gene, *restorer-of-fertility* (*Rf*), together with some modifier genes and is affected by temperature (Peterson, 1958; Shifriss and Guri, 1979; Shifriss 1997).

Genetics of Temperature Sensitive CMS in Chilli

Partial restoration of fertility, producing plants simultaneously produce normal and aborted pollen grains, with most grains stuck to the anther wall, even after dehiscence, resulting in low seed set per fruit (Lee *et al.*, 2008b). The trait was visible only in the presence of Paterson's sterile cytoplasm and was controlled by a recessive nuclear gene, partial restoration (*pr*). These partially restored pepper plants were clearly different from unstable sterile plants, which were usually sterile but whose fertility could temporarily be fully restored by low temperature. The partially fertile plants had abnormally low fruit set and fewer than ten seeds per fruit. This *pr* locus is suspected to be either tightly linked to *Rf* locus or is third allele of *Rf* locus (Lee *et al.*, 2008b). More recently, another commercially utilized hot pepper S-cytoplasm of independent origin has been reported from India (Reddy *et al.*, 2002). Fertility restoration in chilli pepper is controlled in a complex manner by some combination of one major *Rf* gene and sterility modifying genes as well as environmental factors such as temperature (Lee *et al.*, 2008a).

Different Types of Phenotypes

Completely malesterile (*rf/rf*), fully restored (*Rf/Rf*), or partially restored (*pr/pr*), according to the method of Lee *et al.* (2008). The partially restored plants (*pr/pr*) were identified by the simultaneous presence of normal and aborted pollen grains that did not separate easily from the anther wall after dehiscence.

Table 3.1 shows the partial restoration was recessive to the full restoration but dominant to the male sterility. If the *pr* and *Rf* genes are allelic, the *pr* allele may be recessive to the *Rf* allele but dominant to the *rf* allele. On the other hand, if the *pr* and *Rf* are different loci, the *Rf* locus may have dominant epistasis to the *pr* locus, because the partial restoration appeared only in cases of crosses with maintainers.

Selfing as well as crossing of the male sterile individual may be achieved through temperature manipulation because temperature alteration may induce a degree of variation in male sterility, ranging from complete to partial (Gulyas *et al.*, 2006). Hence breeder must be use restorer as well as maintainer lines for various environments to secure a completely sterile female parent but fully fertile hybrids.

Table 3.1: Relationships Between the *pr* and the *Rf* or *rf* Genes

Cross Combination[a]	Genotype of Male Plant[b]	Number of Plants Observed (FR:PR:MS)[c]
(S) *pr/pr* x B1	(N) *rf/rf*	0:25:0
(S) *pr/pr* x B2	(N) *rf/rf*	0:24:0
(S) *pr/pr* x C1	(N) *Rf/Rf*	25:0:0
(S) *pr/pr* x C2	(N) *Rf/Rf*	25:0:0

a: (S) *pr/pr* is the partially restored line that has Peterson's sterile cytoplasm. B1 and B2 are maintainer lines. C1 and C2 are restorer lines

b: (N), normal cytoplasm; *Rf*, restorer allele; *rf*, nonrestorer allele

c: FR, fully-restored pepper; PF, partially-restored pepper; MS, malesterile pepper

Effect of Temperature on CMS in Chilli

The CMS in line PI 164835 can be temporarily restored by a decrease in temperature (under 25°C by day and 17°C at night) (Shifriss 1997). As a result, male fertility was identified as having several phenotypes (Lee *et al.*, 2008a). In chilli low night temperature (LNT) condition (night temperature of 10±2°C) cause a decrease in the number of pollen grains and reduction in their germinability, in comparison with pollen of control plants grown under high night temperature condition (20±2°C) and similar day temperature condition (Shaked *et al.*, 2004). Low night temperature cause a reduction in starch concentration 4 day before anthesis and in the concentration of sucrose and reducing sugar (glucose and fructose) at the time of anthesis. Ishikawa *et al.* (2001) investigated the rate of *in vitro* pollen germination at low temperature was using diploid and tetraploid pepper (*Capsicum annuum* cv. Chigusa) plants. Diploid plants showed high sterility levels at low temperature. The pollen germination of tetraploids at 15°C (22.1 per cent) was higher than that of the diploid plants (6.5 per cent) at the same temperature.

Low temperatures (18°C day/15°C night) had more effect on flower and fruit development in *Capsicum annuum* cv. Vinedale than intermediate (23°C/18°C) or high (28°C/23°C) temperatures. Low temperature caused the formation of abnormal petals, stamens and gynoecium (Polowik and Sawhney 1985). However, in some cases observed deform stamens partly carpel like and produced abnormal nonviable pollen, leading to functional male sterility. In the gynoecium, low temperature grown flowers produced larger ovaries than intermediate temperature or high temperature grown flowers inhibiting the style elongation. Both Intermediate temperature and high temperature produced seeded fruits but larger fruits produced under high temperature. Low temperature grown plants yield small, seedless fruits, but produced normal seeded fruits by pollinating low temperature grown flowers with pollen from intermediate temperature or high temperature grown flowers. Stigma receptivity was found to be affected by fluctuations in temperature and photoperiod. Pollen viability decreased with the increase in storage period (Deogirkar *et al.*, 2007). The greatest fruit set was registered by crossing the male sterile genotypes in the month of July (53.0 per cent) and lowest in June (27.82 per cent) (Saxena *et al.*, 2005). The percentage of sterile pollen registered highest at midday on the day of anthesis (Kato 1989).

Exposure to a temperature of 30/25°C (day/night) induced pollen abortion after meiosis but blocked meiosis, resulting in reduced numbers of pollen grains suggested that cytoplasmic male sterility may be stabilized, provided super optimal field temperatures occur during preanthesis. Single gene segregation occurred for structural differences between normal and male sterile anthers (Shifriss and Guri 1979).

Effect of Superoxide Dismulase, Catalase and Peroxidase Activity

Lower Catalase (CAT) and peroxidase (POD) activity in the pollen of the male sterile line and higher O^2 formulation efficiency and malondialdehyde content was observed than those of the maintainer and restorers during the early stage of pollen mother cell (meiosis to pollen maturity) by Zhang and Hou (2005). However, catalase activity of the maintainer was markedly higher than that of the restorer and SOD and POD activities of the restorer and maintainer did not differ because there were cytoplasmic and nuclear differences among them in pollen development.

Brinjal (*Solanum melongena*)

In brinjal (*Solanum melongena*) stigma receptivity was more pronounced in autumn to winter than in the rainy season in male sterile lines (Hazra *et al.*, 2003). In hybrid seed production, induced male sterility is potentially useful for pollination control, thus reducing the costs and labour involved when manual emasculation of male flowers is done. In Northern India the main flowering season for *S. melongena* occurs during the warmer months of the year (May to November). Normal microsporogenesis followed by flowering and seed set was observed during warm seasons. During off-season flowering (winter) meiosis was abnormal, resulting in sterility and poor or no fruit and seed set due to temperature stress (Karihaloo 1991). Khan and Isshiki (2008) developed the male sterile brinjal utilizing the cytoplasm of *Solanum virginianum* and a biparental transmission of chloroplast DNA in backcrossing. Saito *et al.* (2009) identified the novel source of cytoplasmic male sterility and a fertility restoration gene in eggplant lines. Cytoplasmic male sterility caused by the *S. grandifolium* cytoplasm was stable under a range of environmental conditions. The fertility restoration is controlled by a single dominant nuclear gene (*Rf*).

Low temperature stress during the cool season causes gradual loss of pollen fertility and leads to the development of seedless fruit, but female fertility is not similarly affected. The temperature induced male sterility is transient, and full pollen fertility is regained with normal seed development, as temperature conditions improve (Nothmann and Koller 1975).

Onion (*Allium cepa*)

Presently the genic cytoplasmic male sterility used, worldwide, in onion for commercial exploitation of heterosis was originally derived from Italian Red 13-53. The second source of CMS (T-cytoplasm) in onion was discovered in the French cultivar Jaune paille des venus. This CMS line was found to be different than that from Italian Red 13-53, as three independent segregating restorer loci were identified in this line, responsible for its complex inheritance (Pathak 1997). This male sterility is influenced by the environmental factors, leading to occasional fertile plants in the population. Another male sterility was located in a local cultivar Nasik White Globe. These new lines showed stability in expression of male sterility as it was unaffected by temperature fluctuations.

Genetics of Temperature Sensitive CMS in Onion

Jones and Clarke (1943) established in onion that male sterility is conditioned by the interaction of the male sterile (S) cytoplasm with the homozygous recessive genotype at a single male fertility restoration locus (*Ms*) in the nucleus, written in onion as S *msms*. A dominant allele at *Ms* conditions male fertility in plants possessing S cytoplasm (S Ms–). Plants possessing the normal (N) male fertile cytoplasm are male fertile regardless of the genotype at *Ms*. They described the technique used today to exploit cytoplasmic genic male sterility (CMS) for the production of hybrid seed. A male sterile (S *msms*) inbred line [termed the A line by Jones and Clarke (1943)] is seed propagated by growing it in isolation with a male fertile maintainer (B) line that is N cytoplasmic and homozygous recessive at the nuclear male fertility restoration locus (N *msms*). All seed harvested off of the A line is male sterile (S *msms*); all seed harvested off of the maintainer line is male fertile (N *msrns*). The A line then becomes the female parent for hybrid seed production. This system to seed propagate CMS inbred lines has been widely used for hybrid production of many crops.

CMS in Japanese bunching onion is conditioned by the male sterile cytoplasm and recessive alleles at two nuclear restorer loci (Moue and Uehara 1985). Onion is an example of continued cytoplasmic uniformity in a major crop plant. The majority of hybrid onion seed is produced using S cytoplasm (Havey 2000), which traces back to a single onion plant identified in Davis, CA, in 1925 (Jones and Emsweller 1936). An important consideration for hybrid development using CMS systems is the requirement, or not, for male fertility restoration. For vegetable, fruit, or forage crops, nuclear restoration of male fertility in the hybrid is not necessary. This simplifies the production of hybrids because effort can concentrate on maintainer line development, without concern whether the pollinator restores male fertility in the hybrid. For crops with seeds as the economically important product, such as canola, sunflower, or maize, one or both of the hybrid's parents must bring in male fertility restoration factors or the male sterile hybrid seed must be blended with male fertile hybrid seed. Two sources of CMS are commercially used to produce hybrid onion. The most widely used source of CMS is S cytoplasm, as described by Jones and Clarke (1943). T cytoplasm is a second source of CMS (Berninger, 1965) used to produce hybrids in Europe and Japan (Havey, 2000).

Effect of Temperature on CMS in Onion

The vast majority of onion hybrids are produced using S cytoplasm because it was the first source of onion CMS released in 1951 and was made available both in long and short day germplasms (Goldman *et al.*, 2000). Some lines were predominantly male sterile at 14°C became predominantly male fertile above 17°C. The effect of T cytoplasm was also influenced by temperature (Meer 1978). GA sprayed during bolting proved an effective gametocide, giving complete male sterility with good seed production. An increase in temperature and in length of photoperiod during the formation and ripening of pollen stimulated the production of fertile pollen and led to a reduction in the percentage of plants with male sterility (Khaisin 1975). Barham and Munger (1950) reported the effect of temperature on male sterile onion that the number of stainable pollen grains increase as temperature increased from 5° to 85°F but it decreases with further increase of temperature. Similar effect of temperature were obtained by Meer and Bennekom (1969) on male sterile onion.

Chive (*Allium schoenoprasum* L.)

There are two cytoplasmic male sterility (CMS) systems CMS1 and CMS2 in chives (*Allium schoenoprasum* L.), which can be employed in hybrid breeding.

Genetics of Temperature Sensitive CMS in Chive

Temperature sensitive CMS is controlled by a dominant nuclear gene T; this acts as a restorer gene at high temperatures and is ineffective at low ones, and acts in addition to the restorer gene X/x (Tatlioglu 1987; Turan 1996). In cytoplasmically male sterility some plants showed consistently male sterility, but others were male sterile under 17°C/17°C (day/night) and 20°C/17°C but male fertile at a constant 24°C. The same plants showed reversible temperature induced fertility hence CMS is controlled by genetic component (Tatlioglu 1985). CMS in chive shows unique sensitivity to tetracycline, which restores male fertility (Tatlioglu 1986) and tetracycline susceptibility is conditioned by recessive alleles at a single locus (*aa*) (Tatlioglu and Wricke 1988). The a gene decreased the synthesis of the protein without needing tetracycline treatment, whereas the T gene decreased the amount of protein only after temperature treatment of 24°C.

CMS plants have cytoplasmic factor S for male sterility and recessive alleles of the nuclear restorer gene X for restoration of fertility. Reversion of male sterility to fertility at high temperature is conditioned by the dominant nuclear gene T, and similar reversion after treatment with tetracycline by the recessive allele at another nuclear locus (Aa). Male fertility mediated by these 2 factors is reversible (Ruge *et al.*, 1992). Ultrathin sectioning of anthers from maintainer and restorer genotypes showed a degeneration of the tapetum after meiosis and tetrad formation. During the succeeding pollen mitosis, the uninucleate meiocytes became binucleate pollen grains. Meiocyte genesis was identical in male sterile and fertile plants up to the tetrad stage, but in the male sterile types the tapetum did not degenerate until after meiocyte abortion, suggesting that the abortion may be associated with failure of the tapetum to provide essential nutrients for the meiocytes. Studies of genotypes sensitive to tetracycline and to temperature indicated that flowers become completely male fertile after treatment with (respectively) tetracycline or a constant temperature of 24°C. After 10-14 days' treatment with the appropriate factor, meiocyte genesis in sensitive genotypes showed the normal pattern found in fertile genotypes.

Effect of Temperature on CMS in Chive

The microsporogenesis of CMS plants ((S)xx) and fertile plants ((N)xx or (S)X.) is similar up to the tetrad stage. During the tetrad stage, microspores of CMS plants start to die off. The effect of the (S) cytoplasm on microsporogenesis is overcome by the restorer gene X, the temperature gene T and by tetracycline gene a. Whereas the X gene acts independently of environmental conditions, the gene T is expressed at constantly high temperature of 24°C, and the tetracycline gene a acts after tetracycline treatment. After temperature and tetracycline treatments, respectively, the sensitive genotypes need 10-14 days to change from male sterile into male fertile. This period appears to be associated with the influence of the defined nuclear genes on mitochondrial transcription and translation (Ruge *et al.*, 1993). Genetic control of temperature sensitivity and tetracycline sensitivity of CMS are independent; the recessive allele controlling the latter trait is designated a (Tatlioglu and Wricke 1988). A temperature of 24°C and/or treatment with 2000 p.p.m. of tetracycline induced some male fertility in some male sterile plants. If no further treatment was given, such plants reverted to male sterility (Tatlioglu 1984).

Temperature sensitive genotypes [(S1)xxT_] are able to produce pollen at higher temperatures, and should therefore be excluded from hybrid breeding to avoid self pollination of the maternal parent (Engelke *et al.*, 2004). Allele t, is responsible for the temperature insensitivity, and designated CMS1 maintainer genotypes, leading to the production of temperature insensitive male sterile lines. The incidence of CMS2 maintainers in the German varieties examined was nearly four times lower than CMS1 maintainers. Non restoring allele st2 involved in the CMS2 system. A chimeric mitochondrial gene configuration, mainly derived from sequences associated with the essential genes atp 9 and atp 6, was isolated from the sterility inducing cytoplasm of the CMS1 system in chives (Engelke and Tatlioglu 2004). Schneider and Tatlioglu (1996) investigated the effects of the temperature sensitivity gene T and the tetracycline sensitivity gene a on the CMS specific 18-kDa mitochondrial protein. The CMS specific 18-kDa protein was produced in higher amounts in the mitochondria of tetracycline insensitive CMS genotypes than in tetracycline sensitive CMS genotypes, suggested that there is a critical level of the specific mitochondrial protein required for CMS.

Radish (*Raphanus sativus*)

Recently, the pentatricopeptide repeat protein gene *orf687* was shown to restore fertility in cytoplasmic male sterile Kosena radish (*Raphanus sativus* cv Kosena) (Koizuka *et al.*, 2003). The expression of male sterility is not affected by seasonal influences but reversible temperature effect observed in some population. Most male sterile plants occurred at 10, 14 and 26°C and male fertile plants at 17 and 20°C. Male sterile material at temperatures ranging 17-18°C gave the highest frequency of MS plants (Nieuwhof 1985). Male sterility is probably determined by one dominant and 2 recessive independently acting genes, but minor genes may also be involved (Nieuwhof 1990).

Carrot (*Daucu carota*)

For carrot, two sources of CMS are used to produce hybrid seed. The predominant CMS is the petaloid male sterile cytoplasm, in which the anthers are replaced by a whorl of petals (Eisa and Wallace 1969). The second source of CMS is brown anther, in which complete flowers produce shrivelled anthers with no pollen. Male fertility restoration for these sources of CMS is complexly inherited with up to five loci affecting this trait (Peterson and Simon 1986). The pentaloid CMS is generally preferred because of less frequent reversion to male fertility; however seed yields on the brown anther CMS are generally higher.

Genetics of Temperature Sensitive CMS in Carrot

CMS is conditioned by dominant and recessive genes (Ms1Ms1 ms2ms2) in male sterile S cytoplasm in brown type carrot and three dominant genes (Ms3Ms3 Ms4Ms4 Ms5Ms5) in S cytoplasm of the petaloid type. In both types of CMS, two complementary dominant genes restore male fertility to different degrees (Timin, and Dobrutskaya 1981). Increased temperatures reduced the frequency of plants with CMS and increased the frequency of male fertile plants. Hybrids with CMS of the petaloid type either had all plants with CMS or gave the same frequency of plants with CMS under all cultivation conditions. In artificial environments (higher daytime temperatures and greater variation in temperature per 24 hours), the material showed an increase in male fertility; male sterile plants constituted on average 57 per cent of the total. Line 1025A was the most stable of those studied, while line 1029A reacted most markedly to artificial environments. The S1 from partially fertile lines was 95 per cent male sterile in the field, suggested that the variation in pollen sterility was phenotypic in character (Michalik 1978).

Chinese Cabbage (*Brassica campestris*)

Leaf head of Chinese cabbage (*Brassica campestris* L. ssp. pekinensis) is the main harvesting organ, therefore cytoplasmic male sterility is an effective method in breeding varieties.

Effect of Chemical Composition on Male Sterility

The anther proline, soluble sugar and soluble protein contents, and the alpha-amylase and beta-amylase activities are much lower in the CMS line than in the maintainer line, indicating that these parameters were closely related to microspore development (Shi *et al.*, 2004). There is a similarity for free amino acid content of the CMS line and that of the maintainer line, suggested that this parameter only had a slight effect on microspore development. A new *Brassica chinensis* germplasm, ZS6 (A) 10, was created by asymmetric cell fusion with progeny plants identified in both field and laboratory (Hou *et al.*, 2004). The peroxidase isoenzyme of the radicle and cotyledons and the esterase isoenzyme of the hypocotyl is different with those of the maintainer and the parental sterile line. The soluble protein content and IAA oxidase activity show differences in two lines evaluated. Peroxidase POD activity differed between the two lines and at different temperature treatments. The increase of peroxidase (POD) activity in POL CMS line A3 at 8/4 degrees C for 6 days is probably related to fertility changes. In the sterile line, POD activity at the lower temperature treatment is greater than that at the higher temperature treatment. No differences in esterase banding patterns were noted between the two lines (Yang and Cao 1998).

Effect of Atmospheric Temperature on Male Sterility

Three beta-EST isoenzymes are closely related to the sterility changeover. The beta-EST zymograms changed into maintainer line's (B) during sterile stamen changed into fertile in male sterile line (A) after low temperature treatment, while fertile stamen in male sterile line (A) came back into sterile, the beta-EST zymograms came back into the natural male sterile line's (A) again (Zhang *et al.*, 2008a). The male sterile lines of Chinese cabbage induced to flower at 5°C for 8 days to obtain trace pollens and the relative vigor (Zhang *et al.*, 2008b).

Chinese cabbages with POL cytoplasmic male sterility (CMS) exhibited slightly, moderately and highly sensitive to low temperature at the critical values of the daily lowest temperature 5°C, 8°C and 12°C, respectively (Zhang and Cao 1999). CMS was expressed about 10-12 days after exposure to low temperature. Zhang (1999) evaluated and divided the restoration ability and maintenance of Polima (pol) of non-heading Chinese cabbage CMS into 6 types: complete maintainer lines, complete restorer lines, semi-restorer/maintainer lines, high-maintaining lines, high-restoring lines and unstable maintainer lines. Analysis of segregation ratios in the crosses indicated that pol CMS was conditioned by a pair of dominant genes and a series of minor modifying genes sensitive to low temperature. Restoring genes were dominant over all the categories. The size of floral organs in lines of non-heading Chinese cabbage (*Brassica chinensis*) with polima-type cytoplasmic male sterility (CMS) decreased as the degree of CMS increased; degree of CMS was negatively correlated with stamen length and with stamen length:pistil length ratio. Under temperature conditions which reduced the degree of CMS, produced most of the viable pollen grains, although in highly male sterile plants pollen quantity was low, viability was poor and pollen release was delayed or absent (Yang and Cao 1999).

The male sterile plant produced some pollen under low temperatures; ultrastructurally, but there is some differences between temperature induced pollen and normal fertile pollen at the microspore mother cell or mature pollen stages (Yang and Cao 1997a). Temperature affected fertility before archesporial cell differentiation. A mean daily temperature of 6°C is critical for fertility reversion. The time interval between temperature effect and fertility expression under suitable growth conditions (about 20°C) was 12-16 days, while under natural low temperature conditions it could extend to more than 60-70 days. Fertility could not be restored after application of the low temperature treatment just before flowering (Yang and Cao, 1997b).

Inheritance of Temperature Sensitive Fertility Restoration

The polima cytoplasmic male sterility (pol CMS) three-line and two-line systems have been developed for the production of hybrid seed in *Brassica napus* oilseed rape in China. The pol CMS restorer line FL-204 restores male fertility of hybrid plants in the pol CMS system, but hybrid seed production can only be carried out under autumn sowing in Wuhan in south China under moderate temperatures at flowering. The restorer cannot be used as a male for hybrid seed production in northwestern China (Gansu) under spring sowing conditions, because expression of more or less male sterility is due to high temperatures at flowering.

It was hypothesized that the change of fertility was the result of the interaction between nuclear genes [restoring gene (*Rf*) and temperature sensitive genes (*ts*)] and the cytoplasm. The *Rf* gene in FL-204 was incapable of restoring male fertility of pol CMS lines under spring sowing conditions at Gansu where it is inactivated by the recessive *ts* gene present in FL-204. However, the *ts* gene(s) could be non functional under moderate temperature conditions at flowering which allows full expression of male fertility in FL-204. The recessive *ts* gene(s) can only be expressed in plants containing the polsterile cytoplasm. The male sterile line AB1 was male fertile in autumn and male sterile in summer (Yang *et al.*, 1995). The change in fertility is due to the interaction of cytoplasmic and nuclear gene and environmental factors, probably temperature. Pair of dominant nuclear major genes of the AB1 restorer line masks the effect of the environmentally sensitive genes. It is suggested that the environmentally sensitive genes may be multiple minor genes.

Putative male sterile cybrids of in *B. oleracea* produced by transferring CMS Anand cytoplasm through protoplast fusion showed empty anthers and exhibited stamens and petals varying in size and shape (Cardi and Earle 1995). They reported maternal inheritance of CMS and good female fertility in some lines. In summer rape exposure to day/night temperatures of 30/24°C in a controlled environment for 7 days led to increased pollen production in all of the male sterility based F1 populations in the study. Maximum reversion to male fertility occurred 6-13 days after removal from the high temperature treatment. The restorer lines have only 1 pair of fertility restoring genes (*Rf*) but many temperature sensitive genes (*Ts*) (Yang and Fu 1990). After treating the PolR7 with day/night temperatures of 30°C/24°C for 7-10 days prior to flowering, polypeptides unique to Regent reappeared, while the smaller polypeptides disappeared. PolR7 (TR) produced fertile pollen, while its short stamen filaments resembled those of the male-sterile PolR7 suggested that these changes in protein expression may be causally related to the cytoplasmically male sterile phenotype (Pi *et al.*, 1988).

In CMS under high temperature conditions (32°C/26°C day/night) flower buds remained closed but produced protruding stigmas at maturity. Stamens reduced the size and anthers showed abnormal microsporogenesis. The gynoecia, although normal in appearance but unable to set seed, and ovule development was aberrant (Polowick and Sawhney 1988).

Types of Polima Cytoplasmically Male Sterile Lines

Fu *et al.*, 1990 divided pomila cytoplasmic male sterile line into following three groups according to the sensitivity of their sterility to temperature:

1. High temperature CMS lines
2. Low temperature CMS lines
3. Stable CMS lines

Cabbage (*Brassica oleracea* L.)

In Brassicas, hybrids were historically produced using self-incompatibility (SI). Although the SI system is very effective for hybrid production, some self pollination can occur, allowing competitors to screen for and isolate one, or in some cases both, of the hybrid's parental inbred lines. This has been an impetus for the development of an effective CMS system to produce hybrid Brassicas. Ogura (1968) identified CMS in radish, and this cytoplasm was transferred by backcrossing to *B. oleracea*. Unfortunately the Ogura CMS was associated with cold susceptibility, conditioned by the chloroplast genome. This defect was overcome by several different laboratories after protoplast fusion and organellar sorting to combine the CMS conditioned by the radish mitochondrial genome with cold tolerance conditioned by the *Brassica* chloroplast genome. A cold tolerant form of Ogura CMS was patented by Syngenta and is used to produce hybrid Brassicas.

The main constraint to using Ogura male sterile cytoplasm in Brassica is the induction of leaf yellowing at low temperature and the low seed set (Melo and Giordanow 1994). New types of cytoplasmic male sterility (CMS) in *Brassica oleracea* produced by Card and Earle (1997) in *Brassica oleracea* would be useful for F1 hybrid seed production. The Anand cytoplasm derives from the wild species *B. tournefortii*. Anand mitochondria and the presence of Anand mtDNA fragments were strongly associated with male sterility. The presence of Anand chloroplasts with a *B. oleracea* nucleus did not result in cold temperature chlorosis.

Sugar Beets (*Beta vulgaris*)

The sole source of CMS used to produce hybrid beets was described by Owen in 1945. In beets, male sterility is conditioned by the interaction of the S cytoplasm with recessive alleles at two nuclear loci (xx zz). A second alloplasmic source of CMS from *Beta maritima* has been described (Boutin *et al.*, 1987), but has not been used to date to produce commercial beet hybrids. Partial fertility depended upon some internal condition occurred in plants in which the vegetative generative balance had been shifted towards the generative phase very early as the result of long days and low temperature (Stein 1959). Cortessi (1967) reported that there was an interaction with photoperiod and light intensity which affected the fertility in sugar beet. At 14°C fertility was higher at the 16 hours photoperiod, while at 23°C it was highest at the 24 hours photoperiod than at the16 and 24 hour photoperiods.

Some shoots of Cytoplasmically male sterile sugar beet, subsequently formed fertile pollen when kept for six weeks at 36°C and then transferred to 25°C, the fertility disappeared within three generations. Meristem culture alone was without effect on sterility, but a combination of alternating temperatures (15 hours at 36-37°C; 9 hours at 25°C) for nine weeks plus culture of meristem tissue from shoots developing during that period resulted in the isolation of fully male fertile plants. The possibility of mutation by a gene controlling male sterility is considered, but segregation data obtained on selfing and crossing the converted male fertile plants could not confirm this (Lichter, 1978a). By means of heat treatment (36°C and above) and subsequent meristem culture male fertile plants were obtained from the cytoplasmically male sterile line SL9460M. On the basis of earlier work on other species it is suggested that this new male fertility results from reduction or elimination of particles containing RNA and designated cytoplasmic spherical bodies (Lichter 1978b).

Potato (*Solanum tuberosum*)

Lossl *et al.* (2000) identified the cytoplasm Wy associated with cytoplasmic male sterility by markers in potato. Concilio and Paolini (1986) studied the pollen fertility in relation to temperature in potato clones and suggested that the temperature effects on gametogenesis and meiosis causes sterility.

Cucumber (*Cucumis sativus*)

Male sterility in cucumber involved cytoplasmic and Mendelian factors (Barnes, 1961). Nijs APM den (1984) observed favourable effect of temperature treatment (26°C) on pollen tube growth and fertilization and of pot culture as against outdoor cultivation to overcome sterility barriers in Cucumis species.

Alfalfa (Lucern) (*Medicago sativa* L.)

In lucern an increase in temperature markedly increased pollen fertility in moderately fertile plants while having little or no effect on male sterile plants or on very fertile ones. Depending up on the temperature and genotype there were significant, though slight, differences in the number of ovules per ovary. An increase in temperature resulted in an increase in the number of fertilized ovules per ovary and per flower in the male sterile clones (Blondon *et al.,* 1979). Tereshchenko (1973) reported the degree of pollen sterility in cytoplasmic male sterility is affected by weather conditions (relative humidity and air temperature).

Plantago coronopus

Plantago coronopus is a species of *Plantago* sometimes grown as a leaf vegetable. It is native to Eurasia and North Africa but it can be found elsewhere, including the United States, Australia, and New Zealand as an introduced species. Gynodioecy is a breeding system consisting of male sterile females and hermaphrodites. Within such systems another, often neglected, class of partially male sterility is with intermediate sex expression. These partially male sterile plants constitute a continuous series, linking complete male sterility and hermaphroditism. Temperature appeared to influence sex expression and the degree of male sterility of partially male sterility individuals, but not of male sterile and hermaphroditism individuals (Koelewijn and Damme 1996). At higher temperatures partially male sterile individuals became more male sterile. Difference in cytoplasmic background of the individuals did not influence their response to temperature. It is argued that partial male sterility is a normal feature of gynodioecious species and not an abnormality.

Petunia

Petunia is a widely cultivated genus of flowering plants of South American origin. Seedling and cell culture CMS extracts exhibited a higher sensitivity to high temperature denaturation and the hydrolase activity on mono and triphospho cytosine compounds was significantly higher in CMS than in fertile membranes (Perle *et al.,* 1993).

Cytoplasmic male sterility in petunia was shown to be under plasmon genome control. The male sterile phenotype was variable in that breakdown at different stages of microsporogenesis and gametogenesis was under the control of a multiple genes system. Partial fertility is the effect on microsporogenesis in cytoplasmic male sterile lines, as compared with a normal fertile line, when grown under certain temperature conditions (about 17-25°C) (Izhar 1975). Anthers of normal fertile lines do not get affected by high temperature just prior to the beginning but anthers of CMS are sensitive. Izhar and Frankel (1976) suggested that a single plasmagene is responsible for male sterility due to fertility restoration and asexual transmission of cytoplasmic male sterility through grafts. Variation in cytoplasmic male sterility in petunia may be the result of interaction between a male sterile plasmagene, different numbers of restorer genes and temperature (Izhar 1977). Fertility restoration in petunia can be achieved by 1. A single dominant gene that is not temperature sensitive and 2. A multiple genes system that is temperature sensitive (different amounts of the proper alleles affect breakdown from early meiosis to full fertility) (Izhar 1978).

Conclusion

Cytoplasmic male sterility (CMS) refers to the inability of plants to produce functional pollen, one of the most important traits in crop breeding. CMS is maternally inherited male sterility used for commercial seed production of F1 hybrid cultivars in vegetable crops. Male reproductive development is more sensitive to environmental stresses than female reproductive development and vegetative growth. High temperatures reduce yields in many crops, but the genetics and physiology of heat tolerance in reproductive tissues have received comparatively little attention. Increasing temperature potentially reduced the reproductive fitness of plants in both natural and agro ecosystems, especially in tropical and sub tropical regions. Tapetum and pollen development mechanisms are the most affected male sterility systems. Hybrid seed production of hermaphrodite or monoecious plants requires the maintenance of separate lines of male and female parents. Whether by hand or using chemicals, emasculation is expensive, time consuming, and labour intensive, thereby contributing significantly to the high cost of hybrid seeds. However, most of these lines are genic male sterile mutants (GMS), and hence face the problem of maintenance. The ability to manipulate the restoration of fertility in CMS lines by environmental control is a desirable approach. Breeder must be use restorer as well as maintainer lines for various environments to secure a completely sterile female parent but fully fertile hybrids. Genetic study of cytoplasmic male sterility as affected by temperature is prerequisite to understand the effect of temperature on CMS and their maintainer and restorer.

CMS affected by High and low temperature has been identified in most of the important vegetable crops like tomato, chilli, brinjal, onion, chive, carrot, Chinese cabbage, cabbage, plantago coronopus, petunia, sugar beets, alfalfa (Lucern), potato and cucumber. Selfing as well as crossing of the male sterile individual may be achieved through temperature manipulation because temperature alteration may induce a degree of variation in male sterility, ranging from complete to partial. In tomato low temperature affects the reproductive structure and function, hormonal imbalance, inhibit pollen germination and pollen tube growth and ultimately the male sterile female that restore fertility above 12°C and below 18°C. Partially restored pepper plants were usually sterile but their fertility could temporarily be fully restored by low temperature. In Northern India during off-season flowering (winter) meiosis was abnormal in brinjal, resulting in sterility and poor or no fruit and seed set due to temperature stress. S and T both cytoplasm of some lines in onion were predominantly male sterile at 14°C became male fertile above 17°C. Temperature sensitive genotypes [(S1)xxT_] are able to produce pollen at higher temperatures in chive. In radish most male sterile plants occurred at 10, 14 and 26°C and male fertile plants at 17 and 20°C. Increased temperatures reduced the frequency of plants with CMS and increased the frequency of male fertile plants in carrot. Three beta -EST isoenzymes closely related to the sterility changeover identified in chinese cabbage. These beta-EST zymograms changed into maintainer line's (B) during sterile stamen changed into fertile in male sterile line (A) after low temperature treatment, while fertile stamen in male sterile line (A) came back into sterile, the beta-EST zymograms came back into the natural male sterile line's (A) again. Some shoots of cytoplasmically male sterile sugar beet, subsequently formed fertile pollen when kept for six weeks at 36°C and then transferred to 25°C, the fertility disappeared within three generations. In India temperature fluctuation occurs from region to region which affects the reproductive system of crop plants especially, CMS system which is heart of the economical hybrid seed production. Hence, effective hybrid seed production can be achieved through manipulation of temperature regime by understanding the genetics of CMS as affected by temperature.

References

Barham, W.S. and Munger, H.M., 1950. Stability of male sterility in onion. *Proceeding of American Society of Horticulture Science*, 56: 401–409.

Barnes, W.C. 1961. A male sterile cucumber. *Proceeding of American Society for Horticulture Science*, 77: 415–416.

Berninger, E., 1965. Contribution a letude de La sterilite male de l'oignon (*Alliam cepa L.*). *Annales de I Amelioration des Plantes* 15: 183–199.

Blondon, F., Cambier, B., Dattee, Y. and Guy, P., 1979. Influence of temperature on male and female fertility in lucerne: Controls, male-sterile clones and maintainer clones. *Annales de I Amelioration des Plantes*, 29(1): 89–96.

Boutin, V., Pannenbecker, G., Ecke, W., Schewe, G., Saumitou, L.P. and Jean, R., 1987. Cytoplasmic male sterility and nuclear restorer genes in a natural population of *Beta maritirna*: Genetical and molecular aspects. *Theoretical and Applied Genetics*, 7: 625–629.

Burns, D.R., Scarth, R. and McVetty, P.B.E., 1991. Temperature and genotypic effects on the expression of pol cytoplasmic male sterility in summer rape. *Candian Journal of Plant Science*, 71(3): 655–661.

Cardi, T. and Earle, E.D., 1995. Transfer of CMS 'Anand' cytoplasm from *Brassica rapa* to *B. oleracea* through protoplast fusion. *Cruceiferae Newsletter*, 17: 44–45.

Cardi, T. and Earle, E.D., 1997. Production of new CMS *Brassica oleracea* by transfer of 'Anand' cytoplasm from *B. rapa* through protoplast fusion. *Theoretical and Applied Genetics*, 94(2): 204–212.

Charles, M.R. and E.B. John, E.A., 1967. Temperature-sensitive male-sterile mutant of the tomato. *American Journal of Botany*, 54(5): 601–611.

Charles, W.B. and Harris, R.E., 1972. Tomato fruit-set at high and low temperatures. *Canadian Journal of Plant Science*, 52: 497–506.

Concilio, L. and Paolini, P., 1986. Analysis in pollen fertility in relation to temperature in potato clones (*Solanum tuberosum* sb.sp. *tuberosum*). *Genetica Agraria*, 40(4): 441.

Cortessi, H.A., 1967. Investigations made into male sterile beet. *Euphytica*, 16: 425–432.

Dane, F., Hunter, A.G. and Chambliss, O.L., 1991. Fruit set, pollen fertility and combining ability of selected tomato genotypes under high-temperature field conditions. *Journal of the American Society for Horticultural Science*, 116: 906–910.

Deogirkar, G.V., Kolte, N.M. and Nair, D., 2007. Effect of change in environmental conditions on stigma receptivity and pollen viability in chilli (*Capsicum annuum* L.). *Journal of Soil and Crops*, 17(2): 358–366.

Dhaliwal, M.S., Singh, S., Cheema, D.S. and Singh, P., 2004. Genetic analysis of important fruit characters of tomato by involving lines possessing male-sterility genes. *Acta Horticulture*, 637: 123–132.

Eisa, H.M. and Wallace, D.H., 1969. Morphological and anatomical aspects of petaloidy in the carrot, *Daucus carota* L. I. *American Society of Horticulture Science*, 94: 545–548.

Engelke, T. and Tatioglu, T., 2004. The fertility restorer genes X and T alter the transcripts of a novel mitochondrial gene implicated in CMS1 in chives (*Allium schoenoprasum* L.). *Molecular Genetics and Genomics*, 271(2): 150–160.

Engelke, T., Gera, D. and Tatlioglu, T., 2004. Determination of the frequencies of restorer- and maintainer-alleles involved in CMS1 and CMS2 in German chive varieties. *Plant Breeding*, 123(1): 51–59.

Fu, T.D., Yang, G.S. and Yang, X.N., 1990. Studies on "three line" Polima cytoplasmic male sterility developed in *Brassica napus* L. *Plant Breeding* (http: //abst.uasd.edu: 8595/webspirs/ doLS.ws?ss=Plant–Breeding+in+SO), 104(2): 115–120.

George, R.A.T., 1985. *Vegetable Seed Production*. Longman Company Ltd., New York.

Goetz, M., Godt, D.E., Guivarc'h, A., Kahmann, U., Chriqui, D. and Roitsch, T., 2001. Induction of male sterility in plants by metabolic engineering of the carbohydrate supply. *Proceeding of National Academy of Science*, USA, 98: 6522–6527.

Goldman, I.L., Schroeck, G. and Havey, M.J., 2000. History of public onion breeding programs and pedigree of public onion germplasm releases in the United States. *Plant Breeding Review*, 20: 67–103.

Gomez, P., Jamilena, M., Capel, J., Zurita, S., Angosto, T. and Lozano, R., 1999. Stamenless, a tomato mutant with homeotic conversions in petals and stamens. *Panta*, 209: 172–179.

Gulyas, G., Pakozdi, K., Lee, J.S. and Hirata, Y., 2006. Analysis of fertility restoration by using cytoplasmic male sterile red pepper (*Capsicum annuum* L.) Lines. *Breeding Science*, 56: 331–334.

Hall, A.E., 1992. Breeding for heat tolerance. In: *Plant Breeding Reviews*, (Ed.) J. Janick, 10: 129–168.

Havey, M.J., 2000. Diversity among male sterility inducing and male fertile cytoplasms of onion. *Theoretical and Applied Genetics*, 101: 778–782.

Havey, M.J., 2007. The use of cytoplasmic male sterility for hybrid seed production. In: *Molecular Biology of Plant Organelles*, (Eds.) H. Daniell, C. Chase. Springer, Netherlands, pp. 623–634.

Hazra, P., Mandal, J. and Mukhopadhyay, T.P., 2003. Pollination behaviour and natural hybridization in *Solanum melongena* L. and utilization of the functional male sterile line in hybrid seed production. *Capsicum and Eggplant Newsletter*, 22: 143–146.

Heslop, H.J., 1957. The experimental modification of sex expression in flowering plants. *Biological Review*, 32: 38–90.

Hou, X.L., Cao, S.C. and He, Y.K., 2004. Creation of a new germplasm of CMS non-heading Chinese cabbage. *Acta Horticulturae*, 637: 75–81.

Ishikawa, K., Koboki, H., Mishiba, K. and Nunomura, O., 2001. Tetraploid bell pepper shows high *in vitro* pollen germination at 15°C. *Horticulture Science*, 36(7): 1336.

Izhar, S., 1975. The timing of temperature effect on microsporogenesis in cytoplasmic male sterile petunia. *Journal of Heridity*, 66(5): 313–314

Izhar, S. and Frankel, R., 1976. Cytoplasmic male sterility in petunia. I. Comparative study of different plasmatype sources. *Journal of Heridity*, 67(1): 43–46.

Izhar, S., 1977. Cytoplasmic male sterility in petunia. II. The interaction between the plasmagene, genetic factors, and temperature. *Journal of Heridity*, 68(4): 238–240.

Izhar, S., 1978. Cytoplasmic male sterility in petunia. III. Genetic control of microsporogenesis and male fertility restoration. *Journal of Heridity*, 69(1): 22–26.

Jones, I.I. and Emsweller, S., 1936. A male sterile onion. Proceeding of *American Society for Horticultural Science*, 34: 582–585.

Jones, H. and Clarke, A., 1943. Inheritance of male sterility in the onion and the production of hybrid seed. *Proceeding of American Society for Horticultural Science*, 43: 189–194.

Karihaloo, J.L., 1991. Desynapsis due to temperature stress in three species of *Solanum* L. *Cytologia*, 56(4): 603–611.

Kato, K., 1989. Flowering and fertility of forced green peppers at lower temperatures. *Journal of the Japanese Society for Horticultural Science*, 58(1): 113–121.

Khasin, M.F., 1975. Effect of external factors on changes in cytoplasmic male sterility in onion. *Tezisy doki Konf Selektsiya I genet ovoshch kul tur*, 3: 153–154.

Khan, M.M.R. and Isshiki, S., 2008. Development of a male sterile eggplant by utilizing the cytoplasm of *Solanum virginianum* and a biparental transmission of chloroplast DNA in backcrossing. *Scientica Horticulturae*, 117(4): 316–320.

Koelewijn, H.P. and Damme, J.M.M., 1996. Gender variation, partial male sterility and labile sex expression in gynodioecious Plantago coronopus. *New Phytologist*, 132(1): 67–76.

Koizuka, N., Imai, R., Fujimoto, H., Hayakawa, T., Kimura, Y., Kohno, Murase, J., Sakai, T., Kawasaki, S. and Imamura, J., 2003. Genetic characterization of a pentatricopeptide repeat protein gene, orf687, that restores fertility in the cytoplasmic male-sterile Kosena radish. *Plant Journal*, 34(4): 407–415.

Lasa, J.N. and Bosemark, N.O., 1994. Male-sterility. In: *Plant breeding: Principles and Prospects*, (Eds.) M.D. Hayward, N.O. Bosemark, I. Romagosa and M. Cerezo. Chapman and Hall, London, p. 213–228.

Lee, J., Yoon, J.B. and Park, H.G., 2008a. A CAPS Marker associated with the partial restoration of cytoplasmic male sterility in chilli pepper (*Capsicum annuum* L). *Molecular Breeding*, 21: 95–104.

Lee, J., Yoon, J.B. and Park, H.G., 2008b. Linkage analysis between the partial restoration (*pr*) and the restorer-of-fertility (*Rf*) loci in pepper cytoplasmic male sterility. *Theoretical and Applied Genetics*, 117: 383–389.

Levy, A., Rabinowitch, H.D. and Kedar, N., 1978. Morphological and physiological characters affecting flower drop and fruit set of tomatoes at high temperatures. *Euphytica*, 27: 211–218.

Lichter, R., 1978a. The effect of heat treatment and meristem culture on cytoplasmic male sterility in Beta vulgaris. *Interspecific hybridization in plant breeding. Proceedings of the Eigth Congress of Eucarpia IV Cytopasmic Effects*, p. 215–220.

Lichter, R., 1978b. The restoration of male fertility in cytoplasmic male-sterile sugar beet by heat treatment and meristem culture. *Zeitschrift fur Pflanzenzuchtung*, 81(2): 159–165.

Lossl, A., Gotz, M., Braun, A. and Wenzel, G., 2000. Molecular markers for cytoplasm in potato: male sterility and contribution of different plastid–mitochondrial configurations to starch production. *Euphytica*, 116: 221–230.

Maestro, M.C. and Alvarez, J., 1988. The effects of temperature on pollination and pollen tube growth in muskmelon (*Cucumis melo* L.). *Science of Horticulture*, 36: 173–181.

Masaharu, M., Kenji, K., Kenji, M., Hiroshi, N., Christopher, O.O. and Peter, W.M., 2007. Partial fertility restoration as affected by night temperature in a season-dependent male-sterile mutant tomato, *Lycopersicon esculentum* Mill. *Journal of the Japanese Society of Horticultural Sciences*, 76(1): 41–46.

Masuda, M., Ma, Y., Uchida, K., Kato, K. and Agong, S.G., 2000. Restoration of male-sterility in seasonally dependent malesterile mutant tomato, *Lycopersicon esculentum* cv. First. *Journal of Japanese Society for Horticulture Sciences*, 69: 557–562.

Masuda, M. and Ojiewo, C.O., 2006. Induced mutagensis as a breeding strategy for improvement of solanaceous vegetables. *Gama Field Symposia*, 45: 47–60.

Meer, Q.P.V.D. and Bennekom, J.L.V., 1969. Effect of temperature on the occurrence of male sterility in onion (*Allium cepa* L.). *Euphytica*, 18: 389–394.

Meer, Q.P., 1978. Results of recent investigations on male sterility in onion (*Allium cepa* L.). *Biuletyn warzywniczy*, 22: 73–79.

Mele, P.E. and Giordano, L.B., 1994. Effect of Ogura male-sterile cytoplasm on the performance of cabbage hybrid variety. II. Commercial characteristics. *Euphytica*, 78(1/2): 149–154.

Michalik, B., 1978. Stability of male sterility in carrot under different growth conditions. *Bulletin de I Academie Polonaise des Sciences Sciences Biology*, 26(12): 827–832.

Moore, E.L. and Thomas, W.O., 1952. Some effects of shading and para-chlorophenoxy acetic acid on fruitfulness of tomatoes. *Proceedings of the American Society for Horticultural Science*, 60: 289–294.

Moue, T. and Uehara, T., 1985. Inheritance cytoplasmic male sterility in *Allium flstulosum* L. (Welsh onion). *Journal of Japanese Society for Horticulture Science*, 53: 432–437.

Napoli, C.A., Fahy, D., Wang, H.Y. and Taylor, L.P., 1999. White anther: A petunia mutant that abolishes pollen flavonol accumulation, induces male sterility, and is complemented by a chalcone synthase transgene. *Plant Physiology*, 120: 615–622.

Nieuwhof, M., 1985. Investigations on male sterility in radish at the Institute of Horticultural Plant Breeding. *Zaadbelangen*, 39(4): 96–98.

Nieuwhof, M., 1990. Cytoplasmic–genetic male sterility in radish (*Raphanus sativus* L.): Identification of maintainers, inheritance of male sterility and effect of environmental factors. *Euphytica*, 47(2): 171–177.

Nijs, A.P.M. den, 1984. Interspecific crosses in cucumber. *Bedrifsontwikkeling*, 15(1): 60–63.

Nothmann, J. and Killer, D., 1975. Effects of low-temperature stress on fertility and fruiting of eggplant (*Solanum melongena*) in a subtropical climate. *Experimental Agriculture*, 11(1): 33–38.

Ogura, H., 1968. Studies on the new male sterility in Japanese radish, with special references to the utilization of this sterility towards the practical raising of hybrid seeds. *Mem. Fac. Agriculture Kagoshima University*, 6: 39–78.

Owen, F.V., 1945. Cytoplasmic inherited male sterility in sugar beets. *Journal of Agriculture Research* 71: 423–440.

Pathak, C.S., 1997. A possible new source of male sterility in onion. *Acta Horticulturae*, 433: 313–316.

Peet, M.M. and Bartholomew, M., 1996. Effect of night temperature on pollen characteristics, growth, and fruitset in tomato (*Lycopersicon esculentum* Mill). *Journal of the American Society for Horticultural Science*, 121: 514–519.

Perle, M., Swartzberg, D. and Izhar, S., 1993. Phosphatase and ATPase activities in isonuclear lines of cytoplasmic male-sterile and male-fertile petunia. *Theoretical and Applied Genetics*, 86(1): 49–53.

Peterson, P.A., 1958. Cytoplasmic inherited male sterility in *Capsicum*. *American Naturalist*, 92: 111–119.

Peterson, C.E. and Simon, P.W., 1986. Carrot breeding. In: *Breeding Vegetable Crops*, (Ed.) M.J. Bassett, pp. 321–356.

Pi, P., Hill, R.D. and Scarth, R., 1988. Comparative analysis of stamen polypeptides of a rapeseed cultivar, Regent, a CMS Polima line, and temperature-restored male fertile Polima line. *Sexual Plant Reproduction*, 12: 114–118.

Polowick, P.L. and Sawhney, V.K., 1985. Temperature effects on male fertility and flower and fruit development in *Capsicum annuum* L. *Scientia Horticulturae*, 25(2): 117–127.

Polowick, P.L. and Sawhney, V.K., 1988. High temperature induced male and female sterility in canola (*Brassica napus* L.). *Annals of Botany*, 62(1): 83–86.

Reddy, K.M., Deshpande, A.A. and Sadashiva, A.T., 2002. Cytoplasmic genetic male sterility in chilli (*Capsicum annuum* L.). *Indian Journal of Genetics*, 62(4): 363–364.

Ruge, B., Potz H. and Tatlioglu, T., 1992. Comparative investigations of microsporogenesis using temperature- and tetracycline-sensitive and insensitive CMS genotypes in chives (*Allium schoenoprasum* L.). The genus Allium taxonomic problems and genetic resources. In: *Proceedings of an International Symposium* held at Gatersleben Germany June, 11–13, pp. 281–287.

Ruge, B., Potz, H. and Tatlioglu, T., 1993. Influence of different cytoplasms and nuclear genes involved in the CMS system of chives (*Allium schoenoprasum* L.) on microsporogenesis. *Plant Breeding*, 110(1): 24–28.

Sakata, T. and Higashitani, A., 2008. Male sterility accompanied with abnormal anther development in plants gene and environmental stress with special reference to high temperature injury. *International Journal of Pant Development Biology*, 2(1): 42–51.

Shaked, J.A., Rosenfeld, K. and Pressman, E., 2004. The effect of low night temperature on concentration of carbohydrates metabolism in developing pollen grains of pepper in relation to their number and functioning. *Scientia Horticulturae*, 102(1): 29–36.

Saito, T., Matsunaga, H., Saito, A., Hamato, N., Koga, T., Suzuki, T. and Yoshida, T., 2009. A novel source of cytoplasmic male sterility and a fertility restoration gene in eggplant (*Solanum melongena* L.) lines. *Journal of the Japanese Society for Horticultural Science*, 78(4): 425–430.

Sawhney, V.K., 2004. Photoperiod-sensitive male-sterile mutant in tomato and its potential use in hybrid seed production. *Journal of Horticulture Science and Biotechnology*, 79: 138–141.

Saxena, A., Hundal, J.S. and Dhall, R.K., 2005. Fruit and seed setting studies on male sterile line of chilli. *Indian Journal of Horticulture*, 62(2): 206–209.

Schneider, R. and Tatiliogu, T., 1996. Molecular investigations on tetracycline and temperature sensitivity of cytoplasmic male sterility in *Allium schoenoprasum* L. *Beitrage zur Zuchtungsforschung Bundesanstalt fur Zuchtungsforschung an Kulturpflanzen*, 2(1): 202–205.

Shi, G.J., Hou, X.L. and Zhang, C.W., 2004. Study on the physiological and biochemical characteristics of new OguCMS line and its maintainer line of non-heading Chinese cabbage. *Acta Agriculturae Shanghai*, 20(3): 45–47.

Shifriss, C. and Guri, A., 1979. Variation in stability of cytoplasmic male sterility in *C. annuum* L. *Journal of American Society for Horticulture Science*, 104: 94–96.

Shifriss, C., 1997. Male sterility in pepper (*Capsicum annuum* L.). *Euphytica*, 93: 83–88.

Singh, S. and Sawhney, V.K., 1998. Abscisic acid in a male-sterile tomato mutant and its regulation by low temperature. *Journal of Experimental Botany*, 49(319): 199–203.

Stein, H., Gabelmen, W.H. and Stuck, M.B.E., 1959. Reversion in cytoplasmic male sterile plants of *Beta vulgaris*. *Journal of American Society for Sugar Beet Technology*, 10: 619–623.

Tatlioglu, T., 1984. Influence of chemicals and temperature on the expression of cytoplasmic male sterility in chives (*Allium schoenoprasum* L.). 3rd *Eucarpia Allium* Symposium, Wageninger, 4–6 September, pp. 91–95.

Tatlioglu, T., 1985. Influence of temperature on the expression of cytoplasmic male sterility in chives (*Allium schoenoprasum* L.). *Zeitschrift fur Pflanzenzuchtung*, 94(2): 156–161.

Tatlioglu, T., 1986. Influence of tetracycline on the expression of cytoplasmic male sterility (cms) in chives (*Allium schoenoprasum* L.). *Plant Breeding*, 97: 46–55.

Tatlioglu, T., 1987. Genetic control of temperature-sensitivity of cytoplasmic male sterility (CMS) in chives (*Allium schoenoprasum* L.). *Plant Breeding*, 99(1): 65–76.

Terechchenko, N.M., 1973. Cytoplasmic male sterility in breeding hybrid lucerne. Sb nauch tr Vses elects genet in t 10: 103–109.

Timin, N.I. and Dobrustskaya, E.G., 1981. Frequency and selection of carrots with cytoplasmic male sterility under different environmental conditions. Ekol genet rasti zhivotnykh Tez dok Vses konf 2: 147.

Turan T., 1996. *Allium schoenoprasum* L. A model plant for investigating the genetic and molecular basis of genic (GMS) and cytoplasmic male sterility (CMS). *Beitrage zur Zuchtungsforschung Bundesanstalt fur Zuchtungsforschung an Kulturpflanzen*, 2(1): 146–149.

Yang, G.S. and Fu, T.D., 1990. The inheritance of Polima cytoplasmic male sterility in *Brassica napus* L. *Plant Breeding*, 104(2): 121–124.

Yang, G.S., Fu, T.D. Ya, Yang, X.N. and Ma, C.Z., 1995. Studies on the ecotypical male sterile line of *Brassica napus* L. I. Inheritance of the ecotypical male sterile line *Acta Agronomica Sinica*, 21(2): 129–135.

Yang, X.Y. and Cao, S.C., 1998. Biochemical analysis of temperature effect on fertility of POL CMS in non–heading Chinese cabbage. *Advances in Horticulture*, 2: 539–543.

Yang, X.Y. and Cao, S.C., 1997a. Cytomorphological research on anther development of Pol CMS in non-heading Chinese cabbage (*Brassica campestris* L. ssp. *chinensis* Makino). *Journal of Nanjing Agriculture University*, 20(3): 36–43.

Yang, X.Y. and Cao, S.C., 1997b. Effect of temperature on fertility of Pol cytoplasmic male sterility in Chinese cabbage (*Brassica campestris* ssp. *chinensis* L. Makino). *Journal of Nanjing Agriculture University*, 20(2): 22–27.

Yang, X.Y. and Cao, S.C., 1999. Characteristics of temperature sensitive pol cytoplasmic male sterile non heading Chinese cabbage. *China Vegetable* 4: 15–17.

Yui, R., Iketani, S., Mikami, T. and Kubo, T., 2003. Antisense inhibition of mitochondrial pyruvate dehydrogenase E1alpha subunit in anther tapetum causes male sterility. *Plant Journal*, 34: 57–66.

Zhang, S., Zhang, L.G. and Zhang, M.K., 2008a. Analysis of beta–EST isoenzymes related to sterility changeover in temperature-sensitive CMS of Chinese cabbage (*Brassica campestris* L. ssp. pekinensis) with two-dimensional GEL electrophoresis. *Acta Horticulturae Sinica* 35(5): 681–686.

Zhang, L., Hui, M. and Zhang, M., 2008b. Vigor of trace pollen in temperature-sensitive CMS of Chinese cabbages. *Acta Horticulturae*, 771: 97–102.

Zhang, S.N. and Cao, S.C., 1999a. Effect of atmospheric temperature on male sterility of cytoplasm sensitive to low temperature in non-heading Chinese cabbage. *Jiangsu Journal of Agricultural Sciences*, 15(4): 250–252.

Zhang, S.N., 1999b. Studies on restorative and maintaining ability of different varieties and lines to POL CMS in non-heading Chinese cabbage. *Journal of Jilin Agricultura University*, 21(4): 30–34.

Zhang, Z. and Hou, X., 2005. Relation between cytoplasmic male sterility and reactive oxygen species metabolism in pepper. *Acta Botanica Borealj Occidentalia Sinica*, 25(4): 799–802.

Zheng, Z., Xia, Q., Dauk, M., Shen, W., Selvaraj, G. and Zou, J., 2003 Arabidopsis AtGPAT1, a member of the membrane-bound glycerol-3-phosphate acyltransferase gene family, is essential for tapetum differentiation and male fertility. *Plant Cell*, 15: 1872–1887.

2012, Nutri-Horticulture   *Pages 47–73*
Editor: Professor K.V. Peter
Published by: DAYA PUBLISHING HOUSE, NEW DELHI

Chapter 4

Biointensive Integrated Pest Management in Horticultural Crops

P. Parvatha Reddy

Former Director,
Indian Institute of Horticultural Research, Bangalore – 560 089, INDIA
E-mail: reddy_parvatha@yahoo.com

Through 'Green Revolution' in late 1960's, India achieved self sufficiency in food production, which was hailed as a breakthrough on the farm front by international agricultural experts. But still the country has not achieved self sufficiency in production of horticultural crops. Most of the growth in food production during the green revolution period is attributed to the use of improved crop varieties and higher levels of inputs of fertilizers and pesticides. The modern agricultural techniques like use of synthetic fertilizers and pesticides are continuing to destroy stable traditional ecosystems and the use of high yielding varieties of crops has resulted in the elimination of thousands of traditional varieties with the concurrent loss of genetic resources. The introduction of high yielding varieties changed the agricultural environment leading to numerous pest problems of economic importance. In the process of intensive farming, the environment has been treated in an unfriendly manner.

Swaminathan (2000) emphasized the need for 'Ever Green Revolution' keeping in view the increase in population. The increase in population and diminishing per capita availability of land demand rise in productivity per unit area. In India, annual crop losses due to pests, diseases and weeds are estimated to be about Rs. 600,000 million in 2005. Increasing yields from existing land require effective crop protection to prevent losses before and after harvest. The challenge before the plant protection scientist is to do this without harming the environment and resource base.

Integrated Pest Management (IPM)

Integrated Pest Management is an important principle on which sustainable crop protection can be based. IPM allows farmers to manage pests in a cost effective, environmentally sound and socially acceptable way. According to FAO, IPM is defined as "A pest management system that in the context of the associated environment and the population dynamics of the pest species, utilizes all suitable techniques and methods, in a compatible manner as possible and maintains the pest populations at levels below those causing economic injury".

Biointensive Integrated Pest Management (BIPM)

Biointensive IPM incorporates ecological and economic factors into agricultural system design and decision making and addresses public concerns about environmental quality and food safety. The benefits of implementing biointensive IPM can include reduced chemical input costs, reduced on-farm and off-farm environmental impacts and more effective and sustainable pest management. An ecology-based IPM has the potential of decreasing inputs of fuel, machinery and synthetic chemicals-all of which are energy intensive and increasingly costly in terms of financial and environmental impact. Such reductions will benefit the grower and society.

Over-reliance on the use of synthetic pesticides in crop protection programmes around the world has resulted in disturbances to the environment, pest resurgence, pest resistance to pesticides and lethal and sub-lethal effects on non-target organisms, including humans. These side effects have raised public concern about the routine use and safety of pesticides. At the same time, population increases are placing ever-greater demands upon the "ecological services" *i.e.* provision of clean air, water and wildlife habitat-of a landscape dominated by farms. Although some pending legislation has recognized the costs to farmers of providing these ecological services, it is clear that farmers will be required to manage their land with greater attention to direct and indirect off-farm impacts of various farming practices on water, soil and wildlife resources. With this likely future in mind, reducing dependence on chemical pesticides in favour of ecosystem manipulations is a good strategy for farmers.

Biointensive IPM is defined as "A systems approach to pest management based on an understanding of pest ecology. It begins with steps to accurately diagnose the nature and source of pest problems, and then relies on a range of preventive tactics and biological controls to keep pest populations within acceptable limits. Reduced-risk pesticides are used if other tactics have not been adequately effective, as a last resort, and with care to minimize risks" (Benbrook, 1996).

The primary goal of biointensive IPM is to provide guidelines and options for the effective management of pests and beneficial organisms in an ecological context. The flexibility and environmental compatibility of a biointensive IPM strategy make it useful in all types of cropping systems. Biointensive IPM would likely decrease chemical use and costs even further.

Components of Biointensive IPM

An important difference between conventional and biointensive IPM is that the emphasis of the latter is on proactive measures to redesign the agricultural ecosystem to the disadvantage of a pest and to the advantage of its parasite and predator complex. At the same time, biointensive IPM shares many of the same components as conventional IPM, including monitoring, use of economic thresholds, record keeping and planning.

Planning

Good planning must precede implementation of any IPM programme, but is particularly important in a biointensive programme. Planning should be done before planting because many pest strategies require steps or inputs, such as beneficial organism habitat management, that must be considered well in advance. Attempting to jump-start an IPM programme in the beginning or middle of a cropping season generally does not work.

When planning a biointensive IPM program a few of considerations include:

☆ Options for design changes in the agricultural system (beneficial organism habitat, crop rotations).

☆ Choice of pest-resistant cultivars.

☆ Technical information needs.

☆ Monitoring options, record keeping, equipment, etc.

When making a decision about crop rotation, consider the following questions: Is there an economically sustainable crop that can be rotated into the cropping system? Is it compatible? Important considerations when developing a crop rotation are:

☆ How might the cropping system be altered to make life more difficult for the pest and easier for its natural controls? What two (or three or several) crops can provide an economic return when considered together as a biological and economic system that includes considerations of sustainable soil management?

☆ What are the impacts of this season's cropping practices on subsequent crops?

☆ What specialized equipment is necessary for the crops?

☆ What markets are available for the rotation crops?

Management factors should also be considered. For example, one crop may provide a lower direct return per acre than the alternate crop, but may also lower management costs for the alternate crop, with a net increase in profit.

Pest Identification

A crucial step in any IPM program is to identify the pest. The effectiveness of both proactive and reactive pest management measures depend on correct identification. Misidentification of the pest may be worse than useless; it may actually be harmful and cost time and money. Help with positive identification of pests may be obtained from university personnel, private consultants, the Cooperative Extension Service and books and Web sites.

After a pest is identified, appropriate and effective management depend on knowing answers to a number of questions. These may include:

☆ What plants are hosts and non-hosts of this pest?

☆ When does the pest emerge or first appear?

☆ Where does it lay its eggs?

☆ For plant pathogens, where is the source(s) of inoculum?

☆ Where, how, and in what form does the pest overwinter?

Monitoring (field scouting) and economic injury and action levels are used to help answer these and additional questions.

Monitoring

Monitoring involves systematically checking crop fields for pests and beneficials, at regular intervals and at critical times, to gather information about the crop, pests and natural enemies. Sweep nets, sticky traps and pheromone traps can be used to collect insects for both identification and population density information. Leaf counts are one method for recording plant growth stages. Records of rainfall and temperature are sometimes used to predict the likelihood of disease infections.

The more often a crop is monitored, the more information the grower has about what is happening in the fields. Monitoring activity should be balanced against its costs. Frequency may vary with temperature, crop, growth phase of the crop and pest populations. If a pest population is approaching economically damaging levels, the grower will want to monitor more frequently.

Economic Injury and Action Levels

The economic injury level (EIL) is the pest population that inflicts crop damage greater than the cost of control measures. Because growers will generally want to act before a population reaches EIL, IPM programmes use the concept of an economic threshold level (ETL or ET), also known as an action threshold. The ETL is closely related to the EIL and is the point at which suppression tactics should be applied in order to prevent pest populations from increasing to injurious levels.

ETLs are intimately related to the value of the crop and the part of the crop being attacked. For example, a pest that attacks the fruit or vegetable will have a much lower ETL (*i.e.*, the pest must be controlled at lower populations) than a pest that attacks a non-saleable part of the plant. The exception to this rule is an insect or nematode pest that is also a disease vector. Depending on the severity of the disease, the grower may face a situation where the ETL for a particular pest is zero, *i.e.*, the crop cannot tolerate the presence of a single pest of that particular species because the disease it transmits is so destructive.

BIPM Options

BIPM options may be considered as proactive or reactive.

Proactive Options

Proactive options, such as crop rotations and creation of habitat for beneficial organisms, permanently lower the carrying capacity of the farm for the pest. The carrying capacity is determined by the factors like food, shelter, natural enemy complex and weather, which affect the reproduction and survival of a pest species. Cultural control practices are generally considered to be proactive strategies. Proactive practices include crop rotation, resistant crop cultivars including transgenic plants, disease-free seed and plants, crop sanitation, spacing of plants, altering planting dates, mulches, etc.

The proactive strategies (cultural controls) include:

☆ Healthy, biologically active soils (increasing below-ground diversity).

☆ Habitat for beneficial organisms (increasing above-ground diversity).

☆ Appropriate plant cultivars.

Intercropping

Intercropping is the practice of growing two or more crops in the same, alternate, or paired rows in the same area. This technique is particularly appropriate in vegetable production. The advantage of intercropping is that the increased diversity helps "disguise" crops from insect pests, and if done well, may allow for more efficient utilization of limited soil and water resources.

Strip Cropping

Strip cropping is the practice of growing two or more crops in different strips across a field wide enough for independent cultivation. It is commonly practiced to help reduce soil erosion in hilly areas. Like intercropping, strip cropping increases the diversity of a cropping area, which in turn may help "disguise" the crops from pests. Another advantage to this system is that one of the crops may act as a reservoir and/or food source for beneficial organisms.

The options described above can be integrated with no-till cultivation schemes and all its variations (strip till, ridge till, etc.) as well as with hedgerows and intercrops designed for beneficial organism habitat. With all the cropping and tillage options available, it is possible, with creative and informed management, to evolve a biologically diverse, pest-suppressive farming system appropriate to the unique environment of each farm.

Disease Free Seed and Plants

These are available from most commercial sources, and are certified as such. Use of disease-free seed and nursery stock is important in preventing the introduction of disease.

Resistant Varieties

These are continually being bred by researchers. Growers can also do their own plant breeding simply by collecting non-hybrid seed from healthy plants in the field. The plants from these seeds will have a good chance of being better suited to the local environment and of being more resistant to insects and diseases. Since natural systems are dynamic rather than static, breeding for resistance must be an ongoing process, especially in the case of plant disease, as the pathogens themselves continue to evolve and become resistant to control measures.

Perhaps the greatest single technological achievement is the advance in breeding crops for resistance to pests. Cultivation of resistant varieties is the cheapest and the best method of controlling pests. One of the important components of IPM is the use of resistant cultivars to key pests. Under All India Coordinated Research Projects of Indian Council of Agricultural Research, a large number of highly/moderately resistant varieties are released to the farmers (Table 4.1).

Table 4.1: Horticultural Crop Varieties Resistant to Pests/Diseases

Horticultural Crop	Pest/Disease	Resistant Varieties
Banana	*Radopholus similis*	Kadali, Pedalimoongil, Ayiramkapoovan, Peykunnan, Kunnan, Pisang Seribu, Tongat, Vennettu Kunnan, Anaikomban
	Panama wilt (*Fusarium oxysporum* f. sp. *cubense*)	Robusta, Dwarf Cavendish
Citrus	*Tylenchulus semipenetrans*	Trifoliate Orange, Swingle Citrumelo
Grapevine	Root-knot nematode, *Meloidogyne incognita*	Black Champa, Dogridge, 1613, Salt Creek, Cardinal, Banquabad
	Gummosis, leaf fall, fruit rot (*Phytophthora* spp.)	Cleopatra mandarin, Rangpur lime, Trifoliate orange rootstocks
Papaya	Ring spot virus	Rainbow
Passion fruit	Root-knot nematode, *M. incognita*	Yellow, Kaveri
Potato	Late blight	Kufri Sutlej, Kufri Badshah, Kufri Jawahar (in plains), Kufri Jyothi, Kufri Giriraj, Kufri Kanchan, Kufri Meghad (in hills)
Tomato	Bacterial wilt	Arka Abha, Arka Alok, Arka Shreshta, Arka Abhijit, Megha, Shakthi, Sun 7610, Sun 7611
	Powdery mildew (PM)	Arka Asish
	Fusarium and *Verticillium* wilt	Vaishali, Rupali, Rashmi
	Leaf curl virus	Avinash-2, Hisar Anmol.
	Root-knot nematode	Hisar Lalit, Pusa Hybrid-2, Arka Vardan

Contd...

Table 4.1–Contd...

Horticultural Crop	Pest/Disease	Resistant Varieties
Brinjal	Bacterial wilt	Arka Nidhi, Arka Keshav, Arka Neelkant, Arka Anand, Swarna Shree, Swarna Shyamali, Surya, Ujjwala
	Phomopsis blight	Pusa Bhairav
	Little leaf	Pusa Purple Long, Pusa Purple Cluster (Field resistant)
Chilli	TMV, CMV, Leaf curl	Pusa Sada Bahar, Punjab Lal, Pusa Jwala
	Thrips (T)	NP 46A
	Powdery mildew	Arka Suphala (T)
	Dieback and Powdery mildew (T)	Musalwadi
	Mosaic, Leaf curl	Pant C-1
	Leaf curl and Fruit rot (T)	Jawahar 218
	Viruses	Arka Harita, Arka Meghana
French bean	Angular leaf spot, Mosaic	Pant Anupama
	Rust, Bacterial blight	Arka Anoop
	Rust	Arka Bold, Swarna Priya, Swarna Latha, Arka Anoop
	Rust, *Alternaria* leaf spot	Arka Bold
Pea	Powdery mildew	Pusa Pragati, Jawahar Matar 5, Jawahar Peas 83
	PM, Rust	Arka Ajit, Arka Karthik, Arka Sampoorna
	Fusarium Wilt	JP Batri Brown 3, JP Batri Brown 4
Cowpea	Bacterial blight	Pusa Komal
Pigeon pea	*Fusarium* wilt	Maruti.
Field bean	Viral diseases, jassid, aphid and pod borer	Pusa Sem-2, Pusa Sem-3
Cluster bean	PM, *Alternaria* leaf spot	Gomah Manjari
Okra	YVMV	Pusa Sawani, Arka Abhay, Arka Anamika, Hisar Unnat, DVR-1, DVR-2, IIVR-10, Varsha Upkar, P-7, Pusa A-4, Parbhani Kranti (T), Punjab Kesari, Punjab Padmini, Sun-40, Makhmali.
	YVMV and fruit borer	Pusa A-4
Cucumber	PM	Swarna Poorna
	PM, Downy mildew (DM), Angular leaf spot, Anthracnose	Poinsette
Cabbage	Black rot	Pusa Mukta
	Black leg	Pusa Drum Head
Cauliflower	Black rot	Pusa Snowball K-1
	Black rot and Curd blight	Pusa Shubhra
	Curd blight	Pusa Synthetic
	DM	Pusa Hybrid-2
Onion	Purple blotch, Basal rot, Thrips	Arka Pitamber, Arka Kirtiman, Arka Lalima
	Purple blotch, *Alternaria porri*	Arka Kalyan

Contd...

Table 4.1–Contd...

Horticultural Crop	Pest/Disease	Resistant Varieties
Garlic	Purple blotch, *Stemphylium* disease	Agri-found White
Muskmelon	PM	Arka Rajhans, Pusa Madhuras (MR)
	PM, DM	Punjab Rasila
	Fusarium wilt	Pusa Madhuras, Durgapura Madhu, Arka Jeet, Punjab Sunehari (MR), Harela
Watermelon	PM, DM, Anthracnose	Arka Manik
Pumpkin	Fruit fly	Arka Suryamukhi
Ridge gourd	PM, DM	Swarna Uphaar
Bottle gourd	Blossom end rot	Arka Bahar (T)
	CMV	Punjab Komal
Carrot	PM, Root-knot nematode	Arka Suraj
Amaranth	White rust	Arka Arunima, Arka Suguna (MR)
Palak	*Cercospora* leaf spot	Arka Anupama
China aster	Root-knot nematode, *M. incognita*	Shashank, Poornima (MR)
Tuberose	Root-knot nematode, *M. incognita*	Sringar, Suvasini (T)
Mentha	Root-knot nematode, *M. incognita*	Kukrail, Arka Neera
Black pepper	Root-knot nematode, *M. incognita*	IISR Pournami (T)
	Foot rot, *Phytophthora capsici*	IISR Shakthi
Cardamom	Mosaic	IISR Vijetha
	Rhizome rot	IISR Avinash
Ginger	Root-knot nematodes	IISR Mahima
	Soft rot	Maran
Cumin	Fusarium Wilt	GC-4

MR: Moderately Resistant; T: Tolerant.

Biotech Crops

Gene transfer technology is being used by several companies to develop cultivars resistant to insects, diseases and nematodes. An example is the incorporation of genetic material from *Bacillus thuringiensis* (*Bt*), a naturally occurring bacterium, into brinjal and potatoes, to make the plant tissues toxic to shoot and fruit borer and potato beetle larvae, respectively.

Whether or not this technology should be adopted is the subject of much debate. Opponents are concerned that by introducing *Bt* genes into plants, selection pressure for resistance to the *Bt* toxin will intensify and a valuable biological control tool will be lost. There are also concerns about possible impacts of genetically-modified plant products (*i.e.*, root exudates) on non-target organisms as well as fears of altered genes being transferred to weed relatives of crop plants. Whether there is a market for gene-altered crops is also a consideration for farmers and processors. Proponents of this technology argue that use of such crops decreases the need to use toxic chemical pesticides.

Transgenic crop varieties in horticultural crops (tomato, potato, brinjal, beans, cabbage, cauliflower, musk melon, banana, coffee) have been developed by cloning *Bt* endotoxin genes which are cultivated in large areas. In 2006, India is the 5th largest GM crops growing country (3.8 million ha) in the world only next to USA (54.6 million ha), Argentina (18 million ha), Brazil and Canada. Combining a host gene for resistance with pathogen derived genes or with genes coding for antimicrobial compounds provide for a broad and effective resistance in many host-pathogen combinations (Table 4.2).

Table 4.2: Development of Transgenics in Vegetable Crops

Vegetable Crop	Target Pathogen	Transgene/s	Institute
Potato	Tuber moth	Bt Cry 1Ab	CPRI, Shimla
	Potato virus Y	Coat protein	CPRI, Shimla
Tomato	Leaf curl virus	Leaf curl virus sequence	IIHR, Bangalore IAHS, Bangalore
		Replicase gene	IARI, New Delhi
	Fungal diseases	Chitinase and glucanase	IIHR, Bangalore
		Alfalfa glucanase	IAHS, Bangalore
		OXDC	JNU, New Delhi
	Lepidopteran pests	Bt Cry 1Ab	IARI, New Delhi Proagro PG-S (India) Ltd.
Brinjal	Fungal diseases	Chitinase, glucanase and thaumatin encoding genes	
	Lepidopteran pests	Bt Cry 1Ab	IARI, New Delhi Proagro PG-S (India) Ltd.
Cabbage	Lepidopteran pests	Bt Cry 1Ab	IARI, New Delhi Proagro PG-S (India) Ltd.
		Cry 1H/Cry 9C	Proagro PG-S (India) Ltd.
Cauliflower	Lepidopteran pests	Bt Cry 1Ab	IARI, New Delhi Proagro PG-S (India) Ltd.
		Cry 1H/Cry 9C	Proagro PG-S (India) Ltd.

Sanitation

It involves removing and destroying the overwintering or breeding sites of the pest as well as preventing a new pest from establishing on the farm (*e.g.*, not allowing off-farm soil from farm equipment to spread nematodes or plant pathogens to your land). This strategy has been particularly useful in horticultural and tree-fruit crop situations involving twig and branch pests. If, however, sanitation involves removal of crop residues from the soil surface, the soil is left exposed to erosion by wind and water. As with so many decisions in farming, both the short- and long-term benefits of each action should be considered when tradeoffs like this are involved.

Spacing of Plants

It heavily influences the development of plant diseases. The distance between plants and rows, the shape of beds, and the height of plants influence air flow across the crop, which in turn determines how long the leaves remain damp from rain and morning dew. Generally speaking, better air flow will decrease the incidence of plant disease. However, increased air flow through wider spacing will also allow more sunlight to the ground. This is another instance in which detailed knowledge of the crop ecology is necessary to determine the best pest management strategies. How will the crop react to increased spacing between rows and between plants? Will yields drop because of reduced crop density? Can this be offset by reduced pest management costs or a fewer losses from disease?

Altered Planting Dates

This can at times be used to avoid specific insects or diseases. For example, squash bug infestations on cucurbits can be decreased by the delayed planting strategy, *i.e.*, waiting to establish the cucurbit crop until overwintering adult squash bugs have died. To assist with disease management decisions, the Cooperative Extension Service (CES) will often issue warnings of "infection periods" for certain diseases, based upon the weather.

In some cases, the CES also keeps track of "degree days" needed for certain important insect pests to develop. Insects, being cold-blooded, will not develop below or above certain threshold temperatures. Calculating accumulated degree days, *i.e.*, the number of days above the threshold development temperature for an insect pest, makes the prediction of certain events, such as egg hatch, possible. University of California has an excellent Web site that uses weather station data from around the state to help California growers predict pest emergence.

Some growers gauge the emergence of insect pests by the flowering of certain non-crop plant species native to the farm. This method uses the "natural degree days" accumulated by plants. For example, a grower might time cabbage planting for three weeks after the *Amelanchier* species (also known as saskatoon, shad bush, or service berry) on their farm are in bloom. This enables the grower to avoid peak egg-laying time of the cabbage maggot fly, as the egg hatch occurs about the time *Amelanchier* species are flowering (Couch, 1994). Using this information, cabbage maggot management efforts could be concentrated during a known time frame when the early instars (the most easily managed stage) are active.

Optimum Growing Conditions

Plants which grow quickly and are healthy can compete with and resist pests better than slow-growing, weak plants. Too often, plants grown outside their natural ecosystem range must rely on pesticides to overcome conditions and pests to which they are not adapted.

Mulches

Living or non-living mulches are useful for suppression of insect pests and some plant diseases. Hay and straw, for example, provide habitat for spiders. Research in Tennessee showed a 70 per cent reduction in damage to vegetables by insect pests when hay or straw was used as mulch. The difference was due to spiders, which find mulch more habitable than bare ground (Reichert and Leslie, 1989). Other researchers have found that living mulches of various clovers reduce insect pest damage to vegetables and orchard crops. Again, this reduction is due to natural predators and parasites provided habitat by the clovers.

Mulching minimizes the spread of soil-borne plant pathogens by preventing their transmission through soil splash. Winged aphids are repelled by silver- or aluminum-coloured mulches. Recent springtime field tests at the Agricultural Research Service in Florence, South Carolina indicated that red plastic mulch suppresses root-knot nematode damage in tomatoes by diverting resources away from the roots (and nematodes) and into foliage and fruit (Adams, 1997).

Reactive Options

The reactive options mean that the grower responds to a situation, such as an economically damaging population of pests, with some type of short-term suppressive action. Reactive methods generally include inundative releases of biological control agents, mechanical and physical controls, botanical pesticides and chemical controls.

Biological Controls

Biological control is the use of living organisms - parasites, predators, or pathogens–to maintain pest populations below economically damaging levels, and may be either natural or applied. A first step in setting up a biointensive IPM program is to assess the populations of beneficials and their interactions within the local ecosystem. This determines the potential role of natural enemies in the managed horticultural ecosystem. It should be noted that some groups of beneficials (*e.g.*, spiders, ground beetles, bats) may be absent or scarce on some farms because of lack of habitat. These organisms might make significant contributions to pest management if provided with adequate habitat.

Natural Biological Control

It results when naturally occurring enemies maintain pests at a lower level than would occur without them, and is generally characteristic of biodiverse systems. Mammals, birds, bats, insects, fungi, bacteria and viruses all have a role to play as predators, parasites and pathogens in a horticultural system. By their very nature, pesticides decrease the biodiversity of a system, creating the potential for instability and future problems. Pesticides, whether synthetically or botanically derived, are powerful tools and should be used with caution.

Creation of habitat to enhance the chances for survival and reproduction of beneficial organisms are concepts included in the definition of natural biocontrol. Farmscaping is a term coined to describe such efforts on farms. Habitat enhancement for beneficial insects, for example, focuses on the establishment of flowering annual or perennial plants which provide pollen and nectar needed during certain parts of the insect life cycle. Other habitat features provided by farmscaping include water, alternative prey, perching sites, overwintering sites and wind protection. Beneficial insects and other beneficial organisms should be viewed as mini-livestock, with specific habitat and food needs to be included in farm planning.

The success of such efforts depends on knowledge of the pests and beneficial organisms within the cropping system. Where do the pests and beneficials overwinter? What plants are hosts and non-hosts? When this kind of knowledge informs planning, the ecological balance can be manipulated in favour of beneficials and against the pests.

It should be kept in mind that ecosystem manipulation is a two-edged sword. Some plant pests (such as the tarnished plant bug and lygus bug) are attracted to the same plants which attract beneficials. The development of beneficial habitats with a mix of plants that flower throughout the year can help prevent such pests from migrating *en masse* from farmscaped plants to crop plants.

Applied Biological Control

It is also known as augmentative biocontrol which involves supplementation of beneficial organism populations, for example through periodic releases of parasites, predators, or pathogens. This can be effective in many situations-well-timed inundative releases of *Trichogramma* egg wasps for codling moth control, for instance.

Most of the beneficial organisms used in applied biological control today are insect parasites and predators. They control a wide range of pests from caterpillars to mites. Some species of biocontrol organisms, such as *Eretmocerus californicus*, a parasitic wasp, are specific to one host-in this case the sweet potato whitefly. Others, such as green lacewings, are generalists and will attack many species of aphids and whiteflies.

Information about rates and timing of release are available from suppliers of beneficial organisms. It is important to remember that released insects are mobile; they are likely to leave a site if the habitat is not conducive to their survival. Food, nectar and pollen sources can be "farmscaped" to provide suitable habitat.

The quality of commercially available applied biocontrols is another important consideration. For example, if the organisms are not properly labeled on the outside packaging, they may be mishandled during transport, resulting in the death of the organisms. A recent study by Rutgers University noted that only two of six suppliers of beneficial nematodes sent the expected numbers of organisms, and only one supplier out of the six provided information on how to assess product viability.

While augmentative biocontrols can be applied with relative ease on small farms and in gardens, applying some types of biocontrols evenly over large farms has been problematic. New mechanized methods which may improve the economics and practicality of large-scale augmentative biocontrol include ground application with "biosprayers" and aerial delivery using small-scale (radio-controlled) or conventional aircraft.

Inundative releases of beneficials into greenhouses can be particularly effective. In the controlled environment of a greenhouse, pest infestations can be devastating; there are no natural controls in place to suppress pest populations once an infestation begins. For this reason, monitoring is very important. If an infestation occurs, it can spread quickly if not detected early and managed. Once introduced, biological control agents cannot escape from a greenhouse and are forced to concentrate predation/parasitism on the pest(s) at hand.

An increasing number of commercially available biocontrol products are made up of microorganisms, including fungi, bacteria, nematodes and viruses.

Of late, biological suppression of pests has become an intensive area of research because of environmental concerns. About 60 per cent of the natural control of insect pests are by the natural enemies of pests like parasitoids, predators and pathogens. The Australian lady bird beetle(*Cryptolaemus motrouzieri*) is found very effective against mealy bugs infesting grapes, guava, citrus, mango, pomegranate, ber and custard apple. The encyrtid parasite (*Leptomastix dactylopii*) is effective against mealy bug (*Planococcus citri*) on guava, citrus, pomegranate, ber and custard apple (Mani, 2001). *Bacillus thuringiensis* (*Bt*) is effective against tomato fruit borer, okra fruit borer and diamondback moth on cabbage and cauliflower.

Several methods of enrichment and conservation of natural enemies include providing nesting boxes for wasps and predatory birds; retaining pollen and nectar bearing flowering plants like Euphorbia, wild Clover on bunds to provide supplementary food for natural enemies and placing bundles of paddy straw in fields for attracting predatory spiders. In addition, erecting perching sites, water pans and retaining bushes (Acalypha, Hibiscus, Crotons) help in retention of predatory birds.

The last decade has witnessed a tremendous break through in biological control of fungal diseases and nematodes like *Rhizoctonia, Pythium, Fusarium, Macrophomina, Ralstonia* and *Meloidogyne* in banana, tomato, egg plant, pea, grapes, cucumber, black pepper, cardamom, ginger and turmeric, especially by using species of *Trichoderma, Pochonia, Pseudomonas* and *Bacillus* (Tables 4.3–4.12).

Table 4.3: Biological Control of Fruit Crop Pests

Fruit Crop	Pest	Biocontrol Agent/Dosage
Apple	Woolly aphid (*Eriosoma lanigerum*)	*Aphelinus mali*-1000 adults or mummies/infested tree.
	San Jose Scale (*Quadraspidiotus perniciosus*)	*Encarsia periniciosi*-2000 adults/infested tree.
	Codling moth (*Cydia pomonella*)	*Chilocorus infernalis*-20 adults or 50 grubs/tree; *Trichogramma embryophagum*-2000 adults/tree; *Steinernema carpocapse*
Citrus	Cottony cushion scale (*Icerya purchasi*)	*Rodolia cardinalis* -10 beetles/infested plant.
	Mealy bug (*Planococcus citri*)	*Cryptolaemus montrouzieri*-10 beetles/infested plant; *Leptomastix dactylopii* 3000 adults/ha.
	Red Scale (*Aonidiella aurantii*)	*Chilocorus nigrita*-15 adults/infested tree
	Scale insect (*Coccus viridis*)	*Verticillium lecanii*-16 x 10^4 spores/ml + 0.05 per cent Teepol.
	Leaf miner (*Phyllocnistis citrella*)	*S. carpocapse*
Grapevine	Mealy bug (*Maconellicoccus hirsutus*)	*C. montrouzieri*-2500-3000 beetles/ha or 10 beetles/vine.
Guava	Green shield scale (*Chloropulvinaria psidii*)	*C. montrouzieri*-10-20 beetles/infested plant.
	Aphid (*Aphis gossypii*)	*V. lecanii*-10^9 spores/ml + 0.1 per cent Teepol.

Table 4.4: Biological Control of Fruit Crop Diseases

Fruit Crop	Disease/Pathogen/s	Potential Biocontrol Agent/s
Banana	Panama wilt (*Fusarium oxysporum* f. sp. *Cubense*)	*T. viride, Aspergillus niger, Pseudomonas fluorescens, T. viride + P. fluorescens* – sucker treatment.
Citrus	Root rot (*Phytophthora* spp)	*Trichoderma viride/T. harzianum* at 100 kg/ha; *Penicillium funiculosum, Pythium nunn*-soil treatment.
	Canker (*Xanthomonas campestris* pv. *Citri*)	*Aspergillus niger* AN 27
Strawberry	Grey mold (*Botrytis cineria*)	*T. harzianum*
Mulberry	Leaf spot (*Cercospora moricola*)	*T. viride, T. harzianum, Pseudomonas fluorescens*
	Cutting rot (*F. solani*)	*T. virens, T. harzianum, T. pseudokoningii*
Grapevine	Powdery mildew (*Uncinula necator*)	*Ampelomyces quisqualis*-dispersal from wick cultures at 15 cm of shoot growth and bloom
	Downy mildew (*Plasmopara viticola*)	*Fusarium proliferatum* weekly spray starting from 15 cm of shoot growth – 10^6 spores/ml
Guava	Anthracnose (*Pestalotia psidii*), (*Colletotrichum gloeosporoides*)	*T. harzianum*
	Wilt (*Gliocladium roseum* and *F. Solani*)	*Penicillium citrinum, Aspergillus niger* (AN 17), *T. harzianum*
Mango	Anthracnose (*Colletotrichum gloeosporoides*)	*T. harzianum, Streptosporangium pseudovulgare*
	Powdery mildew (*Oidium mangiferae*)	*S. pseudovulgare*
	Bacterial canker (*Xanthomonas campestris* pv. *Mangiferaeindicae*)	*Bacillus coagulans*

Contd...

Table 4.4–Contd...

Fruit Crop	Disease/Pathogen/s	Potential Biocontrol Agent/s
Apple	Scab (*Venturia inaequalis*)	*Chaetomium globosum, A. pullulans, Microsphaeropsis* sp.; *Chaetomium globosum, Cladosporium* spp., *Trichothecium roseum*-Foliar spray.
	Collar Rot (*Phytophthora cactorum*)	*Enterobacter aerogenes, Bacillus subtilis*-Soil treatment; *T. virens*-soil treatment.
	White root rot (*Dematophora necatrix*)	*T. viride, T. harzianum, T. virens*-soil treatment.
Pear	Blue mold (*Pencillium expansum*); Grey mold (*Botrytis cineria*)	*C. infirmo-miniatus* YY6, *C. laurentii* RR87-108, *R. glutinis* HRB6-fruit spray 3 week or 1 day prior to harvest - 10^8 cfu/ml; *Pantoea agglomerans* CPA-2-post-harvest fruit dipping in 8 x10^8 cfu/ml
	Fire blight (*Erwinia amylovora*)	*Pseudomonas fluorescens*-foliar spray
Peach	Brown rot (*Monilia fructicola*)	*Bacillus subtilis* (B-3)-post-harvest fruit line spray at 5 x 10^8 cfu/g); *Pseudomonas syringae*-post-harvest fruit dipping in 10^7 cfu/ml.
	Twig blight (*Monilia laxa*)	*Pencillium frequentans*-spray shoots in early growing season - 10^{8-9} spores/ml
	Crown gall (*Agrobacterium tumefaciens*)	*Agrobacterium radiobacter* K84, K1026-root dip treat.
Strawberry	Grey mold (*Botrytis cineria*)	*Trichoderma* products (BINAB®TF and BINAB®T), *Bacillus pumilus, Pseudomonas fluorescens, Gliocladium roseum* -spray flowers and fruits – white flower bud to pink fruit-10^6 spores/ml; *G. roseum*-bee vectoring of flowers - 10^9 cfu/g of powder.
Passion fruit	Collar rot (*Rhizoctonia solani*)	*T. harzianum, Trichoderma* sp.
Amla	Bark splitting (*Rhizoctonia solani*)	*Aspergillus niger* AN 27

Table 4.5: Biological Control of Vegetable Crop Pests

Vegetable Crop	Pest	Biocontrol Agent/Dosage
Beans	Mite (*Tetranychus* spp)	*Phytoseiulus persimilis*-10 adults/plant or release 1-6 leaves with predatory mites.
Pigeon pea	Pod borer (*Helicoverpa armigera*)	*Ha* NPV-250 LE/ha
Potato	Cut worm (*Agrotis ipsilon, A. segetum*)	*Steinernema carpocapse, S. bicornutum, Heterorhabditis indica*
Tomato	Fruit borer (*Helicoverpa armigera*)	*Trichogramma brasiliensis/T. chilonis/T. pretiosum*-50,000/ha; *Ha* NPV-250 LE/ha
Brinjal	Fruit and shoot borer (*Leucinodes orbonalis*)	*S. carpocapse, H. indica*
Chilli	Fruit borer (*H. armigera*)	*Ha* NPV-250 LE/ha
Cabbage	Diamondback moth (*Plutella xylostella*)	*S. carpocapse, S. glaseri, S. carpocapse, S. feltiae, S. bicornutum, H. bacteriophora*
Mushroom	*Lycoriella auripila, L. mali, L. solani, Megaselia halterata*	*S. feltiae*

Table 4.6: Biological Control of Vegetable Crop Diseases

Crop	Disease/Pathogen/s	Biocontrol Agent/Mode of Application
French bean	Dry root rot (*Macrophomina phaseolina*)	*Pseudomonas cepacia* UPR5C-seed treatment.
	Wilt (*Fusarium oxysporum* f. sp. *Phaseoli*)	*Streptomyces* spp.- seed treatment.
Pea	Root rot (*Aphanmomyces euteiches*)	*P. fluorescens* PRA25-seed treat., *Pseudomonas cepacia* AMMD-seed treatment.
	Damping-off (*Pythium ultimum*)	*Pseudomonas cepacia* AMMD-seed treatment, *P. putida* NIR-seed treat.
	Wilt (*F. oxysporum* f.sp. *udum*)	*T. viride, T. harzianum, T. koningii* - seed treatment, *Aspergillus niger* AN27
Cluster bean	Bacterial blight (*Xanthomonas axonopodis* pv. *Cyamopsidis*)	*Aspergillus niger* AN27
Cabbage	Damping-of (*R. solani*)	*T. viride, T. harzianum, T. koningii* -seed treatment.
Cauliflower	Blight (*Alternaria brassicola*)	*Streptomyces gresioviridis*-seed treatment.
Okra	*Rhizoctonia solani*	*Bradyrhizobium japonicum, Rhizobium* spp.- seed treatment.
Tomato	Damping-off (*Pythium aphanidermatum*)	*T. viride, T. harzianum, Pseudomonas aeruginosa* 7NSK2
	Wilt (*Fusarium oxysporum* f. sp. *Lycopersici*)	*T. viride, T. harzianum, Aspergillus niger*, non-pathogenic *F. oxysporum, F. oxysporum* f. sp. *dianthi, P. fluorescens* strains Pf1, *P. putida, Pencillium oxalicum, Pythium oligandrum, Bacillus subtilis* strains FZB-G, *Streptomyces* spp.- seed treat., seed and soil treatment.
Potato	Black scurf (*R. solani*)	*T. harzianum, T. viride*- tuber treatment., *Aspergillus niger* AN27, *Verticillium biguttatum*-soil treat., *Laetisaria arvalis*-tuber treatment, Binucleate *Rhizoctonia*.
	Wilt (*Ralstonia solanacearum*)	*Bacillus cereus, B. subtilis*
Bell pepper	*Phytophthora capsici*	*T. viride, T. harzianum*-fruit treat.
	Damping-off (*Pythium aphanidermatum*)	*Streptomyces griseoviridis*-seed and soil treat.
Brinjal	Damping-off, Wilt (*Phytophthora* sp., *Pythium aphanidermatum, F. solani*)	*T. viride, T. harzianum, T. koningii*-seed and soil treatment.
	Collar rot (*Sclerotinia sclerotiorum*)	*T. viride, T. virens, Bacillus subtilis*-soil treatment.
Carrot	Soft rot (*Sclerotinia sclerotiorum*)	*Coniothyrium minitans*-soil treatment.
	Root rot (*R. Solani*)	*T. virens* GL-21
Radish	Wilt (*F. oxysporum* f.sp. *raphani*)	*P. fluorescens* strains WCS374, WCS417r – soil treatment.
	Root rot (*R. solani*)	*Laetisaria rosiepellis, Pythium acanthicum*-soil treatment.
Beet root	Damping-off (*Pythium debarryanum, P. ultimum*)	*Pencillium* spp + *P. fluorescens*-seed treat., *Pythium oligandrum*
Cucumber	Wilt (*Fusarium oxysporum* f. sp. *cucumerinum, R. Solani*)	*Colletotrichum orbiculare, F. oxysporum* f. sp. *niveum, Pseudomonas putida* 89B-27, *Serratia marcescens*, Tobacco necrosis virus.
	Powdery mildew	*Ampelomyces quisqualis*-Foliar spray
	Cucumber mosaic virus	*P. fluorescens* strain 89B-27

Contd...

Table 4.6–Contd...

Crop	Disease/Pathogen/s	Biocontrol Agent/Mode of Application
Watermelon	Wilt (*Fusarium oxysporum* f. sp. *solani, F. o.* f. sp. *Niveum*)	*T. viride, Aspergillus niger*-seed and soil treat., *Pencillium janczewskii.*
Muskmelon	Wilt (*F. oxysporum, F. Solani, R. Solani*)	*T. harzianum, Aspergillus niger* – seed treatment.
Onion	Soft rot (*Sclerotium cepivorum*)	*Chaetomium globosum, Trichoderma* sp. C62 -soil treatment.

Table 4.7: Biological Control of Ornamental Crop Diseases

Crop	Disease/Pathogen/s	Biocontrol Agent/Mode of Application
Rose	Grey mold (*Botrytis cineria*)	*T. viride, T. harzianum*-cutting treatment.
Gladiolus	Yellows and corm rot (*F. oxysporum* f. sp. *gladioli*)	*T. virens, T. harzianum*-corm and soil treatment.
Chrysanthemum	Wilt (*Fusarium oxysporum*	*T. harzianum*- soil appln.-160 kg/ha
	Rhizoctonia solani)	*Aspergillus niger* AN27
Carnation	Wilt (*F. oxysporum* f. sp. *Dianthi*)	*Pseudomonas fluorescens* strain WCS 417r - soil application; *P. putida* WCS 358r-root dip treatment.; *Alcaligenes* sp., *Bacillus* sp.; *Arthrobacter* sp., *Hafnia* sp.; *Serratia liquifaciens.*
Gerbera	*Phytophthora cryptogea*	*Trichoderma* spp.-soil treatment.
Narcissus	Wilt (*F. oxysporum* f. sp. *Narcissi*)	*Streptomyces griseoviridis*-bulb treatment.; *Minimedusa polyspora*-bulb treatment.
Zinnia	*Rhizoctonia solani*	*T. virens* GL-21, *T. virens* GL-20
Marigold	*Pythium ultimum*	*Glomus intraradices, G. mosseae*-soil treatment.

Table 4.8: Biological Control of Medicinal and Aromatic Crop Diseases

Medicinal/ Aromatic Crop	Disease/Pathogen/s	Biocontrol Agent
Opium poppy	Sclerotinia rot and blight (*Sclerotinia sclerotiorum*)	*T. harzianum, T. viride, T.koningii, T. virens*-soil treatment.
	Downy mildew (*Peronospora arborescens*)	*Trichoderma* spp.-seed treatment.
Periwinkle	*Phytophthora parasitica*	*Phytophthora parasitica* var. *nicotianae*- soil treat.
Jasmine	Root rot (*Macrophomina phaseolina*)	*T. viride, T. harzianum*- cutting treatment.
Chinese Rose	Wilt (*Fusarium oxysporum*)	*Aspergillus niger*- soil treatment.
Menthol Mint	Stolon decay (*Sclerotinia sclerotiorum*)	*T. harzianum, T. virens*- sucker treatment.
	Verticillium dahliae	*Verticillium nigrescens*

Table 4.9: Biological Control of Tuber Crop Diseases

Tuber Crop	Disease/Pathogen/s	Biocontrol Agent
Yam	*Botrytis theobromae*	*Trichoderma viride*
Cassava	*Phytophthora dechsleri*	*Trichoderma viride*
Elephant Foot Yam	*Sclerotium rolfsii*	*Trichoderma harzianum, T. pseudokoningii*

Table 4.10: Biological Control of Plantation Crop Pests

Plantation Crop	Pest	Biocontrol Agent/Dosage
Coconut	Black headed caterpillar (*Opisinia arenosella*)	*Goniozus nephantidis*-3000 adults/ha.
	Rhinoceros beetle (*Oryctes rhinoceros*)	Baculovirus-10 infected beetles/tree.
Arecanut	*Ischnaspis longirostris*	*Chilocorus nigrita*-20 to 50 adults/plant.
Coffee	Mealy bugs (*Planococcus* and *Pseudococcus* spp).	*Cryptolaemus montrouzieri*-2-10 beetles/infested plant.

Table 4.11: Biological Control of Plantation Crop Diseases

Crop	Disease/Causal Agent	Biocontrol Agents Reported
Coconut	Stem bleeding (*Thielaviopsis paradoxa; Ceratostomella paradoxa*)	*Trichoderma virens, T. harzianum*
Phosphobacteria	Basal stem rot (*Ganoderma lucidum*)	*G. virens, T. harzianum*
Areca nut	Bud rot (*Phytophthora* spp).	*Trichoderma* spp.
	Fruit rot (*Phytophthora arecae, Colletotrichum capsici*)	*Trichoderma* spp., *P. fluorescens*
	Foot rot/Anabe (*Ganoderma lucidum*)	*T. harzianum*
Tea	Red root rot (*Poria hypolateritia*)	*T. harzianum*
	Brown root (*Fomes noxius*)	*G. virens, T. harzianum*
	Black root rot (*Rosellina arcuata*)	*G. virens, T. harzianum*
Coffee	Black root (*Pellicularia koleroga*)	*G. virens, T. harzianum*
	Brown root (*Fomes noxius*)	*G. virens, T. harzianum*
	Santhaveri wilt (*F. oxysporum* f. sp. *Coffeae*)	*G. virens, T. harzianum*
Rubber	Brown rot (*Phellinus noxius*)	*T. viride, T. harzianum, T. hamatum*
Betelvine	Foot and root rot (*Phytophthora parasitica* pv. *Piperina*)	*T. viride, T. harzianum*-soil treatment.
	Collar rot (*Sclerotium rolfsii*)	*T. harzianum, T. viride, T. koningii, T. virens*-soil treatment.

Table 4.12: Biological Control of Spice Crop Diseases

Spice crop	Disease/Causal Organism	Effective Biocontrol Agents/Mode of Application
Black pepper	Foot rot (*Phytophthora capsici*)	*Trichoderma harzianum, T. virens, Glomus fasciculatum-*soil treat.; *Pseudomonas fluorescens, Bacillus* sp.-foliar spray
	Anthracnose (*Colletotrichum gloeosporoides*)	*P. fluorescens-*foliar spray
	Slow decline (*Radopholus similis, Meloidogyne incognita, Phytophthora capsici*)	*T. harzianum, T. virens, Paecilomyces lilacinus, Pochonia chlamydosporia-* soil treatment.
Cardamom	Damping off (*Pythium vexans*)	*T. harzianum, T. viride-*soil treat. in solarized nursery beds.
	Clump rot/Rhizome rot (*Pythium vexans, Rhizoctonia solani, Meloidogyne incognita*)	*T. harzianum-*soil treatment.
	Capsule rot (*Phytophthora meadii, P. nicotianae* var. *nicotianae*)	*T. harzianum, T. viride, T. virens, T. hamatum-*soil treatment.
Ginger	Rhizome rot (*Pythium aphanidermatum, P. myriotylum*)	*T. harzianum, T. virens-*soil solarization + soil treat.; *T. viride, P. fluorescens-*seed treat.; *Aspergillus niger* AN27-soil treatment.
	Yellows (*Fusarium oxysporum* f. sp. *zingiberi, M. incognita*)	*T. harzianum, T. virens, T. hamatum-*soil solarization + soil treat., Rhizome treatment.
	Bacterial wilt (*Ralstonia solanacearum*)	Avirulent *Ralstonia solanacearum, P. fluorescens,* Endophytic bacteria-soil treatment.
Turmeric	Rhizome rot (*Fusarium* sp., *Pythium graminicolum, P. aphanidermatum, R. similis, M. incognita*)	*T. harzianum, T. viride, T. virens* -soil treatment.
Fenugreek	Root rot (*R. solani*)	*T. viride, P. fluorescens-*seed treatment.
Coriander	Root rot/wilt (*Fusarium oxysporum* f. sp. *Corianderii*)	*T. viride, T. harzianum, Streptomyces* sp.-seed treatment.
Cumin	Wilt (*F. oxysporum* f.sp. *cumini*)	*Trichoderma* spp., *T. virens-*soil treatment.
Vanilla	Root rot (*Phytophthora meadii, F. oxysporum* f.sp. *vanillae*)	*T. harzianum, P. fluorescens-*soil treatment.
Mustard	Damping-off (*P. aphanidermatum*)	*T. viride, T. harzianum-*seed and soil treatment.

Avermectins

Avermectins are the secondary metabolites of an Actinomycete, *Streptomyces avermitilis*, which are highly effective at a very low concentration against mite pests and plant parasitic nematodes. Scientists at the Indian Institute of Horticultural Research, Bangalore, for the first time in India isolated 6 strains of *S. avermitilis* and showed their effectiveness for the management of root-knot nematodes infecting tomato, egg plant, chilli, carnation and gerbera (Parvatha Reddy and Nagesh, 2002) (Table 4.13).

Mechanical and Physical Controls

Methods included in this category utilize some physical component of the environment, such as temperature, humidity, or light, to the detriment of the pest. Common examples are tillage, flaming, flooding, soil solarization and plastic mulches to kill pests.

Table 4.13: Management of Horticultural Crop Nematodes Using Avermectins

Horticultural Crop	Nematode	Avermectin/dose
Citrus	*T. semipenetrans*	A monthly rate of 1.1 kg a.i. per ha
Tomato	*M. incognita*	Aqueous solution of avermectins (250 ml of 0.001 per cent/m² nursery bed); charcoal formulations of *S. avermitilis* at 100 g/m² nursery bed
Carnation and gerbera (commercial polyhouses)	*M. incognita*	Post-plant treatment at 250 ml/m² at two intervals (6 and 12 months after planting)

Heat or steam sterilization of soil is commonly used in greenhouse operations for control of soil-borne pests. Floating row covers over vegetable crops exclude flea beetles, cucumber beetles, and adults of the onion, carrot, cabbage, and seed corn root maggots. Insect screens are used in greenhouses to prevent aphids, thrips, mites, and other pests from entering ventilation ducts. Large, multi-row vacuum machines are used for pest management in strawberries and vegetable crops. Cold storage reduces post-harvest disease problems on produce.

Although generally used in small or localized situations, some methods of mechanical/physical control are finding wider acceptance because they are generally more friendly to the environment.

Chemical Controls (Reduced-Risk Pesticides)

Included in this category are both synthetic pesticides and botanical pesticides.

Synthetic Pesticides

They comprise a wide range of man-made chemicals used to control insects, mites, nematodes, plant diseases and vertebrate and invertebrate pests. These powerful chemicals are fast acting and relatively inexpensive to purchase.

Pesticides are the option of last resort in IPM programs because of their potential negative impacts on the environment, which result from the manufacturing process as well as from their application on the farm. Pesticides should be used only when other measures, such as biological or cultural controls, have failed to keep pest populations from approaching economically damaging levels.

If chemical pesticides must be used, it is to the grower's advantage to choose the least-toxic pesticide that will control the pest but not harm non-target organisms such as birds, fish and mammals. Pesticides that are short-lived or act on one or a few specific organisms are in this class. Examples include insecticidal soaps, horticultural oils, copper compounds (*e.g.*, Bordeaux mixture), sulfur, boric acid and sugar esters.

Biorational Pesticides

Biorational pesticides are generally considered to be derived from naturally occurring compounds or are formulations of microorganisms. Biorationals have a narrow target range and are environmentally benign. Formulations of *Bacillus thuringiensis*, commonly known as *Bt*, are perhaps the best-known biorational pesticide. Other examples include silica aerogels, insect growth regulators, and particle film barriers.

A relatively new technology, particle film barriers are currently available under the trade name Surround® WP Crop Protectant. The active ingredient is kaolin clay, an edible mineral long used as an anti-caking agent in processed foods, and in such products as toothpaste and Kaopectate. There appears to be no mammalian toxicity or any danger to the environment posed by the use of kaolin in pest control. The kaolin in Surround is processed to a specific particle size range, and combined with a sticker-spreader. Non-processed kaolin clay may be phytotoxic. Surround is sprayed on as a liquid, which evaporates, leaving a protective powdery film on the surfaces of leaves, stems and fruit. Conventional spray equipment can be used and full coverage is important. The film works to deter insects in several ways. Tiny particles of the clay attach to the insects when they contact the plant, agitating and repelling them. Even if particles do not attach to their bodies, the insects may find the coated plant or fruit unsuitable for feeding and egg-laying. In addition, the highly reflective white coating makes the plant less recognizable as a host.

Sugar Esters

Sugar esters have performed as well as or better than conventional insecticides against mites and aphids in apple orchards; psylla in pear orchards; whiteflies, thrips and mites on vegetables. However, sugar esters are not effective against insect eggs. Insecticidal properties of sugar esters were first investigated a decade ago when a scientist noticed that tobacco leaf hairs exuded sugar esters for defense against some soft-bodied insect pests. Similar to insecticidal soap in their action, these chemicals act as contact insecticides and degrade into environmentally benign sugars and fatty acids after application.

Inorganic Chemicals

Spray application of K_2HPO_4 or KH_2PO_4 at 3.5 g/litre of water is reported to control powdery mildew in rose and carnation. Similarly, the above treatment was also found effective for management of powdery mildew on mango, grapes and cucurbits.

Strobilurin Fungicides

Strobilurin fungicides are also called Qo inhibitors as they act on cytochrome Qo of the fungi. The Basidiomycetous fungus, *Strobilurus tenacellus*, produces antibiotics to ward off competition from other fungi. Based on this principle, several fungicides have been developed namely Azoxystrobin, Kresoxy methyl, Metominostrobin, Trifloxystrobin, Picoxystrobin, Pyraclostrobin, Famoxadone and Fenomidone during 1996-2001. Within 4 years, sale of these fungicides totaled $620 million, accounting for 10 per cent of total fungicide market in the world. This success is unparalleled in the history of fungicides.

Strobilurin fungicides are naturally occurring compounds and hence eco-friendly, highly systemic, have unique mode of multisite action, hence development of resistance is remote. They have broad spectrum of activity on all groups of fungi and registered in 72 countries on 84 crops representing over 400 crops/diseases systems.

Botanical Pesticides

They can be as simple as pureed plant leaves, extracts of plant parts or chemicals purified from plants. Pyrethrum, neem formulations, and rotenone are examples of botanicals. Some botanicals are broad-spectrum pesticides. Others, like ryania, are very specific. Botanicals are generally less harmful in the environment than synthetic pesticides because they degrade quickly, but they can be just as deadly to beneficials as synthetic pesticides. However, they are less hazardous to transport and in some cases can be formulated on-farm. The manufacture of botanicals generally results in fewer toxic by-products.

Neem products like cake, oil, neem seed kernel extract (NSKE), neem seed powder extract (NSPE), pulverized NSPE and soaps are being used extensively to manage horticultural crop pests [bean fly (*Ophiomyia phaseoli*); serpentine leaf miner (*Liriomyza trifolii*) on several crops; cucurbit fruit fly (*Bactrocera cucurbitae*); tomato fruit borer (*Helicoverpa armigera*); brinjal fruit and shoot borer (*Leucinodes orbonalis*); water melon and chilli thrips (*Thrips* spp.); chilli yellow mite (*Polyphagotarsonemus latus*) and okra leaf hopper (*Amrasca biguttulla biguttulla*) (Krishna Moorthy and Krishna Kumar, 2002).

The soap sprays were highly effective on leaf hoppers, aphids, red spider mites and white flies in many vegetables, but moderately effective on thrips in water melon and chillies (Table 4.14).

Table 4.14: Insect Pests on which Soaps were Found Effective

Cabbage and cauliflower	DBM, leaf webber, aphids, young *Spodoptera* larva
Tomato	White fly, red spider mites, fruit borer (egg laying stage), leaf miner
Okra	Leaf hopper, white fly, aphids
Cucurbits	Fruit fly, leaf miner
Mango	Leaf hopper
Ornamental crops	Mites, white fly

Compost Teas

They are the most commonly used for foliar disease control and applied as foliar nutrient sprays. The idea underlying the use of compost teas is that a solution of beneficial microbes and some nutrients is created, then applied to plants to increase the diversity of organisms on leaf surfaces. This diversity competes with pathogenic organisms, making it more difficult for them to become established and infect the plant.

An important consideration when using compost teas is that high-quality, well-aged compost be used, to avoid contamination of plant parts by animal pathogens found in manures that may be a component of the compost. There are different techniques for creating compost teas. The compost can be immersed in the water, or the water can be circulated through the compost. An effort should be made to maintain an aerobic environment in the compost/water mixture.

Case Studies

Mango IPM (Uttarakhand)

☆ Spraying of copper oxychloride (0.3 per cent) for control of die-back, anthracnose and red rust diseases wherever they appeared during September-October.

☆ Ploughing of orchard in November-December to expose pupae of fruit flies, midges, leaf hoppers and eggs of mealy bugs to natural enemies.

☆ Polythene banding of tree trunk in December-January and application of 5 per cent NSKE and *Beauveria bassiana* in January.

☆ Spraying of sulfex (0.2 per cent) for the control of powdery mildew disease.

☆ Spraying of *Verticillium lecanii* on orchards for control of hoppers.

☆ Fixing methyl eugenol traps (wooden blocks impregnated with methyl eugenol) to control fruit flies from April to August.

☆ Mechanical removal of mango leaf webber larvae and webs by leaf web removing device (developed by the Central Institute of Sub-tropical Horticulture, Lucknow) from April to September-October.

The IPM package was successfully validated in 16.8 ha of mango orchards in Gulabkhera, Habibpur, Budhadia, Pathakganj, Rehmankhera and Kanar villages in Malihabad and Kakori belt of mango near Lucknow on Dashehari variety during 2000-04. As a result of adoption of IPM, yield of mango increased from 6.0-9.0 tonnes/ha as it was 3.5-7.0 tonnes/ha earlier in non-IPM orchards. By adopting IPM, the mango growers in that area earned a profit of Rs.30,000/- to Rs.55,000/- while the farmers who did not adopt IPM, earned a profit of Rs.17,000/- to 35,000/- per ha only (Table 4.15) (Amerika Singh *et al.*, 2004).

Table 4.15: Economics of IPM in Mango

Parameters	IPM Plots	Non-IPM Plots	Per cent Increase
Yield tonnes/ha	6.0-9.0	3.5-7.0	28.57-71.43
Net profit (Rs/ha)	30,000-55,000	17,000-35,000	57.14-76.47

Apple IPM (Himachal Pradesh)

☆ Use of urea (5 per cent) at leaf shedding stage for early decomposition of the infested leaves and to encourage the population of antagonists in the plant rhizosphere.

☆ Use of Bordeaux paint during autumn on the naked plant stem to overcome direct effect of UV rays on the plant skin to reduce sun burn and cankers disease complex.

☆ Overwinter spray of Bordeaux mixture as eradicative action to pathogens and total disinfection of plant surface.

☆ Use of Neemarin at pink bud stage *i.e.* pre-bloom stage to manage blossom thrips population.

☆ Use of *Bacillus thurungiensis* at fruit development stage for the management of fruit scrapper insect pests.

☆ Use of *Trichoderma viride* (Bioderma) for the control of root rot fungus.

☆ Use of Bordeaux mixture for the control of root and collar rots.

During 2001-04, IPM for apple crop was validated and promoted in 30 ha of orchards in Kotkhai, Jhubbal, Theneder and Rohroo villages of Himachal Pradesh. By adopting IPM package, farmers were able to harvest 580 boxes (4.99 tonnes/ha) of apple in IPM plots as compared to 380 boxes (4.12 tonnes/ha) in case of non-IPM farmers. Apple growers who adopted IPM earned a profit of Rs. 145733/- while the farmers who did not adopt IPM earned a profit of Rs.110889/-. Average benefit cost ratio of IPM to non-IPM was 4.07 to 3.01 (Table 4.16) (Amerika Singh *et al.*, 2004).

Table 4.16: Economics of Apple IPM

Parameters	IPM Plots	Non-IPM Plots	Per cent Increase
Yield (tonnes/ha)	4.99	4.12	21.11
Net profit (Rs/ha)	145733	110889	31.42
Benefit: cost ratio	4.07	3.01	35.21

Tomato Fruit Borer IPM

Use of African marigold (*Tagetes erecta*) as a trap crop for the management of fruit borer on tomato involves planting one row of 45 days-old marigold seedlings after every 16 rows of 25 days-old tomato seedlings and spraying of *Ha* NPV at 250 LE/ha or 4 per cent NSKE or 4 per cent pulverized NSPE, 28 and 45 DAP coinciding with peak flowering (Srinivasan *et al.*, 1994) (Figure 4.1).

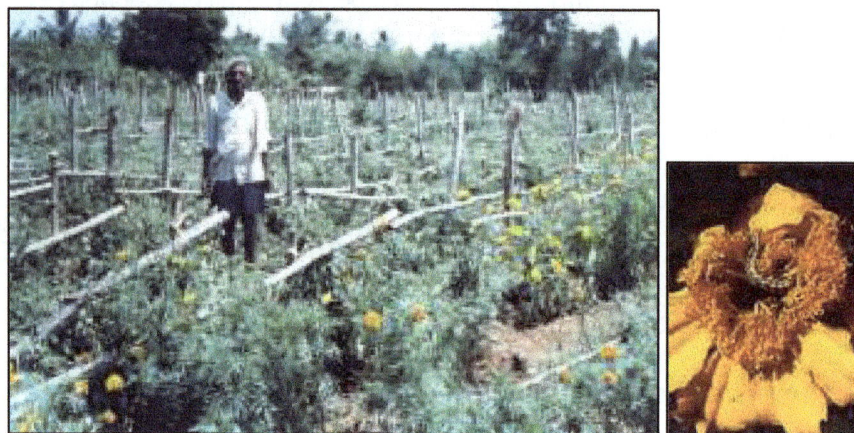

Figure 4.1: Integrated Management of Tomato Fruit Borer Using African Marigold as a Trap Crop

Table 4.17: Economics of Tomato Fruit Borer IPM

Centre	Yield (tonnes/ha)	Net Returns (Rs.)	Benefit : Cost Ratio
Bangalore - IPM	74.03	249721	4.82
Bangalore - Non IPM	45.05	69704	0.61
Varanasi - IPM	14.25	39917	3.30
Varanasi - Non IPM	13.00	38167	2.02
Ranchi - IPM	22.29	56705	1.87
Ranchi - Non IPM	18.77	41776	1.32

Figure 4.2: IPM in Cabbage Using Indian Mustard as a Trap Crop

During 2001-04 IPM technology in tomato was validated and promoted in more than 40 ha in 42 villages covering 88 families located 40 km from Bangalore. Similarly, near Varanasi also IPM technology was validated in 8 villages in about 40 ha area covering 100 families in tomato. Near Ranchi, IPM technology was validated and promoted in 20 villages with the support of 100 farming families covering an area of 40 ha together. In IPM validation studies conducted at three locations (Bangalore, Varanasi and Ranchi), IPM fields recorded higher tomato fruit yields of 74.038, 14.250 and 22.293 tonnes/ha as compared to 45.056, 18.772 and 18.772 tonnes/ha in non-IPM fields, respectively (Amerika Singh *et al.*, 2004) (Table 4.17).

Cabbage IPM

IPM using Indian mustard as a trap crop involves planting of paired rows of mustard after every 25 rows of cabbage/cauliflower (Figure 4.2) and spraying of NSKE(4 per cent) at primordial formation. Two more sprays of NSKE(4 per cent) may be given at 10-15 days interval after the first spray. The IPM gave 152 per cent more returns than pure cabbage crop. IPM controls diamondback moth (*Plutella xylostella*); leaf webber (*Crocidolomia binotalis*); stem borer (*Hellula undalis*); aphids (*Brevicorne brassicae, Hyadaphis erysimi*) and bug (*Bagrada cruciferarum*) (Srinivasan and Krishna Moorthy, 1991).

The IPM gave 60 per cent more yield and 152 per cent more returns than pure cabbage crop (Table 4.18) (Khaderkhan *et al.*, 1998b).

Table 4.18: Economics of Cabbage IPM

Practice	Yield (tonnes/ha)	(Per cent) Increase	Net Teturns (Rs.)	(Per cent) Increase
Farmers' practice	20	--	19,817	--
IPM practice	32	60	49,251	152

In another study, the benefit-cost ratio in IPM and non-IPM plots was 2.42 and 0.83, respectively (Table 4.19) (Krishna Moorthy *et al.*, 2003).

Table 4.19: Economics of Cabbage IPM

Treatment	Yield (tonnes/ha)	Net Returns (Rs.)	Benefit : Cost Ratio
IPM plots	55	30085	2.42
Non-IPM plots (farmers' practice)	35	5090	0.83

Okra Fruit Borer IPM

☆ Planting of yellow vein mosaic virus resistant (YVMV) hybrids *viz.* Sun-40 and Pusa Makhmali.

☆ Sowing of sorghum/maize as border crop.

☆ Installation of yellow sticky polythene traps smeared with castor oil and delta traps set up for white fly and other small sucking pests.

☆ Erection of bird perches at 25/ha for facilitating predation of borer larvae.

☆ Installation of pheromone traps at 5/ha for monitoring *Earias vitella*.

☆ Three sprays of NSKE (5 per cent) for hopper, white fly and mites starting at 28 DAS.

☆ Five releases of *Trichogramma chilonis* at 1 lakh/ha starting from 42 DAS at weekly interval.

☆ Rouging out YVMV affected plants from time to time.

In okra crop, IPM technology was validated in about 3 ha area in Raispur village near Ghaziabad during 2003-04. IPM fields gave higher yields of 10.305 tonnes/ha as compared to 7.246 tonnes/ha in non-IPM fields (Amerika Singh *et al.*, 2004) (Table 4.20).

Table 4.20: Yield and Economics of IPM in Okra

Parameters	IPM Plots	Non-IPM Plots	Per cent Increase
Yield (tonnes/ha)	10.30	7.24	42
Net returns (Rs/ha)	64797	34678	86
Benefit: cost ratio	1.28	0.72	77

Black Pepper Foot Rot and Nematodes Disease Complex IPM

Integrated management of foot rot (*Phytophthora capsici*) and nematodes (*Meloidogyne incognita* and *Radopholus similis*) on black pepper was achieved by:

☆ Mixing VAM and *Trichoderma harzianum* in solarized nursery mixture to raise healthy and robust seedlings.

☆ Application of *T. harzianum* and FYM in planting pit.

☆ Field application of neem cake at 1 kg/vine mixed with 50 g of *T. harzianum* during August (Sarma, 2003).

Cardamom Capsule Rot IPM

Management of capsule rot (*Phytophthora* spp.) of cardamom was achieved by two applications of *T. harzianum* at 1 kg/plant (grown on decomposed coffee pulp and FYM in 1:1 ratio) during May and July integrated with foliar spray of Akomin (potassium phosphonate) (Anandraj and Eapen, 2003).

Banana Burrowing Nematode IPM

Integration of neem cake at 200 g/plant with *Glomus mosseae* at 100 g/plant (containing 25-30 chlamydospores/g of inoculum) was the most effective in reducing *R. similis* population both in soil and roots, while karanj cake with *G. mosseae* gave maximum increase in fruit yield of banana. Mycorrhizal root colonization and number of chlamydospores of *G. mosseae* were maximum in neem cake amended soil (Table 4.21) (Parvatha Reddy *et al.*, 2002).

Tomato Root-Knot Nematode IPM

In nursery, integration of *Pasteuria penetrans* (at 28×10^4 spores/m^2), *P. lilacinus* (at 10 g/m^2 with 19×10^9 spores/g) and neem cake (at 0.5 kg/m^2) gave maximum increase in plant growth and number of seedlings/bed. Parasitization of *M. incognita* females was the highest when neem cake was integrated with *P. penetrans*, while parasitization of eggs was the highest when neem cake was integrated with *P. lilacinus*. In field, planting of tomato seedlings (raised in nursery beds amended with neem cake + *P. penetrans*) in pits incorporated with *P. lilacinus* (at 0.5 g/plant) gave the least root galling and nematode multiplication rate and increased fruit weight and yield of tomato (Table 4.22) (Parvatha Reddy *et al.*, 1997).

Table 4.21: Effect of *Glomus mosseae* and Oil Cakes on Population of
Radopholus similis and Yield of Banana

Treatment	Dose (g)/plant	Population of R. similis		Yield(kg)/plant
		Roots (10 g)	Soil (250 ml)	
G. mosseae	200	112	122	8.64
Castor cake	400	146	132	8.18
Karanj cake	400	118	128	10.34
Neem cake	400	118	112	8.91
G. mosseae + Castor cake	100 + 200	90	108	12.68
G. mosseae + Karanj cake	100 + 200	76	80	16.61
G. mosseae + Neem cake	100 + 200	48	62	14.80
Control	--	218	184	5.45
CD (P = 0.05)		11.97	8.31	0.84

Table 4.22. Effect of Neem Cake, *Pasteuria penetrans* and *Paecilomyces lilacinus* on
Root Galling, Nematode Multiplication Rate and Yield of Tomato

Treatment		Root-knot Index	Yield (kg)/ 6 m²
Nursery (m²)	Main Field (per plant)		
Neem cake -1 kg	P. lilacinus – 0.5 g	3.4	9.168
Neem cake -1 kg	P. penetrans (28 x 10⁴ spores)	3.2	9.312
P. lilacinus – 20 g	P. penetrans (28 x 10⁴ spores)	3.0	9.504
P. penetrans (28 x 10⁷ spores)	p. lilacinus – 0.5 g	2.9	9.624
Neem cake – 0.5 kg + P. lilacinus – 10 g	P. penetrans (28 x 10⁴ spores)	2.5	9.672
Neem cake – 0.5 kg + P. penetrans (28 x 10⁴ spores)	P. lilacinus – 0.5 g	2.0	9.984
Neem cake – 0.5 kg + P. lilacinus – 10 g + P. penetrans (28 x 10⁴ spores)	—	2.6	9.600
Control	—	4.6	8.352
CD (P = 0.05)		0.14	0.100

Transfer of Technology

The constraint for wider adoption of eco-friendly crop protection technologies is transfer of technology. Crop protection technologies developed have not reached the small and marginal farmers. Unless these technologies are assessed in farmer's fields and refined to suit local conditions, the fruits of research will not benefit farmers. Researchers and extension personnel should work hand in hand for successful transfer of crop protection technologies. Communication media like radio, TV, audio and video cassettes, agri portals, farmer's field schools and KVK's should be used for effective transfer of technologies.

Conclusions

Globalization driven by WTO is opening up fantastic opportunities for export of horticultural products and processed food from India. It is a revolution, which is taking place and our farmers' will miss this golden opportunity if they are not equipped with the right crop protection technologies to produce horticultural products of international standards without pesticide residues. The challenge before the plant protection scientists is to prevent crop losses due to pests before and after harvest without harming the environment. There is need to develop low input and eco-friendly crop protection technologies so as to be very competitive in the international market.To quote Swaminathan(2000)"The ever-green revolution will be triggered by farming systems that can help produce more from the available land, water and labour resources without either ecological or social harm".

References

Adams, S., 1997. See in' red: colored mulch starves nematodes. *Agricultural Research*, October, p. 18.

Anandraj, M. and Eapen, S.J., 2003. Achievements in biological control of diseases of spice crops with antagonistic organisms at Indian Institute of Spices Research, Calicut. In: *Current Status of Biological Control of Plant Diseases Using Antagonistic Organisms in India*, (Eds.) B. Ramanujam and R.J. Rabindra. Project Directorate of Biological Control, Bangalore, pp. 189–215.

Benbrook, C.M., 1996. *Pest Management at the Crossroads*. Consumers Union, Yonkers, NY. 272 p.

Couch, G.J., 1994. The use of growing degree days and plant phenology in scheduling pest management activities. *Yankee Nursery Quarterly*. Fall. p. 12–17.

Krishna Moorthy, P.N. and Krishna Kumar, P.N., 2002. Advances in the use of botanicals for the IPM of major vegetable pests. *Proc. of the Int. Conf. on Vegetables,* Bangalore. Dr. Prem Nath Agri. Sci. Foundn., Bangalore, pp. 262–272.

Mani, M., 2001. Biological control of fruit crop pests. In: *Integrated Pest Management in Horticultural Ecosystems*, (Ed.) P. Parvatha Reddy, A. Verghese and N.K. Krishna Kumar. Capital Publishing Co., New Delhi, pp. 93–107.

Marsh, S., 2000. Mechanized delivery of beneficial insects. *The IPM Practitioner*, April, p. 1–5.

Parvatha Reddy, P. and Nagesh, M., 1998. Integrated nematode management in horticultural crops: Eco–friendly approaches for the future. In: *Advances in IPM for Horticultural Crops*, (Eds.) P. Parvatha Reddy, N.K. Krishna Kumar and A. Verghese. Association for Advancement of Pest Management in Horticultural Ecosystems, Bangalore, pp. 275–300.

Parvatha Reddy, P. and Nagesh, M., 2002. *Avermectins: Isolation, Fermentation, Preliminary Characterization and Screening for Nematicidal Activity*. Tech. Bull. – 17, Indian Inst. of Hort. Res., Bangalore, 28 pp.

Parvatha Reddy, P., Nagesh, M. and Devappa, V., 1997. Effect of integration of *Pasteuria penetrans, Paecilomyces lilacinus* and neem cake for the management of root-knot nematode infecting tomato. *Pest Managmt. Hort. Ecosystems*, 3: 100–104.

Parvatha Reddy, P., Rao, M.S. and Nagesh, M., 2002. Integrated management of burrowing nematode (*Radopholus similis*) using endomycorrhiza (*Glomus mosseae*) and oil cakes. In: *Banana*, (Eds.) H.P. Singh and K.L. Chadha. AIPUB, Trichy, pp. 344–348.

Reichert, S.E. and Leslie, B., 1989. Prey control by an assemblage of generalist predators: Spiders in garden test systems. *Ecology*. Fall. p. 1441–1450.

Sarma, Y.R., 2003. Recent trends in the use of antagonistic organisms for the disease management in spice crops. In: *Current Status of Biological Control of Plant Diseases Using Antagonistic Organisms in India*, (Eds.) B. Ramanujam and R.J. Rabindra. Project Directorate of Biological Control, Bangalore, pp. 49–73.

Singh, Amerika, Trivedi, T.P., Sardana, H.R., Sabir, N., Krishna Moorthy, P.N., Pandey, K.K., Sengupta, A., Ladu, L.N. and Singh, D.K., 2004. Integrated pest management in horticultural crops: A wide area approach. In: *Crop Improvement and Production Technology of Horticultural Crops*, (Eds.) K.L. Chadha, B.S. Ahluwalia, K.V. Prasad and S.K. Singh. Hort. Soc. of India, New Delhi, pp. 621–636.

Srinivasan, K. and Krishna Moorthy, P.N., 1991. Indian mustard as a trap crop for management of major lepidopterous pests on cabbage. *Trop. Pest Mangmt.*, 37: 26–32.

Srinivasan, K., Krishna Moorthy, P.N. and Raviprasad, T.N., 1994. African marigold as a trap crop for the management of the fruit borer *Helicoverpa armigera* on tomato. *Int. J. Pest Mangmt.*, 40: 56–63.

Swaminathan, M.S., 2000. For an evergreen revolution. *The Hindu Survey of Indian Agriculture 2000*, pp. 9–15.

2012, Nutri-Horticulture
Editor: Professor K.V. Peter
Published by: DAYA PUBLISHING HOUSE, NEW DELHI

Pages 75-97

Chapter 5

In vitro Breeding Strategies in the Development of Australian Native Plants

Zul Zulkarnain[1], Tanya Tapingkae[2] and Acram Taji[3]
[1]*Agricultural Faculty, University of Jambi, Indonesia*
[1]*Faculty of Agricultural Technology, Chiang Mai Rajabhat University, Thailand*
[3]*Queensland University of Technology, Brisbane, Australia*
E-mail: acram.taji@qut.edu.au

Plant tissue culture is a technique that exploits the ability of many plant cells to revert to a meristematic state. Although originally developed for botanical research, plant tissue culture has now evolved into important commercial practices and has become a significant research tool in agriculture, horticulture and in many other areas of plant sciences.

Plant tissue culture is the sterile culture of plant cells, tissues or organs under aseptic conditions leading to cell multiplication or regeneration of organs or whole plants. The steps required to develop reliable systems for plant regeneration and their application in plant biotechnology are reviewed in countless books. Some of the major landmarks in the evolution of *in vitro* techniques are summarised in Table 5.1.

In this chapter the current applications of this technology to agriculture, horticulture, forestry and plant breeding are briefly described with specific examples from Australian plants where applicable.

Clonal Micropropagation

The general term tissue culture is used to refer to the *in vitro* culture of various types of plant parts including stems, leaves, roots, flowers, callus, cells, protoplasts and embryos. These parts, known as explants, are isolated from the *in vivo* condition and cultured on an artificial sterile medium so that they regenerate and differentiate into new intact plants (Street, 1973). A more specific term, micropropagation, is applied to indicate the use of tissue culture techniques to the plant propagation system started with very small plant parts (explants) grown aseptically in a test tube or other similar containers (Hartmann *et al.*, 1990). However, it is often found in practice that these two terms are used interchangeably to describe any plant propagation technique involving aseptic culture.

Table 5.1: Historical Landmarks in the Evolution of Plant Cell and Tissue Culture Technology

---

☆ Gautheret, Nobecourt and White achieved the first success in developing plant tissue culture. Gautheret (1934) obtained callus formation from cultured explants of tree cambium and phloem tissue.

☆ After the discovery of cytokinins by Skoog and co-workers, Skoog and Miller (1957) observed that the shoot and root formation are controlled by the auxin/cytokinin balance.

☆ *In vitro* somatic embryogenesis was first described by Steward *et al.* (1958b) and Reinert (1958a).

☆ Anther culture and production of haploid plants was achieved by Guha and Maheshwari (1964; 1966) and by Bourgin and Nitsch (1967).

☆ Protoplast culture, fusion and development of somatic hybrids were described in 1960s and 1970s (Cocking, 1960; Belliard *et al.*, 1979; Gleba and Sytnik, 1984).

☆ During 1980s, recombinant DNA technology and production of transgenic plants were achieved (Schell,1987 and Schell and Vasil,1989).

Micropropagation is widely used by many private and publicly funded companies around the world for mass production of plants. Many ornamental plants (orchids, gerberas, ferns, roses, carnations, lilies, etc), vegetables (tomato, carrot, celery, etc) food crops (cassava, potato, sugarcane), fruits (banana, pineapple, apple, strawberry, cherries), plantation crops (coconut, tea, cocoa) and spices (clove, cinnamon, ginger, turmeric) are successfully propagated using tissue culture techniques. Generally speaking success is relatively rapid with herbaceous plants. Many Australian native plants, including numerous rare woody plant species, are successfully propagated using *in vitro* micropropagation techniques (Speer, 1993; Johnson and Burchett, 1996; Taji and Williams, 1996; Taji *et al.*, 1997)

In vitro multiplication involves three main pathways. These are axillary shoot formation, adventitious shoot production and somatic embryogenesis. Axillary shoot formation is the true-to-type multiplication of plants from pre-existing meristems (axillary and apical). This type of tissue culture involves an expanded shoot of terminal and lateral growing points where axillary shoot proliferation is promoted and the growth of terminal shoot is suppressed. Hartmann *et al.* (1990) stated that this condition enables the multiplication of microshoots, which can be excised and rooted *in vitro* to produce microplants, or which can be cut into microcuttings to be rooted outside the *in vitro* system.

The advantage of using axillary shoot proliferation from meristem, shoot tip or bud as a means of regeneration is that the incipient shoots have already been differentiated *in vivo*. Only shoot elongation and root differentiation are required to establish a complete plant. *In vitro* organogenesis and embryogenesis, on the other hand, must undergo developmental changes which usually involve callus formation (Hu and Wang, 1983) that frequently causes genetic mutation in regenerated propagules.

The term adventitious shoot refers to the shoots arising from any plant parts other than the leaf axil or shoot apex (Bhojwani and Razdan, 1983). In other words, adventitious shoot production is *de novo* bud formation on structures without a pre-existing meristem, for example leaf segments of *Begonia rex*, leaf petiole of *Saintpaulia sp.* or root segments of many other species. Hartmann *et al.* (1990) stated that adventitious shoot induction includes initiation of adventitious shoot development either directly from the explant or indirectly from the callus that is produced on the explant as a result of wounding and growth regulator treatment.

Although adventitious shoot induction offers a great potential for rapid plant propagation, Boulay (1987) claimed that there are several problems associated with this technique. The first is the problem of obtaining true-to-type genetic copies which is the main purpose of clonal propagation. This particularly appears when the adventitious shoot proliferation passes through an intervening callus stage. There is also variation in growth behaviour of resulted plants. It is found that some copies are not always juvenile. Some clones manifest plagiotropic growth and poor vigour, whereas others exhibit orthotropic growth and vigour typical of young seedlings. These problems may arise from damage to meristems of clonal stock during the propagation process.

Somatic Embryogenesis as a Mean of Plant Regeneration

Somatic embryogenesis was first demonstrated by Steward *et al.* (1958c; 1958b) and Reinert (1959) in cultures of carrot tissues. Somatic embryogenesis is a remarkable developmental process by which non-sexual cells undergo a developmental sequence, similar to that seen in zygotic embryos, without the need for sexual reproduction. Somatic embryogenesis is influenced by the genotype, explant source, developmental stage, culture medium and other inductive factors, like the mineral composition of the media, type and concentration of carbon sources, amino acids, heavy metal ions and manipulation of environmental conditions (Tapingkae *et al.*, in press).

Somatic embryogenesis has become one of the most desired pathways in the regeneration of plants via tissue culture because it bypasses the necessity of time-consuming and costly manipulation of individual explants, which is a problem with organogenesis (Folta and Dhingra, 2006; Carneros *et al.*, 2009). To date somatic embryoids are obtained in a large number of monocotyledonous and dicotyledonous plant species including some Australian plants (Lakshmanan and Taji, 2000; Taji and Williams, 2005).

Application of somatic embryogenesis to Australian plants was reported for *Eucalyptus* species (Watt *et al.*, 1991; McComb *et al.*, 1996; Prakash and Gurumurthi, 2010), Australian cotton cultivars (Zhang *et al.*, 2009; Yan *et al.*, 2010), sorghum (Sargent *et al.*, 1997), papaya (Sieler *et al.*, 1993; Ernawati *et al.*, 1997), sedges and rushes (Sieler *et al.*, 1993).

Somatic embryogenesis provides a possible *in vitro* system for a number of difficult to propagate Australian native species including *Lysinema ciliatum* (Senaratna, 2000), *Leucopogon verticillatus* (Anthony *et al.*, 2004), *Swainsona formosa* (Sudhersan and AboEl-Nil, 2002), fanflower *Scaevola aemula* (Wang and Bhalla, 2004; 2006), Koala Fern *Baloskion tetraphyllum* (Panaia *et al.*, 2004; Ma *et al.*, 2006) and Black Kangaroo Paw *Macropidia fuliginosa, Stirlingia latifolia* and *Lepidosperma squamatum* (Panaia *et al.*, 2005).

Embryogenic callus was established for Australian monocots genera like *Anigozanthos, Blandfordia* and *Thysanotus* which are used as cut flowers or pot plants. However, the process of somatic embryogenesis was not fully utilised and applied to these plants (Johnson, 1997). In the study on waxflowers, a plant regeneration system via somatic embryogenesis of leaf and immature seed tissues was developed for *Chamelaucium repens* and *C. uncinatum* (Ratanasanobon, 2007).

Extensive *in vitro* research and their success are achieved for a number of hardwoods including the *Pinus, Eucalyptus,* and *Acacia* species (Midgley and Turnbull, 2003; Merkle and Nairn, 2005; Nehra *et al.*, 2005). If somatic embryos are encapsulated to produce artificial or synthetic seeds, it has enormous potential to reduce the cost of production, reduce loss of biodiversity, increase the efficiency of rehabilitation programs and increase the supply of many Australian native species to the horticultural industry.

Somaclonal Variation in Micropropagation

Genetic stability during micropropagation is controlled by many factors including genotype, presence of chimeral tissue, explant type and origin, culture medium, culture conditions and duration of culture (reviewed by Debnath and Teixeira da Silva, 2007). The occurrence of variation in plants regenerated from *in vitro* cultures was named as 'somaclonal variation' by Larkin and Scowcroft (1981). Somaclonal variation is one of the most important concerns in commercial micropropagation. It can be distinguished by their morphological, biochemical, physiological and genetic characteristics. Molecular markers are powerful tools in genetic identification of somaclonal variation.

Rapid and efficient *in vitro* regeneration methods which minimise somaclonal variation are critical for the genetic transformation and mass propagation of commercial varieties. Although somaclonal variation is undesirable for commercial micropropagation, it is useful to crop breeders, where variation in tissue culture regenerated plants from somatic cells can be used in the development of crops with novel traits.

The mechanism of somaclonal variation is still not completely understood; however, both genetic and epigenetic mechanisms are believed to play a role (Gao *et al.*, 2009). Somaclonal variation has been associated with changes in chromosome number and structure, point mutations and alteration in DNA methylation (reviewed by Baenziger *et al.*, 2006; Bartoszewski *et al.*, 2007; Santana-Buzzy *et al.*, 2007). Somaclonal variation frequency is determined by a number of factors, including genotype, explant source, medium composition and duration of culture (reviewed by Predieri, 2001). Turner *et al.* (2001) found that genetic fidelity and shoot apex viability (for cryopreserved material) of *Anigozanthos viridis* subsp. *terraspectans* (Haemodoraceae), a threatened plant from south west Australia, were maintained following tissue culture, cold storage and cryostorage of *A. viridis* subsp. *terraspectans* for 12 months. Lakshmanan *et al.* (2006) developed the rapid and efficient *in vitro* regeneration methods which minimise somaclonal variation for the genetic transformation and mass propagation in commercial varieties of sugarcane (*Saccharum* spp. interspecific hybrids).

In vitro Micrografting

There is also scope for blending the traditional horticultural practice of grafting with micropropagation. Grafting enables the combination of selected stocks or scions which are not successful on their own roots as well as providing various biotic stresses and pest resistances or growth controlling properties. Reduction in vigour, yield and quality are usually attributed to various virus, bacteria and fungi infections leading to loss of productivity and commercial usage of plants. There are methods like thermotherapy and meristem culture available to overcome pathogen infection problems. However, their involvement in recovering pathogen free in tree species is very limited. Micrografting, on the other hand, offers an alternative strategy of producing pathogen free plants (Tangolar *et al.*, 2003; Naz *et al.*, 2007).

Micrografting may involve either scions or rootstocks taken from *in vitro* stock plant banks utilising the phenomenon of rejuvenation that occurs *in vitro* to enable adventitious rooting of mature rootstock selections. Micrografting may be carried out aseptically using *in vitro* rootstocks with either *in vitro* or disinfested *ex vitro* scion material or on micropropagated rootstocks after transfer to the nursery (Taji *et al.*, 2002). Since it was first introduced by Murashige *et al.* (1972) and followed by the success of Navarro *et al.* (1975), this technique is now routinely used by several commercial fruit nurseries for eliminating graft-transmitted viral diseases, for rejuvenation of parent stock and for mass production. Standardized procedures exist for citrus, cherry, apple, grape and avocado (Obeidy and Smith, 1991; Tangolar *et al.*, 2003; Suarez *et al.*, 2005). Micrografting applied to woody species shows rather slow progress (Detrez, 1994; Reid *et al.*, 2001), the kinds of plants which can be propagated by micrografting will diversify in the future. In our own laboratory, Kawaguchi and Taji (2005) and Kawaguchi *et al.* (2008) were successful in micrografting *Swainsona formosa* to a number of rootstocks improving the plant's resistance to a number of soil borne diseases.

There are a number of benefits in *in vitro* micrografting system. It may be of great value in studying stock-scion relationship, because with *in vitro* system it is easier to isolate and identify the special compounds produced from grafted plants and to screen new scion-rootstock combinations. All types of *in vitro* grafting can be used to evaluate incompatibility, since symptoms of incompatibility under *in vivo* conditions are the same as under *in vitro* conditions. In addition, virus elimination in fruit trees has become the major application of *in vitro* micrografting (Taji *et al.*, 2002; Suarez *et al.*, 2005; Naz *et al.*, 2007).

Factors affecting the success of *in vivo* micrografting are also known to influence the success of *in vitro* micrografting. Obeidy and Smith (1991) suggested that variables associated with proper attachment of stock and scion, protection of the union during healing and removal of grafting implements after healing had significant influence on success of the procedure. In addition, Tangolar *et al.* (2003) claimed that different factors like the age of plant materials and grafting technique could affect the success rate of micrografting. There are also many reports on the success of *in vitro* micrografting under different environmental culture conditions and growth hormones used (Naz *et al.*, 2007).

Table 5.2: A Few Examples of Species which are Successfully Used as Rootstocks and Scions in *in vitro* Micrografting

Rootstocks	Scions	References
Acacia tortilis subsp. *raddiana*	*Acacia tortilis* subsp. *raddiana*	Detrez (1994)
Citrus jambheri Lash	*Citrus nobilis* Lour x *C. deliciosa* Tenora	Naz *et al.* (2007)
Citrus limon L. Burm. f.	*Citrus paradisi* Macf.	Obeidy and Smith (1991)
Citrus limon L. Burm. f.	*Citrus limon* L. Burm. f.	Obeidy and Smith (1991)
Citrus limon L. Burm. f.	*Citrus sinensis* L. Osbeck	Obeidy and Smith (1991)
Citrus paradisi Macf.	*Citrus limon* L. Burm. f.	Obeidy and Smith (1991)
Citrus sinensis L. Osbeck	*Citrus limon* L. Burm. f.	Obeidy and Smith (1991)
Malus domestica Borkh.	*Malus domestica* Borkh.	Obeidy and Smith (1991)
Persea americana Mill.	*Persea americana* Mill.	Suarez *et al.* (2005)
Prunes cerasus L.	*Prunes cerasus* L.	Obeidy and Smith (1991)
Swainsona formosa (G. Don) J. Thompson	*Clianthus puniceus*	Kawaguchi and Taji (2005)
Vitis vinifera L.	*Vitis vinifera* L.	Tangolar *et al.* (2003)

The benefits of micrografting are reported by many authors, particularly on the elimination of viral diseases (Murashige *et al.*, 1972; Navarro *et al.*, 1975; Suarez *et al.*, 2005; Naz *et al.*, 2007). To date, *in vitro* micrografting is extensively used to recover disease-free plants from the commercial citrus varieties and is becoming a potential tool for separating virus and virus like agents (Navarro, 1981). The apical meristem contains little or no virus titre and meristematic cells grow faster than all viruses. Therefore, production of disease-free foundation plants by micrografting remains the only means to supply disease-free bud stocks to the growers. The challenge of successfully grafting the very small and delicate materials from *in vitro* plants has warranted development of novel techniques to facilitate the process. Furthermore, this technique is useful in shortening the juvenile phase in tree crops, and considerably reduces the time in breeding programs (Taji *et al.*, 2002).

On the other hand, however, *in vitro* micrografting is a cumbersome, tedious and extremely time-consuming exercise. The refinements in media composition and techniques as well as growth factors in micrografting process may result in viable means of producing high yielding and pathogen-free plants. More studies are required to evaluate the degree of juvenility re-established by such techniques with the objectives of obtaining true-to-type propagation.

Haploid Plant Production

Obtaining homozygous lines from highly outcrossing species would take a long time (Williams and Taji, 1992). The process starts with cross-pollination to combine desirable parental traits resulting in heterozygous but genetically uniform offspring. Reproduction of these offspring is frequently accompanied by the separation of homologous chromosomes and genes from different parents at meiosis and production of genetic variability within the population of the next generation (Croughan, 1995). Therefore, the overall process of producing homozygous lines as predicted by Ferrie and Keller (1995) may take 10 years or even more, depending on the plant species. In contrast, use of microspore embryogenesis *via* anther or microspore culture is valuable for the detection of recessive gene traits and the exploitation of gametoclonal plants (Christou, 1992). More importantly, this technique offers the opportunity to generate pure homozygous lines more rapidly and efficiently than with conventional ways (Tomasi *et al.*, 1999). Taji *et al.* (2002) claimed that such lines could be generated in only one generation, while conventional methods need at least five generations.

The purpose of anther and pollen culture is to produce haploid plants by induction of embryogenesis from repeated division of monoploid spores, either microspores or immature pollen grains. Plant breeders are especially interested in haploid plants as fertile double haploid homozygous plants could be obtained either by spontaneous doubling of the chromosome or by the application of ploidy inducing chemicals like colchicine or oryzalin. Through chromosome doubling treatment, it is possible to produce homozygous, fertile, doubled-haploid and pure breeding lines (Ferrie and Keller, 1995). Thus, microspore embryogenesis makes mutational breeding and selection of beneficial traits possible.

Since it was first demonstrated by Guha and Maheshwari (1964) on *Datura inoxia* and Nitsch and Nitsch (1969) on tobacco, haploids were used to produce homozygous genotypes in a number of economically important monocotyledonous and dicotyledonous species (Table 5.3). Production of haploid plants through anther culture has not been exploited extensively with Australian native plants. However, in our own laboratory Tade (1992), Olde (1994) and Zulkarnain (2003) have some success in anther culture of Sturt's Desert Pea (*Swainsona formosa*).

Haploid technology is of significant interest for developmental and genetic research, as well as for plant breeding and biotechnology. Haploid plants are useful in understanding cellular totipotency because they develop from single male or female gametes without fertilization (Powell, 1990). Haploid individuals also provide an excellent example when studying induced mutagenesis, where recessive traits can be easily detected (Seguí-Simarro and Nuez, 2008b). Tolerance to unfavourable conditions like drought, cold, heavy metals or low nutrients, are amongst recessive traits which can be detected promptly in haploid plants. The problems associated with outcrossing and self-incompatibility in some species may also be solved by microspore embryogenesis (Taji *et al.*, 2002).

Table 5.3: Examples of Haploid Plants which are Successfully Regenerated *via* Microspore Embryogenesis

Species	References
Albizzia lebbeck	Crosser *et al.* (2006)
Anemone sp.	Custers *et al.* (2001)
Brassica napus	Maluszynsky *et al.* (2003)
Cajanus cajun	Crosser *et al.* (2006)
Delphinium sp.	Custers *et al.* (2001)
Glycine max	Crosser *et al.* (2006)
Hordeum vulgare	Maraschin *et al.*(2005)
Lupinus spp.	Bayliss *et al.* (2002)
Malus domestica	Höfer *et al.* (1999)
Medicago sativa	Zagorska *et al.* (1997)
Nicotiana tabacum	Maluszynsky *et al.* (2003)
Oryza sativa L.	Aryan (2002),
Peltophorum pterocarpum	Crosser *et al.* (2006)
Populus sp.	Hyun *et al.* (1986)
Psophocarpus tetragonolobus	Crosser *et al.* (2006)
Triticum aestivum	Maluszynsky *et al.* (2003)
Vigna unguiculata	Crosser *et al.* (2006)
Zantedeschia sp.	Custers *et al.* (2001)

However, haploid individuals tend to be smaller in size, less vigorous, more sensitive to diseases and environmental stresses and most importantly, they are sterile (Seguí-Simarro and Nuez, 2008b). Therefore, for practical purposes, it is usually desirable to obtain doubled haploids. Regardless of the mode of development, doubling of haploid plants either spontaneously via endoreduplication (DNA duplication without mitosis), nuclear fusion (merging of coalescing nuclei into a larger nuclei, mixing both DNA contents), endomitosis (mitosis in the absence of mitotic spindle and nuclear envelope breakdown) or by chemical means like colchicine or oryzalin treatments, leads to a homozygous doubled haploid individual with two identical copies of each chromosome (Crosser *et al.*, 2006). For many families of angiosperms, doubled haploidy is used as a routine tool in breeding programs because this technique provides pure lines in a single generation, which may save considerable time in the breeding of new cultivars (Tuvesson *et al.*, 2000; Maluszynsky *et al.*, 2003).

Embryo Rescue Technology

Embryo rescue techniques like embryo culture, ovule culture and ovary culture, are used in interspecific and intergeneric hybridisation programs. Embryo rescue holds great promise not only for effecting distant crosses, but also for obtaining plants from inherently weak and immature embryos, obtaining haploid plants as well as for shortening the breeding cycle. Moreover, embryo rescue can provide a means of overcoming dormancy of recalcitrant seeds (Raghavan, 2003; Taji and Williams, 2005).

A good example of the application of embryo rescue is the work of Drew *et al.* (2006a; 2006b) who developed a successful embryo-rescue and culture protocol for use with the intergeneric hybrids between *Carica papaya* L. and *Vasconcellea quercifolia*, which is papaya ringspot virus type P resistant. Palmer *et al.* (2002) obtained many of the new hybrid plants from the crosses between indigenous *Vigna* species and mungbean cultivars grown in Australia. By employing embryo rescue, a hybrid between *Sorghum bicolor* and *S. macrospermum* was retrieved and cultured *in vitro* by Dillon *et al.* (2007).

The successful production of plants from the cultured embryos largely depends on parental genotype, maturation stage and medium composition (Palmer *et al.*, 2002). Liu *et al.* (2008) reported that culture medium, genotype and year of cross affected the embryo development and recovery from *in vitro* cultured ovules in breeding stenospermocarpic seedless grape varieties for the Australian table and dried grape industries. Early embryo rescue of wax flower hybrids is the major research activity in the tissue culture laboratory in Western Australia (Shan, 2007). It is impossible to imagine modern plant breeding without embryo rescue technique.

In vitro Flowering

The transition from the vegetative to reproductive phase in plants is an important developmental process with considerable practical importance in plant breeding. Despite decades of research and rapid advances in technology, our understanding of this important developmental process is still fragmentary. From the results of previous research, it is evident that the majority of plants use environmental cues to regulate flowering. Environmental variables with regular seasonal patterns such as temperature, photoperiod and irradiance are the key signals in floral induction. These factors are perceived by different plant parts, and strong and diverse interactions between the environmental variables are required for floral induction to occur in many species.

Classical physiological, genetic and grafting experiments, though invaluable in deciphering various aspects of flowering have failed to unravel the true nature of the flowering stimulus or the mechanism(s) by which various environmental cues induce flowering. The *in vitro* technique is a useful strategy for studying flowering physiology in vascular plants (Van Staden and Dickens, 1991). Novel approaches involving *in vitro* flowering and molecular techniques offer unique opportunities to investigate flowering process from new perspectives especially in species which are difficult to flower or produce flowers only once in several years (Taji *et al.*, 2002; Lakshmanan and Taji, 2004).

Factors controlling *in vivo* flowering are also found to be important in the induction of *in vitro* flowering. In addition, studies on the *in vitro* flowering have unravelled many aspects of flowering more than *in vivo* investigations have. The involvement of factors like photoperiod (Singh *et al.*, 2006), light intensity (Sheeja and Mandal, 2003), light quality (Victorio and Lage, 2009) and plant growth regulators (Singh *et al.*, 2000; Britto *et al.*, 2003; Amutha *et al.*, 2008; Kanchanapoom *et al.*, 2009) are clearly critical for the induction of flowering *in vitro*.

To date, the significant aspect of *in vitro* flowering is that the capability of producing fertile flowers is more quickly than *in vivo* system. This is important in breeding of plants which are difficult to flower during *in vivo* culture like orchids, or produce flowers once in several years like bamboo. An example of the success of producing orchid fertile flower under *in vitro* system was reported in *Dendrobium* hybrids (Hee *et al.*, 2007; Sim *et al.*, 2007; Sim *et al.*, 2008). Using one of the hybrids, *Dendrobium* Chao Praya Smile, as a model system, flowers were produced following 5-6 months *in vitro* seed sowing by appropriate benzyladenine treatment. This duration is much shorter than the 2-

3 years required using conventional growing method. As a result, early evaluation of the characteristics of flowers is possible. In addition, following *in vitro* flowering, pollination is also performed *in vitro* with formation of viable seeds. Hence, the breeding cycles of orchid were reduced significantly.

Another example illustrating the significant contribution of *in vitro* flowering is the breeding strategy of bamboos. Bamboos are one of groups of plants which have a peculiar behaviour of flowering and seeding at the end of a very long vegetative growth phase. In nature, flowering of bamboo takes place after 12 to 24 years of growth and then the plants dying at the end of fruiting season (Taji *et al.*, 2002). Because of the peculiar flowering habits, it was almost impossible to breed for superior traits in bamboos. This situation, however, was changed following Nadgauda *et al.* (1990) reports on the first success of *in vitro* flowering in bamboos *Bambusa arundinacea* and *Dendrocalamus brandisii*, and subsequently in a few more species of bamboos (Chambers *et al.*, 1991; Rout and Das, 1994). Since then *in vitro* flowering is observed in many types of bamboos (Gielis *et al.*, 2002).

Table 5.4: Examples of Species which are Successfully Induced to Flower *in vitro*

Species	References
Ammi majus L.	Pande *et al.* (2002)
Bambusa edulis	Lin *et al.* (2004)
Basilicum polystachyon (L.) Moench	Amutha *et al.* (2008)
Ceropegia bulbosa Roxb. var. *bulbosa*	John Britto *et al.* (2003)
Citrus limon	Tisserat *et al.* (1990)
Citrus nobilis x *C. deliciosa* Tenora	Singh *et al.* (2006)
Cymbidium ensifolium var. *misericors*	Chang and Chang (2003)
Dendrobium hybrids	Hee *et al.* (2007), Sim *et al.* (2007,2008)
Dendrocalamus strictus	Singh *et al.* (2000)
Eulophia graminea Lindl.	Chang *et al.* (2010)
Gentiana triflora	Zhang and Leung (2000), Zhang and Leung (2002)
Hypericum brasiliense	Abreu *et al.* (2003)
Kniphofia leucocephala	Taylor *et al.* (2005)
Lycopersicon esculentum Mill.	Sheeja and Mandal (2003)
Murraya paniculata (L.) Jack.	Jumin and Ahmad (1999)
Nicotiana tabacum L.	Peeters *et al.* (1991)
Pharbitis nil	Galoch *et al.* (2002)
Psygmorchis pusilla	Vaz *et al.*(2004)
Rosa hybrida L.	Kanchanapoom *et al.* (2009)
Streptocarpus nobilis	Floh and Handro (2001)
Swainsona formosa (G. Don) J. Thompson	Tapingkae *et al.* (2009)

More than 10 years after the first report on *in vitro* flowering in bamboo, some positive results are obtained on a wide range of monocotyledonous and dicotyledonous species (Table 5.4), but practical and commercially exploitable results were not reported yet. Many hurdles still need to be overcome before the methods really become applicable at agricultural and commercial scales. The major challenge at the moment is to establish reliable methods which allow continuous production of fertile flowers and viable seeds of desired species. Taji *et al.* (2002) suggested that the physiological state of donor plant, the properties of culture medium, as well as environmental and endogenous signals are among other factors which need to be considered to induce *in vitro* flowering reliably. Undoubtedly, the *in vitro* flowering system capable of producing viable seeds would be a valuable approach in enhancing the breeding programs of those species with a long juvenile growth phase.

In vitro Pollination and Fertilisation

In vitro pollination and fertilisation is a method in which male and female gametes are isolated and introduced to each other under optimum conditions for zygote development. It involves pollen tube penetration of the embryo sac by manipulation of maternal tissue and by methods other than the normal *in situ* process. Initially developed to bypass pre-zygotic incompatibility barriers, this technique was used for the production of hybrids, the induction of haploid plants, overcoming sexual self-incompatibility and in the study of pollen physiology and fertilization (Taji and Williams, 2004).

Various *in vitro* methods are developed to overcome incongruity barriers in a number of plant species (Taji *et al.*, 2002; Taji and Williams, 2005). Sexual barriers preventing interspecific hybridisation are divided into pre- and post-fertilization barriers (Stebbins, 1958). Pre-fertilization barriers are bypassed using *in vitro* pollination and fertilization. *In vitro* pollination enhances the possibility of pollen tube penetration into the egg apparatus and central cell. Once fertilization has occurred, hybrid embryo growth is restricted by post-fertilization barriers. Post-fertilization barriers could be overcome by using embryo rescue followed by *in vitro* culture techniques (Van Tuyl *et al.*, 1997). The technique of *in vitro* pollination and fertilization was first reported in 1962 in poppy, *Papaver somniferum* (Kanta *et al.*, 1962). Since that time, several species were pollinated successfully and fertilized *in vitro* including *Swainsona laxa*, a rare Australian plant (Taji and Williams, 1987).

Interspecific hybridisation is an important research area in the genus *Sorghum* because of the possibility of transferring the genes for resistance to important insects and pathogens from wild Australian *Sorghum* species to the grain sorghum (*Sorghum bicolor*) genome (Price *et al.*, 2006). Several diseases limit pawpaw (papaya) production in Australia but the main concern is papaya ringspot virus-type P (PRSV-P). Protocols were developed to produce large number of intergeneric hybrids between *Carica papaya* L. and *Vasconcellea quercifolia*, which is PRSV-P resistant. A very efficient protocol was developed for both the rescuing and the germination of the embryos. Most of the rescued embryos produce embryogenic callus and multiple plantlets. The results demonstrate that efforts in wide hybridization to transfer PRSV-P resistance to *C. papaya*, are better directed towards crosses between *C. papaya* and *V. quercifolia* than with other *Vasconcellea* species (Drew *et al.*, 2006a; Drew *et al.*, 2006b).

The success of *in vitro* pollination, fertilization and subsequent production of viable seed depend on a number of exogenous and endogenous factors (Dusi *et al.*, 2010; Skalova *et al.*, 2010). The main exogenous factor is the composition of culture media. The most important endogenous factor seems to be ovule and pollen grain maturity.

Protoplast Technology

The plant protoplast consisting of cytoplasm and nucleus with the cell wall removed, provides a unique single cell system to underpin several aspects of modern biotechnology. Protoplasts can be isolated mechanically by cutting or breaking the cell wall, and by digesting it away with enzymes or by a combination of mechanical and enzymatic separation (Davey *et al.*, 2005). Isolated protoplasts serve as the field of somatic cell cloning and development of mutant lines. Teulieres and Boudet (1991) reported preliminary attempts to isolate protoplasts from *Eucalyptus globulus*.

Reliable procedures are available to isolate and culture protoplasts from a range of plants (Debnath and Teixeira da Silva, 2007). Several factors especially the source tissue, culture medium and environmental factors, influence the ability of protoplasts and protoplast-derived cells to express their totipotency and to develop into fertile plants. Recently, direct differentiation of globular embryo structures from mesophyll protoplast cultures of *Scaevola aemula* (fan flower) was reported (Wang, 2010). This is the first achievement of direct differentiation of embryo structure in protoplast culture of Australian native plants. It will enhance the study of embryo development in Australian native plants which are well known for their low seed viability and germination.

Protoplast fusion comprises of removal of cell wall and then fusion of cell contents. Somatic hybridization by protoplast fusion and plant transformation could overcome many of the barriers to interspecific hybridization among plant species or genera. The hybrid plants are then assessed for desirable new traits. Traits of interest range from disease resistance, increased nutritional value (in the case of food crops), seedlessness, improved vigour to other stress factors, etc.

Plant cell and protoplast isolation and fusion techniques have become important and well-accepted methods for studying the physiology, biochemistry and breeding of plants. Since the first successful report on somatic hybridization with tobacco (Carlson *et al.*, 1972), the potentials of somatic fusions are exploited for crop improvement in many crops including rice, wheat, rapeseed, canola, tobacco, tomato, potato and citrus (Guo *et al.*, 2004; Tapingkae *et al.*, in press).

Successes with somatic hybridization have mainly involved the production of useful and fertile hybrid plants as a result of simple additive combination of the complete genomes of two unrelated species. However, production of cybrids which contain the nuclear genome of one parent and either the cytoplasmic genome of the other parent or a combination of both parents, and their potential use has been a common approach in plant improvement (Guo *et al.*, 2004; Grosser and Gmitter, 2005). Successful somatic cell hybridization depends totally on development of the optimum conditions to regenerate whole plants from hybrid protoplasts. Each species requires its own unique regeneration conditions which need to be determined (Ratanasanobon, 2007).

Within Australia, breeding programs have focused upon producing new varieties exhibiting increased adaptations to the Australian environment, with additional objectives towards overcoming a range of undesirable traits. Australian native plant species like *Grevillea* (family Proteaceae) are of increasing commercial importance in the horticulture and revegetation industries. Tissue degradation and low protoplast yields were investigated to provide a basis for breeding with the aim of creating a salt tolerant plant (Kennedy and De Filippis, 2004). Chikkala *et al.* (2009) studied adventitious shoot regeneration and protoplast isolation and culture of several cauliflower (*Brassica oleracea* var. *botrytis*) cultivars, sourced from Europe and Australia to develop improved nuclear and plastid transformation protocols. He *et al.* (1997) obtained transgenic wheat (*Triticum aestivum*) through protoplast electroporation.

In forest trees, protoplast fusion and gene transfer are methods which potentially can be used to enhance wood quality and oil, along with tolerance to salt, drought, temperature and disease (Sartoretto *et al.*, 2008).

There is great demand for Australian native flowers of different colours. Fused protoplasts from two different *Chamelaucium* species were achieved at the floriculture laboratory at the Department of

Agriculture and Food, Western Australia (Shan, 2007). Employing protoplast technology to create new colour variants may enable the industry to compete in the domestic and international markets.

Plant Genetic Engineering and Transformation

The isolation and addition of specific foreign genes to a plant species to enhance its properties is a technique with almost unlimited potential. This technique allows us to investigate the evolution, structure, function or regulation of a particular gene, which can be manipulated for production of clones with enhanced economic characteristics (Daggard, 1996). Karp *et al.* (1998) introduced the term "molecular breeding" to describe the development and application of molecular genetic techniques to introduce novel and desirable characteristics with high value to plant breeding program.

Genetic engineering may also involve the transfer of genes - and thus the characteristics governed by those genes - from one species to another. The objectives of plant genetic engineering include improving flower quality in ornamentals and crop production in agriculture as well as introducing new traits like enhanced nutrient intake, resistance to environmental stress or better post harvest quality. Certain characteristics allow a genetically modified crop to be grown, harvested, or shipped at a lower cost.

There are several approaches to inserting foreign genes into plants. These include direct DNA uptake by protoplasts -(electroporation, polyethylene glycol (PEG)- induced uptake, micro-injection or sonication)-, direct DNA delivery by biolistic methods or indirectly via vectors (viral or bacterial). The system which has proved the most successful is based on the T_i plasmid of a soil bacterium, *Agrobacterium tumefaciens*. Whilst vectors based on viral genome have also been extensively studied, none have yet been developed for general use in plant transformation. This is mainly due to the pathogenic nature of the virus, the restrictions on genome size and the fact that virus DNA is not stably transmitted to the progeny of infected plants. Another recent advanced tool for plant genetic engineering, developed by Krichevsky (2008), is the modular satellite (pSAT) vector system. This molecular tool was first developed to provide N- and C-terminal fusions of the genes under investigation to five different autofluorescent tags, EGFP, EYFP, Citrine-YFP, ECFP, and DsRed2. However, this plasmid system also allows cloning of untagged ORFs, or genes marked with different tags, for simultaneous expression in the plant cell.

Whilst extensive research was undertaken in the area of genetic transformation of dicotyledonous species since the production of the first transgenic plants in 1983, only a limited number of agronomically important genes were successfully transferred to crop plants. These genes confer resistance to certain herbicides, insects or viruses and have all been stably integrated into the genome of several species, including maize, soybean, oilseed rapes (canola), cotton and tomato (Vasil, 1991). Due to the high cost, this technology is presently limited to crop species with potential for high economic returns (Table 5.5). It is anticipated that these molecular tools will be used routinely for the improvement of ornamental species and in the development of crops for biofuel production in the foreseeable future.

The application of molecular biology technique in improving plant quality is increasingly important in both commercial and research purposes in Australian native species. Though publication on the application of molecular based biotechnology in Australian native plants is still limited, there is no doubt that access to this technology will be of increasing importance in studies of plant systematics, conservation biology as well as commercialisation of Australian native species.

Table 5.5: Examples of Transgenic Crop Plants with their Advantages and Modification Methods

Plants	Advantages	Modification Methods	References
Rice	Rice plants containing high provitamin A (β-carotene).	Genes from narcissus, maize and *Erwinia* bacteria were inserted into rice chromosome.	Gupta (2004)
Maize, cotton, potato	Resistance to pests	Toxic gene Bt was transferred from *Bacillus thuringiensis*	Gupta (2004)
Tobacco	Resistance to cold climate	Genes regulating the resistance to cold climate from *Arabidopsis thaliana* or cyanobacteria (*Anacyctis nidulans*) were inserted into tobacco chromosomes.	Gupta (2004)
Tomato	Fruits remain firm and fresh for a long time	Antisenescens gene was transferred into tomato chromosome to inhibit the production of polygalacturonase enzyme.	Gupta (2004)
Soybean	Plants containing high oleic acid and resistant to herbicide glyphosate	Herbicide resistant gene from *Agrobacterium* line CP4 was introduced to soybean chromosome. Molecular technology was used to increase oleat acid formation	Gupta (2004); Heller (2007)
Sweet potato	Plant resistance to viral disease	Gene from certain virus was transferred into sweet potato by the aid of gene silencing technology	Loebenstein and Thottappilly (2009)
Canola	Plants producing canola oil high in lauric acid content	Gene *FatB* from *Umbellularia californica* was transferred into canola chromosome to increase laurate acid content.	Scarth and Tang (2006)
Papaya	Plant resistance to certain virus such as *Papaya Ringspot Virus* (PRSV)	Gene encoding PRSV was transferred into papaya chromosome	Gonsalves (2004)
Rockmelon	Slow ripening of fruits	New gene from bacteriophage T3 was taken out to reduce the production of ethylene in rockmelon.	ILSI Research Foundation (2010)
Sugar beet	Plant resistance to glyphosate and gluphosinate herbicides	Genes from *Agrobacterium* line CP4 and *Streptomyces viridochromogenes* were transferred into sugar bit chromosomes	Haim D. Rabinowitch and Currah (2002)
Plum	Plant resistance to plum pox virus	Gene encoded plum pox virus was transferred into plum chromosomes.	Scorza *et al.* (1994)
Wheat	Plant resistance to *Fusarium graminearum*	Gene encoding chitinase enzyme from barley was transferred into wheat chromosome.	Shin *et al.* (2008)

Current efforts in genetic engineering in Australian native plants are focussed primarily on species of eucalyptus, because of their commercial importance (Chandler, 1995). Field trials of genetically engineered *Eucalyptus grandis* carrying selectable marker genes was reported by Edwards *et al.* (1995) following the availability of lasting tomato (Flavr Savr, *Lycopersicon esculentum*) on market in 1994. This genetically modified tomato produces less of the substance that causes tomatoes to rot, so remains firm and fresh for longer time. Strawberries, pineapples, sweet peppers and bananas have also been genetically modified by scientists to remain fresh for longer. In 2008, the first transgenic canola was harvested in Australia.

In future, research and development of transgenic crop should focus on creating species with improved production stability; better nutritional value; resistant to environmental impacts of intensive and extensive agriculture; as well as developing protocols and regulations which ensure transgenic crops designed for purposes other than food, such as pharmaceuticals, industrial chemicals, biofuels, vaccines, etc.

Conclusion

Development of Australian plants using the *in vitro* technology has come a long way since the first comprehensive publication of Ron deFossard in 1976. His recent book (de Fossard, 1993) provides protocols developed for a large number of plants including many Australian native species.

Biotechnology is a very powerful tool to further advance the various fields of plant sciences. In practice it should be combined with "classical" breeding strategies and with conventional plant propagation practices. These tools aid the domestication and development of Australian native plant species as potential ornamental crops. Further advances and applications of plant biotechnology require more basic research. Indeed biotechnology provides some powerful tools by which we can extend our comprehension of the physiology, metabolism and developmental biology of plants. This is particularly important with Australian plants for which we still know very little about their unique biology. Furthermore, the endless possibilities for improvements in plants through biotechnology have the potential to help solve world hunger and problems in agriculture as a result of climate change.

References

Abreu, I. N., M. T. A. Azevede, V. M. Solferini and P. Mazzafera. 2003. *In vitro* propagation and isozyme polymorphism of the medicinal plant *Hypericum brasiliense. Biologia Plantarum* 47: 629–632.

Amutha, R., M. Jawahar, and S. R. Paul. 2008. Plant regeneration and *in vitro* flowering from shoot tip of *Basilicum polystachyon* (L.) Moench – An important medicinal plant. *Journal of Agricultural Technology* 4: 117–123.

Anthony, J. M., T. Senaratna, K. W. Dixon, and K. Sivasithamparam. 2004. Somatic embryogenesis for mass propagation of *Ericaceae* – a case study with *Leucopogon verticillatus. Plant Cell Tissue and Organ Culture* 76: 137–146.

Aryan, A. P. 2002. Production of double haploids in rice: anther vs. microspore culture. *In* A. Taji and R. Williams [eds.], The Importance of Plant Tissue Culture and Biotechnology in Plant Sciences, 201–208. University of New England Press, Armidale, Australia.

Baenziger, P. S., W. K. Russell, G. L. Graef, and B. T. Campbell. 2006. Improving lives: 50 years of crop breeding, genetics, and cytology (C–1). *Crop Science* 46: 2230–2244.

Bartoszewski, G., M. J. Havey, A. Ziólkowska, M. Dlugosz, and S. Malepszy. 2007. The selection of mosaic (MSC) phenotype after passage of cucumber (*Cucumis sativus* L.) through cell culture – a method to obtain plant mitochondrial mutants. *Journal of Applied Genetic* 48: 1–9.

Bayliss, K. L., J. M. Wroth, and W. A. Cowling. 2002. Production of multicellular microspores of *Lupinus* species: first step toward haploid lupin embryos. *In* A. Taji and R. Williams [eds.], The Importance of Plant Tissue Culture and Biotechnology in Plant Sciences, 145–157. University of New England Press, Armidale, Australia.

Belliard, G., F. Vedel, and G. Pelletier. 1979. Mitochondrial recombination in cytoplasmic hybrids of *Nicotiana tabacum* by protoplast fusion. *Nature* 281: 401–402.

Bhojwani, S. S., and M. K. Razdan. 1983. Plant Tissue Culture: Theory and Practice. Development in Crop Science 5. Elsevier Press, Amsterdam.

Boulay, M. 1987. *In vitro* Propagation of Tree Species. *In* C. E. Green, D. A. Somers, W. P. Hacket and D. D. Biesboer [eds.], Plant Biology Volume 3: Plant Tissue and Cell Culture, 367–382. Alan R. Liss, Inc, New York.

Bourgin, J. P., and J. P. Nitsch. 1967. Obtention de Nicotiana haploides a' partir de'etamines cultivees in vitro. *Annales de Physiologie Vegetale* 9: 377–382.

Britto, S. J., E. Natarajan, and D. I. Arockiasamy. 2003. *In Vitro* Flowering and Shoot Multiplication from Nodal Explants of *Ceropegia bulbosa* Roxb. var. *bulbosa. Taiwania* 48: 106–111.

Carlson, P. S., H. H. Smith, and R. D. Dearing. 1972. Parasexual interspecific plant hybridisation. *Proceedings of National Academy of Science, USA* 69: 2292–2294.

Carneros, E., C. Celestino, K. Klimaszewska, Y. S. Park, M. Toribio, and J. Bonga. 2009. Plant regeneration in Stone pine (*Pinus pinea* L.) by somatic embryogenesis. *Plant Cell, Tissue and Organ Culture* 98: 165–178.

CERA. 2010. GM Crop Database. Center for Environmental Risk Assessment (CERA), ILSI Research Foundation, Washington D.C. Retrieved December 29, 2010, from http://cera–gmc.org/index.php?action=gm_crop_database.

Chambers, S. M., J. H. R. Heuch, and A. Pirrie. 1991. Micropropagation and *in vitro* flowering of the bamboo *Dendrocalmus hamiltonii* Munro. *Plant Cell, Tissue and Organ Culture* 27: 45–48.

Chandler, S. F. 1995. Commercialisation of Genetically Engineered Trees. *In* B. M. Potts, N. M. G. Borralho, J. B. Reid, R. N. Cromer, W. N. Tibbits and C. A. Raymond [eds.], Eucalyptus Plantations: Improving Fibre Yield and Quality, 381–385. CRC for Temperate Hardwood Forestry, Hobart.

Chang, C., and W. C. Chang. 2003. Cytokinins promotion of flowering in *Cymbidium ensifolium* var. *misericors* in vitro. *Plant Growth Regulators* 39: 217–221.

Chang, C., H. Wei–Hsin, C. Ying–Chun, S. Yu–Ling, and C. Yi Tien. 2010. In vitro flowering and mating system of *Eulophia graminea* Lindl. *Botanical Studies* 51: 357–362.

Chikkala, V. R. N., G. D. Nugent, P. J. Dix, and T. W. Stevenson. 2009. Regeneration from leaf explants and protoplasts of *Brassica oleracea* var. botrytis (cauliflower). *Scientia Horticulturae* 119: 330–334.

Christou, P. 1992. Genetic Engineering and *In Vitro* Culture of Crop Legumes. Technomic Publishing Co. Inc., Lancaster, Pennsylvania.

Cocking, E. C. 1960. A method for the isolation of plant protoplast and vacuoles. *Nature* 1987: 927–929.

Crosser, J. S., L. L. Lülsdorf, P. A. Davies, H. J. Clarke, K. L. Bayliss, N. Mallikarjuna, and K. H. M. Siddique. 2006. Toward doubled haploid production in the Fabaceae: progress, constraints, and opportunities. *Critical Review in Plant Sciences* 25: 139–157.

Croughan, T. P. 1995. Anther Culture for Doubled Haploid Production. *In* O. L. Gamborg and G. C. Phillips [eds.], Plant Cell, Tissue and Organ Culture: Fundamental Methods, 143–152. Springer Verlag, Berlin.

Custers, J., M. Visser, R. Snijder, K. Litovkin, and L. v. d. Geest. 2001. Model plants pave the way to haploid technology; microspore embryogenesis in ornamentals. Plant Research International B.V. Poster, Wageningen, The Netherlands.

Daggard, G. E. 1996. Aspects of Molecular Biotechnology with Reference to Australian Plants. *In* A. Taji and R. Williams [eds.], Tissue Culture of Australian Plants, 284–306. University of New England, Armidale, Australia.

Davey, M. R., P. Anthony, J. B. Power, and K. C. Lowe. 2005. Plant protoplasts: status and biotechnological perspectives. *Biotechnology Advances* 23: 131–171.

de Fossard, R. A. 1993. Plant Tissue Culture Propagation. Xarma Pty. Ltd, Eagle Heights, Queensland, Australia.

Debnath, S. C., and J. A. Teixeira da Silva. 2007. Strawberry culture *in vitro*: applications in genetic transformation and biotechnology. *Fruit, Vegetable and Cereal Science and Biotechnology* 1: 1–12.

Detrez, C. 1994. Shoot production through cutting culture and micrografting from mature tree explants in *Acacia tortilis* (Forsk.) Hayne subsp. *raddiana* (Savi) Brenan. *Agroforestry Systems* 25: 171–179.

Dillon, S. L., P. K. Lawrence, R. J. Henry, and H. J. Price. 2007. *Sorghum* resolved as a distinct genus based on combined ITS1, *ndhF* and *Adh1* analyses. *Plant Systematics and Evolution* 268: 29–43.

Drew, R. A., S. V. Siar, C. M. O'Brien, and A. G. C. Sajise. 2006a. Progress in backcrossing between *Carica papaya* x *Vasconcellea quercifolia* intergeneric hybrids and *C. papaya*. *Australian Journal of Experimental Agriculture* 46: 419–424.

Drew, R. A., S. V. Siar, C. M. O'Brien, P. M. Magdalita, and A. G. C. Sajise. 2006b. Breeding for papaya ringspot virus resistance in *Carica papaya* via hybridisation with *Vasconcellea quercifolia*. *Australian Journal of Experimental Agriculture* 46: 413–418.

Dusi, D., E. Alves, M. Willemse, R. Falcao, C. do Valle, and V. Carneiro. 2010. Toward in vitro fertilization in *Brachiaria spp*. *Sexual Plant Reproduction* 23: 187–197.

Edwards, G. A., N. W. Fish, K. J. Fuell, M. Keil, J. G. Purse, and T. A. Wignal. 1995. Genetic Modification of Eucalypts: Objectives, Strategies and Progress In B. M. Potts, N. M. G. Borralho, J. B. Reid, R. N. Cromer, W. N. Tibbits and C. A. Raymond [eds.], Eucalypt Plantation: Improving Fibre Yield and Quality, 389–391. CRC for Temperate Hardwood Forestry, Hobart.

Ernawati, A., R. A. Drew, S. W. Adkins, and I. D. Godwin. 1997. Multiplication of *Carica papaya* L. x *C. parviflora* (A.DC.) Solms. hybrids through somatic embryogenesis. *In* A. M. Taji and R. R. Williams [eds.], Tissue culture: Towards the next century, 235–237. University of New England Press, Armidale, Australia.

Ferrie, A. M. R., and W. A. Keller. 1995. Microspore Culture for Haploid Plant Production. *In* O. L. Gamborg and G. C. Phillips [eds.], Plant Cell, Tissue and Organ Culture: Fundamental Methods, 155–164. Springer Verlag, Berlin.

Floh, E. I. S., and W. Handro. 2001. Effect of photoperiod and chlorogenic acid on morphogenesis in leaf discs of *Streptocarpus nobilis*. *Biologia Plantarum* 44.

Folta, K., and A. Dhingra. 2006. Transformation of strawberry: the basis for translational genomics in *Rosaceae*. *In Vitro Cellular and Developmental Biology – Plant* 42: 482–490.

Galoch, E., J. Czaplewska, E. Burkacka–Laukajtys, and J. Kopcewicz. 2002. Induction and stimulation of *in vitro* flowering of *Pharbitis nil* by cytokinin and gibberellin. *Plant Growth Regulators* 37: 199–205.

Gao, X., D. Yang, D. Cao, M. Ao, X. Sui, Q. Wang, J. Kimatu, *et al.*, 2009. *In vitro* micropropagation of *Freesia hybrida* and the assessment of genetic and epigenetic stability in regenerated plantlets. *Journal of Plant Growth Regulation* Published online: 30 December 2009: DOI 10.1007/s00344-00009-09133-00344.

Gautheret, R. J. 1934. Culture du tissus cambial. *Comptes Rendus Hebdomadaires des Seances de l Academie des Sciences* 198: 2195–2196.

Gielis, J., H. Peeters, K. Gillis, J. Oprins, and P. C. Debergh. 2002. Tissue culture strategies for genetic improvement of bamboo. *Acta Horticulturae* 552: 195–203.

Gleba, Y. Y., and K. M. Sytnik. 1984. Protoplast Fusion: Genetic Engineering of Higher Plants. Springer–Verlag, Heidelberg.

Gonsalves, D. 2004. Transgenic papaya in Hawaii and beyond. *AgBioForum* 7: 36–40.

Grosser, J. W., and F. G. Gmitter. 2005. Thinking outside the cell: Applications of somatic hybridization and cybridization in crop improvement, with citrus as a model. *In Vitro Cellular and Developmental Biology – Plant* 41: 220–225.

Guha, S., and S. C. Maheshwari. 1964. *In vitro* production of embryos from anthers of *Datura*. *Nature* 204: 497.

Guha, S., and S. C. Maheshwari. 1966. Cell division and differentiation of embryos in the pollen grains of *Datura in vitro*. *Nature* 212: 97–98.

Guo, W. W., X. D. Cai, and J. W. Grosser. 2004. Somatic Cell Cybrids and Hybrids in Plant Improvement. *In* H. Daniell and C. D. Chase [eds.], Molecular Biology and Biotechnology of Plant Organelles, 635–659. Springer, The Netherlands.

Gupta, P. K. 2004. Biotechnology and Genomics. Rastogi Publications, Meerut – New Delhi, India.

Hartmann, H. T., D. E. Kester, and F. T. Davis–Jr. 1990. Plant Propagation: Principles and Practices. Prentice–Hall International, Inc, Englewood Clifts, New Jersey.

He, D. G., Y. Yang, and K. J. Scott. 1997. Embryogenic clones of wheat (*Triticum aestivum* L.). *In* A. M. Taji and R. R. Williams [eds.], Tissue culture: Towards the next century, 259–262. University of New England Press, Armidale, Australia.

Hee, K. H., C. S. Loh, and H. H. Yeoh. 2007. *In vitro* flowering and rapid in vitro embryo production in *Dendrobium* Chao Praya Smile (Orchidaceae). *Plant Cell Reports* 26: 2055–2062.

Heller, K. 2007. Genetically Engineered Food: Methods and Detection. Wiley–VCH.

Höfer, M., A. Touraev, and E. Heberle–Bors. 1999. Induction of embryogenesis from isolated apple microspores. *Plant Cell Reports* 18: 1012–1017.

Hyun, S. K., J. H. Kim, E. W. Noh, and J. I. Park. 1986. Induction of haploid plants of *Populus* species. *In* L. A. Withers and P. G. Alderson [eds.], Plant Tissue Culture and its Agricultural Application, 413–418. Butterworths, London.

John Britto, S., E. Natarajan, and D. I. Arockiasamy. 2003. *In vitro* flowering and shoot multiplication from nodal explants of *Ceropegia bulbosa* Roxb. var. *bulbosa*. *Taiwania* 48: 106–111.

Johnson, K., and M. Burchett. 1996. Native Australian Plants – Horticulture and Uses. University of New South Wales Press, Sydney, Australia.

Johnson, K. A. 1997. Induction of somatic embryogenesis in three australian monocots. *In* A. M. Taji and R. R. Williams [eds.], Tissue culture: Towards the next century, 165–168. University of New England Press, Armidale, Australia.

Jumin, H. B., and M. Ahmad. 1999. High–frequency *in vitro* flowering of *Murraya paniculata* (L.) Jack. *Plant Cell Reports* 18: 764–768.

Kanchanapoom, K., N. Posayapisit, and K. Kanchanapoom. 2009. *In Vitro* Flowering from Cultured Nodal Explants of Rose (*Rosa hybrida* L.). *Notulae Botanicae Horti Agrobotanici Cluj–Napoca* 37: 261–263.

Kanta, K., N. S. Rangaswamy, and P. Maheshwari. 1962. Test–tube fertilization in a flowering plant. *Nature* 194: 1214–1217.

Karp, A., P. G. Isaac, and D. Ingram. 1998. Molecular Tools for Screening Biodiversity. Chapman and Hall, New York.

Kawaguchi, M., and A. Taji. 2005. Anatomy and physiology of graft incompatibility in Sturt's desert pea (*Swainsona formosa*), an Australian native plant. *Acta Horticulturae* 683: 249–257.

Kawaguchi, M., A. Taji, D. Backhouse, and M. Oda. 2008. Anatomy and physiology of graft incompatibility in solanaceous plants. *Journal of Horticultural Science and Biotechnology* 83: 581–588.

Kennedy, B. F., and L. F. De Filippis. 2004. Tissue degradation and enzymatic acitivity observed during protoplast isolation in two ornamental *Grevillea* species. *In Vitro Cellular and Developmental Biology – Plant* 40: 119–125.

Krichevsky, A. 2008. Advances in plant genetic engineering. *SciTopics*: Retrieved December 24, 2010, from http://www.scitopics.com/Advances_in_plant_ genetic_ engineering.html.

Lakshmanan, P., and A. Taji. 2000. Somatic embryogenesis in leguminous plants. *Plant Biology* 2: 136–148.

Lakshmanan, P., and A. Taji. 2004. *In Vitro* Flowering. *In* R. M. Goodman [ed.], Encyclopedia of Plant and Crop Science. 576–578 Marcel Dekker, Inc, New York.

Lakshmanan, P., R. J. Geijskes, L. F. Wang, A. Elliott, C. P. L. Grof, N. Berding, and G. R. Smith. 2006. Developmental and hormonal regulation of direct shoot organogenesis and somatic embryogenesis in sugarcane (*Saccharum* spp. interspecific hybrids) leaf culture. *Plant Cell Reports* 25: 1007–1015.

Larkin, P. J., and W. R. Scowcroft. 1981. Somaclonal variation a novel source of variability from cell cultures for plant improvement. *Theory and Application Genetic* 60: 197–214.

Lin, C. S., C. C. Lin, and W. C. Chang. 2004. Effect of thidiazuron on vegetative tissue–derived somatic embryogenesis and flowering of bamboo *Bambusa edulis*. *Plant Cell, Tissue and Organ Culture* 76: 75–82.

Liu, S. M., S. R. Sykes, and P. R. Clingeleffer. 2008. Effect of culture medium, genotype, and year of cross on embryo development and recovery from *in vitro* cultured ovules in breeding stenospermocarpic seedless grape varieties. *Australian Journal of Agricultural Research* 59: 175–182.

Loebenstein, G., and G. Thottappilly. 2009. The Sweetpotato. Springer, Berlin.

Ma, G. H., E. Bunn, K. Dixon, and G. Flemati. 2006. Comparative enhancement of germination and vigor in seed and somatic embryos by the smoke chemical 3–methyl–2H–furo[2,3–c]pyran–2–one in *Baloskion tetraphyllum* (Restionaceae). *In Vitro Cellular and Developmental Biology–Plant* 42: 305–308.

Maluszynsky, M., K. J. Kasha, B. P. Forster, and I. Szarejko. 2003. Doubled Haploid Production in Crop Plants. Kluwer Academic Publishers, Dordrecht, The Netherlands.

Maraschin, S. D., M. Vennik, G. E. M. Lamers, H. P. Spaink, and M. Wang. 2005. Time–lapse tracking of barley androgenesis reveals position–determined cell death within pro–embryos. *Planta* 220: 531–540.

McComb, J. A., I. J. Bennett, and C. Tonkin. 1996. *In vitro* propagation of *Eucalyptus* species. *In* A. M. Taji and R. R. Williams [eds.], Tissue culture of Australian plants, 112–156. University of New England Press, Armidale, Australia.

Merkle, S. A., and C. J. Nairn. 2005. Hardwood tree biotechnology. *In Vitro Cellular and Developmental Biology – Plant* 41: 602–619.

Midgley, S. J., and J. W. Turnbull. 2003. Domestication and use of Australian acacias: case studies of five important species. *Australian Systematic Botany* 16: 89–102.

Murashige, T., W. P. Bitters, E. M. Rangan, E. M. Naue, C. N. Roistacher, and P. B. Holliday. 1972. A technique of shoot apex grafting and its utilization towards recovering virus–free citrus clones. *HortScience* 7: 118–119.

Nadgauda, R. S., V. A. Parasharami, and A. F. Mascarenhas. 1990. Precocious flowering and seeding behaviour in tissue cultured bamboos. *Nature* 344: 335–336.

Navarro, L. 1981. Shoot–tip grafting in vitro (STG) and its application: a review. *Proceedings of The International Society of Citriculture* 1: 452–456.

Navarro, L., C. N. Roistacher, and T. Murashige. 1975. Improvement of shoot–tip grafting in vitro for virus–free citrus. *Journal of American Society of Horticultural Science* 100: 471–479.

Naz, A. A., M. J. Jaskani, H. Abbas, and M. Qasim. 2007. *In vitro* studies on micrografting technique in two cultivars of citrus to produce virus free plants. *Pakistan Journal of Botany* 39: 1773–1778.

Nehra, N. S., M. R. Becwar, W. H. Rottmann, L. Pearson, K. Chowdhury, S. Chang, H. D. Wilde, *et al.*, 2005. Forest biotechnology: . *In Vitro Cellular and Developmental Biology – Plant* 41: 701–717.

Obeidy, A. A., and M. A. L. Smith. 1991. A versatile new tactic for fruit tree micrografting. *Hoechnology* October–December 1991: 91–95.

Olde, D. A. 1994. Towards Haploid Plant Production of Sturt's Desert Pea (*Swainsona formosa*) Using *in vitro* Pollen Culture Techniques. Graduate Diploma in Horticultural Science *Thesis*. Agronomy and Soil Science, University of New England, Armidale, Australia.

Palmer, J. L., R. J. Lawn, and S. W. Adkins. 2002. An embryo–rescue protocol for *Vigna* interspecific hybrids. *Australian Journal of Botany* 50: 331–338.

Panaia, M., E. Bunn, and K. Dixon. 2005. Somatic embryogenesis as an efficient method for the clonal propagation of Koala Fern, Black Kangaroo Paw and Blue Boy – Important species for rehabilitation of disturbed habitats and horticultural utilisation, The Plant Tissue Culture and Biotechnology Conference, 21–24 September 2005, Perth, Western Australia.

Panaia, M., T. Senaratna, K. W. Dixon, and K. Sivasithamparam. 2004. High–frequency somatic embryogenesis of Koala Fern (*Baloskion tetraphyllum*, Restionaceae). *In Vitro Cellular and Developmental Biology–Plant* 40: 303–310.

Pande, D., M. Purohit, and P. S. Srivastava. 2002. Variation in xanthotoxin content in *Ammi majus* L. cultures during in vitro flowering and fruiting. *Plant Science* 162: 583–587.

Peeters, A. J. M., W. Gerards, G. W. M. Barendse, and G. J. Wullems. 1991. *In vitro* flower bud formation in tobacco: interaction of hormones. *Plant Physiology* 97: 402–408.

Prakash, M. G., and K. Gurumurthi. 2010. Effects of type of explant and age, plant growth regulators and medium strength on somatic embryogenesis and plant regeneration in *Eucalyptus camaldulensis*. *Plant Cell, Tissue and Organ Culture* 100: 13–20.

Predieri, S. 2001. Mutation induction and tissue culture in improving fruits. *Plant Cell, Tissue and Organ Culture* 64: 185–210.

Price, H. J., G. L. Hodnett, B. L. Burson, S. L. Dillon, D. M. Stelly, and W. L. Rooney. 2006. Genotype dependent interspecific hybridization of *Sorghum bicolor*. *Crop Science* 46: 2617–2622.

Rabinowitch, H. D., and L. Currah. 2002. Allium Crop Science: Recent Advances. CAB International, Oxon, UK.

Raghavan, V. 2003. One hundred years of zygotic embryo culture investigations. *In Vitro Cellular and Developmental Biology – Plant* 39: 437–442.

Ratanasanobon, K. 2007. Somatic hybridisation for Australian cut flowers: protoplast fusion. *Floriculture News* 69, November 2007: 7.

Reid, M. S., W. P. Hackett, and J. S. Julian. 2001. Rootstocks for "difficult" plants: rhododendrons, azaleas and grevilleas. *Slosson Report 2000–2001*: 1–3.

Reinert, J. 1958a. Morphogeneses und ihre kontrolle an gewebe kulturen aus karotten. *Naturwissenschaft* 45: 344–345.

Reinert, J. 1958b. Morphogenese und ihre kontrolle an gewebekulturen aus carotten. *Naturwissenchaften* 45: 344–345.

Reinert, J. 1959. Ueber die kontrolle der morphogenese und die induktion von adventiveembryonen an gewebekulturen aus karotten. *Planta* 53: 318–333.

Rout, G. R., and P. Das. 1994. Somatic embryogenesis and in vitro flowering of 3 species of bamboo. *Plant Cell Reports* 13: 683–686.

Santana–Buzzy, N., R. Rojas–Herrera, R. M. Galaz–Avalos, J. R. Ku–Cauich, J. Mijangos–Cortes, L. C. Gutierrez–Pacheco, A. Canto, *et al.*, 2007. Advances in coffee tissue culture and its practical applications. *In Vitro Cellular and Developmental Biology–Plant* 43: 507–520.

Sargent, H. R., I. D. Godwin, and S. W. Adkins. 1997. The effects of putrescine, spermidine and spermine on somatic embryogenesis of *Sorghum bicolor*. *In* A. M. Taji and R. R. Williams [eds.], Tissue culture: Towards the next century, 75–79. University of New England Publications Unit, Armidale, NSW.

Sartoretto, L. M., C. W. Saldanha, and M. P. M. Corder. 2008. Genetic transformation: strategies for forest species breeding. *Ciencia Rural* 38: 861–871.

Scarth, R., and J. Tang. 2006. Modification of brassica oil using conventional and transgenic approaches. *Crop Science* 46: 1225–1236.

Schell, J. 1987. Transgenic plants as tool to study the molecular organisation of plant genes. *Science* 237: 1176–1183.

Schell, J., and K. Vasil [eds.]. 1989. Molecular Biology of Plant Nuclear Genes, vol. 6. Academic Press, New York.

Scorza, R., M. Ravelonandro, A. M. Callahan, J. M. Cordts, M. Fuchs, J. Dunez, and D. Gonsalves. 1994. Transgenic plums (*Prunus domestica* L.) express the plum pox virus coat protein gene. *Plant Cell Reports* 14: 18–22.

Senaratna, T. 2000. Mass propagation of difficult Australian plants. *Australian Horticulture* November 2000: 59–60.

Shan, F. 2007. Floriculture laboratory activities. *Floriculture News* 69, November 2007: 6.

Sheeja, T. E., and A. B. Mandal. 2003. In vitro flowering and fruiting in tomato (*Lycopersicon esculentum* Mill.). *Asia Pacific Journal of Molecular Biology and Biotechnology* 11: 37–42.

Shin, S., C. A. Mackintosh, J. Lewis, S. J. Heinen, L. Radmer, R. Dill–Macky, G. D. Baldridge, *et al.*, 2008. Transgenic wheat expressing a barley class II chitinase gene has enhanced resistance against *Fusarium graminearum*. *Journal of Eexperimental Botany* 59: 2371–2378.

Sieler, I., K. Dixon, and I. R. Dixon. 1993. Horticultural development of rushes and sedges. *Australian Horticulture* 91: 20–21.

Sim, G. E., C. S. Loh, and C. J. Goh. 2007. High Frequency early in vitro flowering of *Dendrobium* Madame Thong–In (Orchidaceae). *Plant Cell Reports* 26: 383–393.

Sim, G. E., C. J. Goh, and C. S. Loh. 2008. Induction of in vitro flowering in *Dendrobium* Madame Thong–In (Orchidaceae) seedlings is associated with increase in endogenous N^6–(Δ^2–isopentenyl)–adenine (iP) and N^6–(Δ^2–isopentenyl)–adenosine (iPA) levels. *Plant Cell Reports* 27: 1281–1289.

Singh, B., S. Sharma, G. Rani, G. S. Virk, A. A. Zaidi, and A. Nagpal. 2006. *In vitro* flowering in embryogenic cultures of Kinnow mandarin (*Citrus nobilis* Lour x *C. deliciosa* Tenora). *African Journal of Biotechnology* 5: 1470–1474.

Singh, M., U. Jaiswal, and V. S. Jaiswal. 2000. Thidiazuron–induced in vitro flowering in *Dendrocalamus strictus* Nees. *Scientific Correspondence* 79: 1529–1530.

Skalova, D., B. Navratilova, V. Ondrej, and A. Lebeda. 2010. Optimizing culture for *in vitro* pollination and fertilization in *Cucumis sativus* and *C. melo*. *Acta Biologica Cracoviensia Series Botanica* 52/1: 111–115.

Speer, S. S. 1993. Micropropagation of some Myrtaceae species which show potential as "new" ornamental plants. *Australian Journal of Experimental Agriculture* 33: 385–391.

Stebbins, G. L. 1958. The inviability, weakness, and sterility of interspecific hybrids. *Advances in Genetics* 9: 147–215.

Steward, F. C., M. O. Mapes, and K. Meats. 1958a. Growth and organized development of cultured cells, II. Organization in cultures grown from freely suspended cells. *American Journal of Botany* 45: 704–708.

Steward, F. C., M. O. Mapes, and K. Mears. 1958b. Growth and organised development of cultured cells. *American Journal of Botany* 45: 693–713.

Steward, F. C., M. O. Mapes, and K. Mears. 1958c. Growth and organized development of cultured cells II. Organization in cultures grown from freely suspended cells. *American Journal of Botany* 45: 705–708.

Street, H. E. 1973. Plant tissue and cell cultures. Blackwell Scientific Publications, Oxford, London.

Suarez, I. E., R. A. Schnell, D. N. Kuhn, and R. E. Litz. 2005. Micrografting of ASBVd–infected Avocado (*Persea americana*) plants. *Plant Cell, Tissue and Organ Culture* 80: 179–185.

Sudhersan, C., and M. AboEl–Nil. 2002. Somatic embryogenesis on Sturt's desert pea (*Swainsona formosa*). *Current Science* 83: 1074–1076.

Tade, E. 1992. Anther and ovule culture of *Clianthus formosus*. Master of Rural Science *Thesis*. Agronomy and Soil Science, University of New England, Armidale, Australia.

Taji, A., and R. Williams. 2005. Use of *in vitro* breeding strategies in the development of Australian native plants. *Acta Horticulturae* 683: 87–94.

Taji, A., P. Kumar, and P. Lakshmanan. 2002. *In Vitro* Plant Breeding. Haworth Press, Inc., New York.

Taji, A. M., and R. R. Williams. 1987. Perpetuation of the self–incompatible rare species of *Swainsona laxa* R. Br. by pollination in vitro and in situ. *Plant Science* 48: 137–140.

Taji, A. M., and R. R. Williams. 1996. Overview of plant tissue culture. *In* A. M. Taji and R. R. Williams [eds.], Tissue Culture of Australian Plants, 1–15. University of New England Press, Armidale, Australia.

Taji, A. M., and J. S. Williams. 2004. *In Vitro* Pollination and Fertilisation. *In* R. M. Goodman [ed.], Encyclopedia of Plant and Crop Science, 584–587. Marcel Dekker, Inc, New York.

Taji, A. M., W. A. Dodd, and R. R. Williams. 1997. Plant Tissue Culture Practice (third edition). University of New England Press, Armidale, Australia.

Tangolar, S. G., K. Ercik, and S. Tangolar. 2003. Obtaining plants using in vitro micrografting method in some grapevine varieties (*Vitis vinifera* L.). *Biotechnology and Biotechnology Equipment* 17: 50–55.

Tapingkae, T., P. Kristiansen, and A. Taji. 2009. Influence of carbohydrate source on the *in vitro* flowering of Sturt's Desert Pea (*Swainsona formosa*). *Acta Horticulturae* 829: 225–230.

Tapingkae, T., Z. Zulkarnain, M. Kawaguchi, T. Ikeda, and A. Taji. in press. Somatic (Asexual) Procedures (Haploids, Protoplasts, Cell Selection) and Their Applications. *In* A. Altman and M. Hasegawa [eds.], Plant Biotechnology and Agriculture: Prospects for The 21st Century, in press. Elsevier.

Taylor, N. J., M. E. Light, and J. Van Staden. 2005. In vitro flowering of *Kniphofia leucocephala*: influences of cytokinins. *Plant Cell, Tissue and Organ Culture* 83: 327–333.

Teulieres, C., and A. M. Boudet. 1991. Isolation of protoplasts from different *Eucalyptus* species and preliminary studies on regeneration. *Plant Cell Tissue and Organ Culture* 25: 133–140.

Tisserat, B., P. D. Galletta, and D. Jones. 1990. In vitro flowering from *Citrus limon* lateral buds. *Journal of Plant Physiology* 136: 56–60.

Tomasi, P., D. A. Dierig, R. A. Backhaus, and K. B. Pigg. 1999. Floral bud and mean petal length as morphological predictors of microspore cytological stage in Lasquerella. *HortScience* 34: 1269–1270.

Turner, S., S. L. Krauss, E. Bunn, T. Senaratna, K. Dixon, B. Tan, and D. Touchell. 2001. Genetic fidelity and viability of *Anigozanthos viridis* following tissue culture, cold storage and cryopreservation. *Plant Science* 161: 1099–1106.

Van Staden, J., and C. W. S. Dickens. 1991. *In vitro* induction of flowering and its relevance to micropropagation. *In* Y. P. S. Bajaj [ed.], Biotechnology in Agriculture and Forestry: High–Tech and Micropropagation, vol. 17, 85–115. Springer–Verlag, Berlin.

Van Tuyl, J. M., M. J. De Jeu, and V. K. Sawhney. 1997. Methods for Overcoming Interspecific Crossing Barriers. *In* K. R. Shivanna and V. K. Shawney [eds.], Pollen Biotechnology for Crop Production and Improvement, 273–292. Cambridge University Press Cambridge.

Vasil, I. K. 1991. Plant tissue culture and molecular biology as tools in understanding plant development and in plant improvement. *Current Opinion in Biotechnology* 2: 158–163.

Vaz, A. P. A., L. de Cassia, F. Riberiro, and G. B. Kerbauy. 2004. Photoperiod and temperature effects on *in vitro* growth and flowering of *P. pusilla*, an epiphytic orchid. *Plant Physiology and Biochemistry* 42: 411–415.

Victorio, C. P., and C. L. S. Lage. 2009. *In vitro* flowering of *Phyllanthus tenellus* Roxb. cultured under different light qualities and growth regulators. *General and Applied Plant Physiology* 35: 44–50.

Wang, Y. H. 2010. An efficient protocol for stimulating cell development in protoplast culture of *Scaevola*. *Plant Growth Regulation* Published online: 28 November 2010: DOI 10.1007/s10725–10010–19544–z.

Wang, Y. H., and P. L. Bhalla. 2004. Somatic embryogenesis from leaf explants of Australian fan flower, *Scaevola aemula* R. Br. *Plant Cell Reports* 22: 408–414.

Wang, Y. H., and P. L. Bhalla. 2006. Plant regeneration from cell suspensions initiated from leaf– and root–derived calli of the Australian ornamental plant *Scaevola aemula* R.BR. *Propagation of Ornamental Plants* 6: 55–60.

Watt, M. P., F. Blackway, C. F. Cresswell, and B. Herman. 1991. Somatic embryogenesis in *Eucalyptus grandis*. *South African Forestry Journal* 157: 59–65.

Williams, R. R., and A. M. Taji. 1992. Stock plant management for *Clianthus* cutting production. *Acta Horticulturae* 314: 317–322.

Yan, S. F., Q. Zhang, J. E. Wang, Y. Q. Sun, M. K. Daud, and S. J. Zhu. 2010. Somatic embryogenesis and plant regeneration in two wild cotton species belong to G genome. *In Vitro Cellular and Developmental Biology – Plant* 46: 298–305.

Zagorska, N., B. Dimitrov, P. Gadeva, and P. Robeva. 1997. Regeneration and characterisation of plants obtained from anther culture of *Medicago sativa* L. *In Vitro Cellular and Developmental Biology – Plant* 33: 107–110.

Zhang, B. H., Q. L. Wang, F. Liu, K. B. Wang, and T. P. Frazier. 2009. Highly efficient plant regeneration through somatic embryogenesis in 20 elite commercial cotton (*Gossypium hirsutum* L.) cultivars. *Plant Omics* 2: 259–268.

Zhang, Z., and D. W. M. Leung. 2000. A comparison of *in vitro* and *in vivo* flowering in gentian. *Plant Cell, Tissue and Organ Culture* 63: 223–226.

Zhang, Z., and D. W. M. Leung. 2002. Factors influencing the growth of micropropagated shoots and *in vitro* flowering of gentian. *Plant Growth Regulators* 26: 245–251.

Zulkarnain. 2003. Breeding Strategies in Sturt's Desert Pea (*Swainsona formosa*) Using *In Vitro* and *In Vivo* Techniques. Ph.D *Thesis*. Agronomy and Soil Science, University of New England, Armidale, Australia.

2012, Nutri-Horticulture
Editor: Professor K.V. Peter
Published by: DAYA PUBLISHING HOUSE, NEW DELHI

Pages 99–122

Chapter 6

Science of Polyembryony

R.K. Roshan[1] and Nongallei Pebam[2]
[1]*Krishi Vijyan Kendra, Churachandrapur, Personmun Village,*
Churachandrapur Distt. Manipur – 795 128, INDIA
E-mail: rk_ryan2000@yahoo.com
[2]*Institute of Bioresources and Sustainable Development (IBSD),*
Takyelpat Institutional Area, Imphal – 795 001, INDIA

Polyembryony is the formation of more than one embryo in a seed. The phenomenon of polyembryony was discovered by A. Leeuwenhoek in 1719, who found two germlings developing from the same seed in Citrus. Later, Strasburger (1878) described the formation of adventives embryos from nucellus cells in Citrus. He called this phenomenon nucellar or adventive polyembryony. Embryological studies in 19–20th centuries demonstrated that adventive embryos present in the seed in addition to the sexual embryo may be formed on the basis of different ovule structures and embryo sac structures. The available materials on polyembryony called for the necessity of systematizing these data. The first system was proposed by Braun (1859), who described four possible ways of formation of adventive embryos as a result of fusion of two or more ovules, development of several embryo sacs in the same ovule, or as a result of the proembryo division. This phenomenon was also recorded by Schacht in 1859. Subsequently, Horn (1943) independently observed polyembryony in mango. In polyembryonic seeds, one embryo is produced as a result of fertilization and several adventitious embryos arise from the cells of nucellus. These adventitious embryos are known to originate from the epidermal cells of the nucellus situated close to the micropolar end of the side opposite to the funicle (Juliano, 1937 and Maheshwari *et al*,1955). The polyembryony is one of the apomictic processes which have been described to occur in the ovules of angiospermic species (Koltunow, 1993). Polyembryony in horticultural crops occurs as facultative apomixis wherein simultaneous growth of multiple embryos of somatic origin co-exist in the same seed containing sexual embryo resulting from either selfing or cross pollination. Polyembryony in citrus is referred as nucellar embryony because of the origin of plural embryos from nucellus and the resultant seedlings are called as "nuclear seedlings". Nucellar embryos develop by ordinary mitotic division, as such the genetic constitution of the embryos is identical to that of seed parent, except in cases of probable bud variation. Nucellar embryony is of wide occurrence in the genera *Citrus*, *Poncirus* and *Fortunella* of Rutaceae family.

Occurrence of Polyembryony

The polyembryony is a type of apomixis that may occur by following ways (Bhatnagar and Bhojwani, 1999).

1. Cleavage of proembryos
2. Formation of embryos by cells of embryo sac other than egg
3. Development of more than one embryo sac within the same ovule or multiple polyembryony
4. Formation of embryo by some sporophytic/maternal tissue of ovule

Cleavage of Proembryo

The simplest method is the cleavage polyembryony. In this case, splitting of zygote or proembryo occurs and each split part develops into an embryo. This type of polyembryony is more common in gymnosperms than in angiosperms. *Erythronium americanum* (Figure 6.1), *Nymphaea advena*, orchids, Crotalaria etc. are some of the angiosperm showing cleavage polyembryony. Swamy (1943) recorded three different modes of super numerary embryo formation.

1. The zygote divides irregularly to form a mass of cells of which those lying towards the chalazal end grow simultaneously and give rise to many embryos.
2. The proembryo gives out small buds or outgrowths which may themselves function as embryos
3. The filamentous embryo becomes branched and each branch gives rise to an embryo. Unlike the cleavage embryony, arising during seed development very common in orchids, the formation of plural embryos during seed germination is exclusively observed in *Vandas*

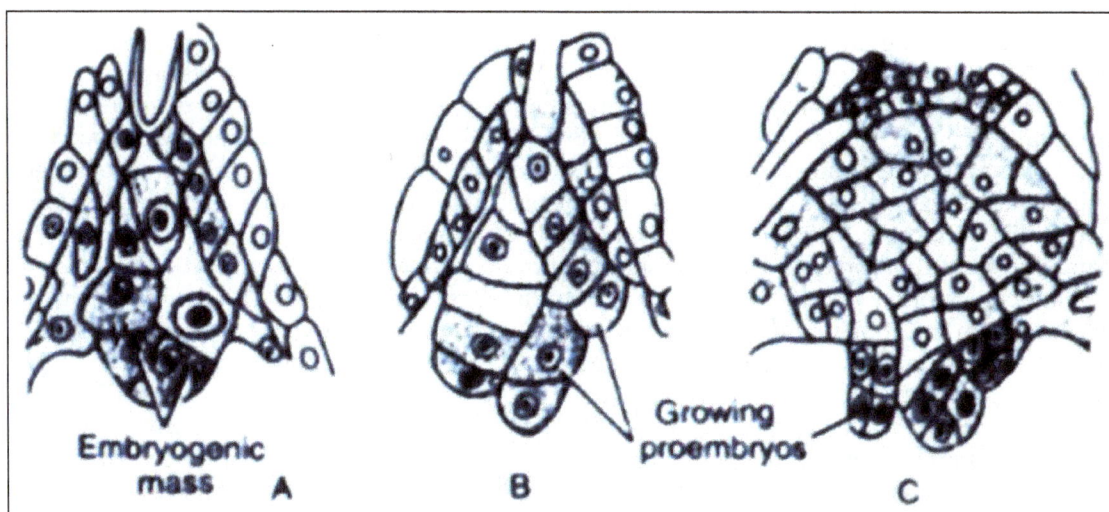

Figure 6.1A–C: Cleavage Polyembryony

(A) Embryonic mass form by the basal cell of the zygote in *Erythronium americanum;*
B–C: Differentiation of embryos from the cells of the embryonic mass.

only. The apical promeristem of embryo in this genera divides into a number of primordial (3-9) and each of which further organized into an embryo.

Suspensor Polyembryony

It is common in some of the genera belonging to family Santalaceae, where as many as six embryo may develop simultaneously in an ovule by the proliferation of the suspensor cells. But only one of them fully develops into full maturity.

Embryos from Cells of the Embryo Sac Other than the Egg

Polyembryony also arises due to formation of embryos from other cells of embryo sac eg. Synergids (as in *Argemone maxicana*) and antipodals (as in *Ulmus americana*). The cells outside the embryo sac develop into embryo which later penetrate the embryo sac to obtain nourishment. These embryos are known as adventives embryos and the process as adventives polyembryony. Such embryos may developed from nucellus, *e.g. Mangifera, Citrus, Opuntia* etc. In Eugenia both nucellus and integumentary embryos (those developing from the cells of the integument) are known to occur.

The ploidy of embryo depends on the origin. If it arises from unfertilized synergid, the embryo will be haploid and if it arises from fertilized synergid the resultant embryo will be diploid. The formation of diploid embryo from synergid is brought about by the entry of more than one pollen tube into the embryo sac or by the presence of additional male gamete in the same pollen tube. Embryos arising from unfertilized synergids are known in French bean (*Phaseolus vulgaris*). In this case, the synergid embryo (haploid) and zygotic embryo (diploid) can be distinguished on the basis of ploidy level. The development of embryo from antipodal cells and endosperm cells is rather rare. The embryo from antipodal cell may be haploid or diploid depending up on the fusion with male gamete but generally fails to produce germinating embryo and gets degenerated at an early stage whereas, endosperm produces triploid embryo and seedling.

More than One Embryo Sac in the Same Ovule

Multiple embryo sac may arise in an ovule either from the same megaspore mother cell or from other megaspore mother cell or from any other sporophytic tissue of ovule. Although many embryos develop inside an ovule but hardly one or in few cases two embryos survive and produce seedlings. This condition is also called multiple polyembryony.

Multiple embryosacs in an ovule emerge from:

1. Derivatives of the same megaspore mother cells
2. Derivatives of two or more megaspore mother cells
3. Nucellar cells.

Formation of twin embryosac within an ovule is common in *Casuarina equisetifolia* and *Citrus* sp. In *Pennisetum ciliare*, 22 per cent seed contain twin embryos. The normal embryosac develops only upto the 4 nucleate stage and multiple embryos are formed by aposporous embryo sacs.

Activation of some Sporophytic Cells of the Ovule

The process of development of embryos from tissues other than the zygote or endosperm is generally refered to as adventitious embryony. Maternal sporophytic tissues outside the embryosac. *i.e.,* nucellus and integument form the resource tissues in this situation. In addition to the well known examples of *Mangifera* and *Citrus* genera, nucellar polyembryony is observed in some species under the genera *Opuntia* and *Trillium*. Species variation with respect to manifestation of polyembryony is noted in citrus in which some are monoembryonic (*C. grandis, C. limon*) while others are polyembryonate (*C. microcarpa, C. reticulata*). As many as 40 embryos are observed in seeds of *C. unshiu*. In the polyembryonate species, the adventives embryos arise by proliferation of nucellar cells with rare exceptions. Nucellar embryos arise from the micropylar half of the nucellus. In *Mangifera*, the nucellar cells destined to form adventives embryos can be distinguished from other cells of the nucellus by their dense cytoplasm and starchy contents. The inception of nucellar embryos takes place outside the embryosac, but they are gradually pushed into the embryosac cavity where they divide and differentiate into mature embryos. The adventive embryos do not show synchronous development. A single seed may show embryo at various stages of development. Nuceller embryos can be differentiated from zygotic embryos by their lateral position on the embryosac, irregular shape and lack of suspensor. But in *C. microcarpa*, the nucellar embryos possess a distinct suspensor from very early stages. Mostly, for the initiation of nucellar cells, the pollination stimulus is essential.

Classification of Polyembryony

Different polyembryony classifications have shown that the main criteria for classification include the origin of initial cells, ways of formation of embryos and their genetic characteristics (Figure 6.2). Based on these criteria, polyembryony is divided into two main types: gametophytic and sporophytic (Yakovlev, 1967; Bouman and Boesewinkel, 1969).

Gametophytic Polyembryony

This type of polyembryony unites the phenomena related to the formation of adventive embryos from the gametophyte cells: synergids and antipodals (apogamety), as well as from the egg cells, when additional embryo sacs develop.

The term "apogamety" was introduced by Renner (1916) to designate formation of embryos from the gametophyte cells (synergids and antipodals), except the egg cell. Two forms of apogamety are distinguished in the literature: synergidal and antipodal (Figure 6.3).

The development of synergidal embryos begins, as a rule, after egg cell fertilization and zygote division. In this respect and also with an account of their haploid nature, such embryos develop at a slower rate than the sexual embryo and are, in most cases, not competitive, as result of which die at the early developmental stages. In some cases, when many pollen tubes penetrate into the embryo sac, one of synergids may be fertilized by an additional sperm cell (*Najas major* –Guignard, 1901; *Cuscuta reflexa* –Johri and Tiagi, 1952; *Tamarix ericoides* – Johri and Kak, 1954; *Pennisetum squamulatum* – Sindhe *et al.*, 1980) and give rise to an embryo. In this case, both embryos characterized by biparental inheritance reach, as a rule, the mature state.

The formation of apogametic embryos was described in both haploid embryo sacs in amphimicts (*Lilium martagon* –Cooper, 1943; *Arabis lyallii* –Lebègue, 1948; *Bergenia delavai* – Lebègue, 1949; *Argenone Mexicana* –Sachar, 1955; *Fritillaria meleagroides* –Smirnov and Grakhantseva, 1982; *Ulmus Americana* –Shattuck, 1905; *U. glabra* –Ekdahl, 1941) and diploid ones

Table 6.1: Gametophytic Polyembryony Described in Different Species and Families

Gametophytic Polyembryony		
Synergidal Apogamety		
	Species	Families
Dicotyledonous	140	56
Monocotyledonous	83	19
Antipodal apogamety		
Dicotyledonous	10	5
Monocotyledonous	17	9

Solntseva, 1999.

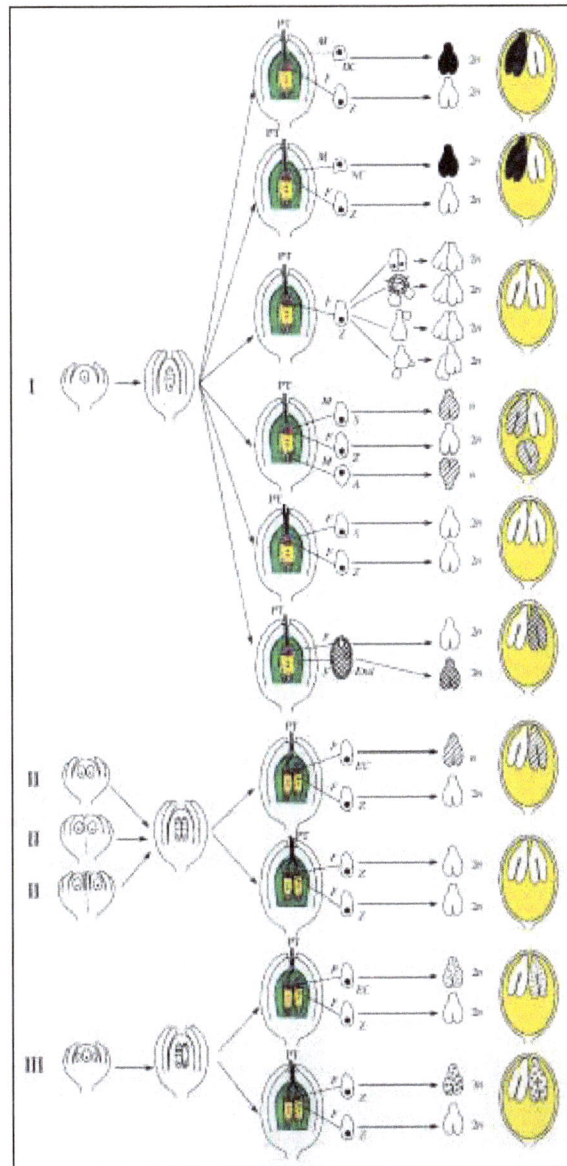

Figure 6.2: Possible Pathways of Formation of Several Embryos in the Same Seed: (I) From cells of one embryo sac and surrounding tissues; (II) As a result of development of two embryo sacs; (III) As a result of development of additional aposporic embryo sac. M, mitosis; F, fertilization; P, parthenogenesis; A, antipode; Z, zygote; IIC , inner integument cells; NC - nuclleus cells; PT, pollen tube; S, synergid; End, endosperm; EC, egg cell. Embryo developing from: ( ) zygote, 2 n; ( ) cells of nucellus and inner integument, 2 n; ( ) unfertilized synergids and antipodals, n; ( ) endosperm cells, 3 n; ( ) unfertilized egg cell of apoaporic embryo sac, 2 n; ( ) fertilized egg cell of aposporic embryo sac, 3 n.

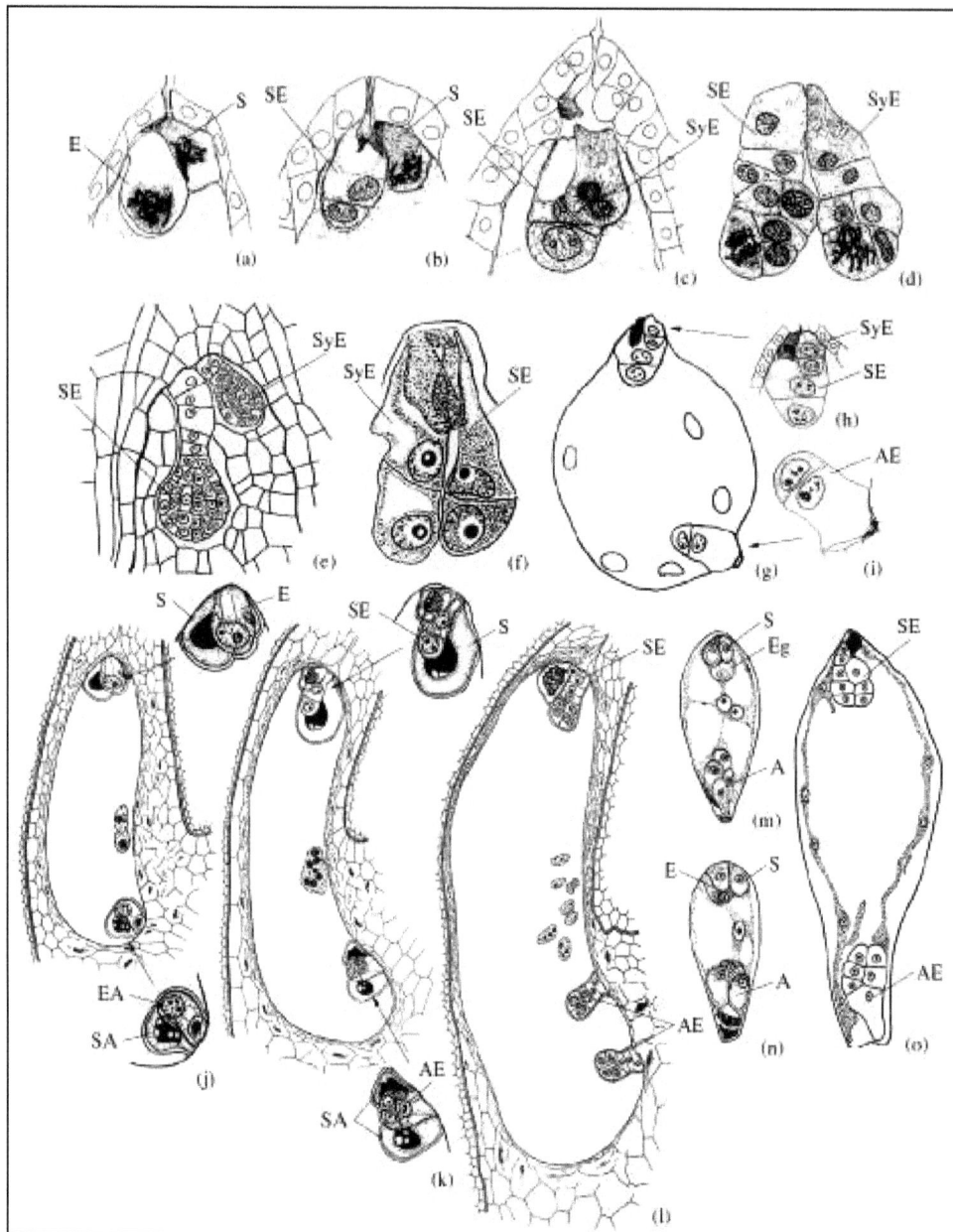

Figure 6.3: Gametophytic polyembryony

(a–h) formation of synergidal embryos in: (a–d) *Lilium martagon* (after Cooper, 1943),
(e) *Bergenia delavayi* (after Lebegue, 1949), (f) *Tamarix ericoides* (after Johri and Kak, 1954),
(g–h) *Fritillaria meleagroides* (after Smirnov and Grakhantseva, 1982); (i–o) development of
antipodal embryos in: (i) *Fritillaria meleagroides* (after Smirnov and Grakhantseva, 1982),
(j–l) *Allium ramosum* (after Vinogradova, unpublished data), (m–o) *Ulmus glabra* (after Ekdahl, 1941).
Embryo: antipodal (AE), sexual (SE), synergidal (SyE) antipodals: synergid-like (SA), egg-like (EA).
For other designations see Figure 6.2.

in apomicts (*Rudbeckia laciniata* – Solntseva, 1973). In most cases, synergidal and antipodal embryos develop without fertilization and, hence, apogamety in haploid embryo sacs may be considered as a possible source of the origin of haploids.

Sporophytic Polyembryony

When multiple embryos arise either from zygote or from sporophytic cells of ovule (nucellus, integument) and the resulting embryo will be diploid. Hence citrus, mango and jamun exhibit true and sporophytic polyembryony.

This type is characterized by the development of adventive embryos from the sporophyte cells: mother (nucellar and integumental polyembryony) or daughter (monozygotic cleavage polyembryony) (Figure 6.4).

Nucellar and integumental polyembryonies are widespread among angiospermous plants. In this type of polyembryony, somatic cells of nucellus (*Citrus trifoliate* –Osawa, 1912; *Nigritella nigra* – Afzelius, 1928; *Zeuxine sulcata* –Swamy, 1946; *Eugenia jambos* –Roy, 1953; *Mangifera indica* –Sachar and Chopra, 1957; *Syzygium caryophyllifolium* –Roy and Sahai, 1962) or inner integument (*Spiranthes cernua* –Swamy, 1948) act as initial cells of adventive embryos. Adventive embryos developing from somatic cells of the mother tissues, nucellus and integument, are somatic (embryoids) and have uniparental inheritance, thus repeating the mother genotype.

Nucellar embryos can develop from one cell (*Poa pratensis* –Batygina and Freiberg, 1979; *Sarcococca humilis* –Naumova and Villemse, 1983; *Poncirus trifoliata, Opuntia elata* –Naumova, 1992) or so-called embryonic cell complex is preliminarily formed from nucellar cells (P. *pratensis* –Batygina and Freiberg, 1979). Such embryos develop, as a rule, in the micropilar ovule part, but can also be formed in the lateral and chalasal parts. They are usually competitive and reach the mature state; in some cases (*Opuntia elata* – Archibald, 1939; Naumova, 1992) they can even replace the sexual embryos, as a result of which the mature seed contains only adventive embryos.

Integumental embryos are usually formed from internal epiderm cells of inner integument (for example, in *Spiranthes cernua* –Swamy, 1948). In some cases, they are formed not only in the micropylar ovule part, but also in its middle parts. This is inherent in the species with "endothelium" (integumental tapetum), the innermost layer of inner integument consisting of large cells with dense cytoplasm (*Carthamus tinctorius* –Maheshwari Devi and Pullaiah, 1977; *Melampodium divaricatum* –Maheshwari Devi and Pullaiah, 1976). Usually, such embryos do not reach maturity and begin to degenerate at the early developmental stages. Monozygotic cleavage polyembryony is a way of formation of somatic embryos on the basis of daughter sporophyte, zygote or multicellular embryo.

Ernst (1918) divided polyembryony into two categories on the basis of embryogenesis

True Polyembryony

Two or more embryos arising in the same embryo sac, from the zygote or embryo (Eulophia: vanda), from synergids (Sagittaria), from antipodal cell (Ulmus) or from nucellus or integument (citrus, mango, jamun)

Table 6.2: Sporophytic Polyembryony Described in Different Species and Families

Sporophytic Polyembryony		
Nucellar Polyembryony		
	Species	Families
Dicotyledonous	162	35
Monocotyledonous	33	11
Integumental Polyembryony		
Dicotyledonous	49	9
Monocotyledonous	9	2

Naumova, 1992.

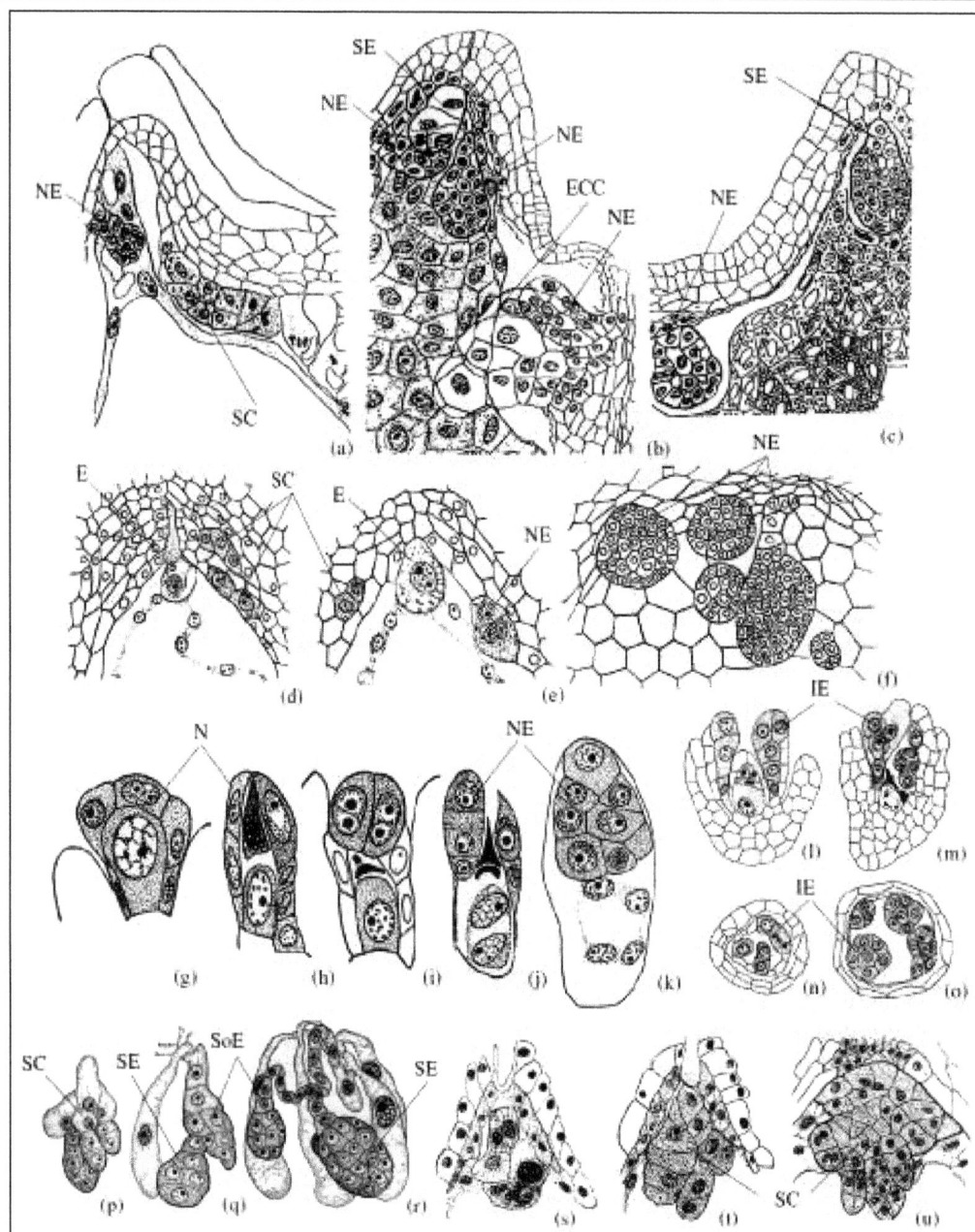

Figure 6.4: Sporophytic polyembryony

(a–k) development of nucellar embryos in: (a–c) *Poa pratensis* (after Batygina and Reiberg, 1979),
(d–f) *Citrus trifoliate* (after Osawa, 1912), (g–k) *Nigritella nigra* (after Afzelius, 1928); (l–o) formation of
integumental embryos in *Spiranthes cernua* (after Swamy, 1948); (p–u) monozygotic cleavage
polyembryony in: (p–q) *Eulophia epidendraea* (after Swamy, 1943), (r–u) *Erythronium americanum*
(after Jeffrey, 1895). Embryo: integumental (IE), nucellar (NE), somatic (SoE). N, nucellus;
SC, slem cells; ECC, embryonic cell complex. For other designations see Figure 6.3.

False Polyembryony

In this case, more than one embryo sac is formed in an ovule (*Fragaria* sp.) which is followed by the formation of multiple embryos.

Bouman and Boesewinkel (1969) put forth another classification based on origin of additional embryos. According to them, spontaneous polyembryony should split into four categories:

1. Supernumerary embryos arising from sporophytic cells of generation (integuments and nucellus)
2. Supernumerary embryos arising from cells of the gametophyte (by one or more embryo sacs in the same ovule).
3. Supernumerary embryos arising from the new sporophyte (fertilizer egg or proembryo).
4. Supernumerary embryos arising from the male gametophyte (disputed group).

On the basis of degree of polyembryony species/varieties can be divided into three categories; slightly polyembryonic (polyembryony up to 25 per cent), moderately polyembryonic (polyembryony up to 50 per cent) and highly polyembryonic (polyembryony > 50 per cent).

Factors Affecting Polyembryony

Several factors, like environment, food supply, pollination, pollen source, hormones etc, are reported to influence polyembryony.

Geographical location influences polyembryony in citrus. Nasharty (1945) observed more number of embryos in citrus when grown at Los Angels than at Riverside in the same season. Nakatami *et al.* (1978) reported that temperature also influences the production of extra embryos. The reduction in the number of embryos per seed is reported due to cessation of division in nucellar cells in the glass house due to higher temperature than at the field. He also suggested that cross pollination of polyembryonate seed parent under high temperatures may be beneficial for citrus breeding to overcome the hurdle of polyembyrony.

The evidence suggests strongly that pollination and fertilization are necessary for nucellar embryony (Furusato, 1953, Hodgson, 1961). Polyembryony is also affected by the type of pollinator (Soares *et al.*, 1995), amount and pollen viability, plant nutrition, air temperature, environmental and soil humidity and speed of the wind. Dhillon *et al.* (1993) reported that adventitious embryos developed into globular or early cotyledon stages in the absence of pollination, but the embryos required the endosperm in order to grow, making pollination and fertilization necessary for development of polyembryonic seeds. The growth of nucellar embryo depends upon fertilization for endosperm formation. During fertilization, one male gamete meets with egg cell to form zygotic embryo while other male gamete fuses with polar nuclei to form endosperm. Since endosperm supplies food to developing embryos, the fertilization is a prerequisite for production of matured embryo. The growth of nucellar embryo is arrested at later stage of development if endosperm is not formed (Koltunow *et al.*, 1995). Therefore any factor that affects pollination, fertilization or seed development will also affect the percentage of polyembryony and embryos/seed. The difference in results obtained in various locations shows the importance of determining the characteristics of polyembryony.

Type of pollination also plays a role, particularly in the degree of polyembryony. Cross-pollination gives higher percentage of polyembryony than self-pollination (Cekvava, 1968).

Seeds from the northern side of the tree have higher mean number of embryos than those from southern side (Furusato, 1953). Reduction in food supply to the fruit reduces number of seeds producing more than one embryo by 51-100 per cent in several varieties of Citrus (Traub, 1936). Harvesting season also influences the number of embryos per seed. Sarkar (1961) reported higher number of embryos per seed in lemon and citron fruits harvested in October than from fruits harvested in March.

Table 6.3: Horticulturally Important Polyembryonic Crops and their Varieties

Sl.No.	Crops	Species	Varieties/Cultivar
1.	Citrus and related species	(a) *Citrus reticulata* L.	Nagpur Mandarin, Khasi Mandarin, Coorg Mandarin,
		(b) *C. reshni*	Cleopatra
		(c) *C. deliociosa* Ten	Willow Leaf Mandarin
		(d) *C. unshu* Marc.	Satsuma Orange
		(e) *C. medurensis* Lour.	Calamondin
		(f) *C. sinensis* Osbeck	Hamlin, Jaffa, Mosambi, Malta Blood Red, Sathgudi, Valentia, Pineapple
		(g) *C. paradisii* Macfad	Duncan, Foster, Marsh, Paradise Navel
		(h) *C. aurantium* L.	Seville orange, Bergamot orange, Bergamot orange
		(i) *C. aurantifolia* Swingle	Jai Devi, Sai Sarbati, Vikram
		(j) *C. limon* Burm	Assam Lemon, Eureka, Baramasi Lemon, Galgal
		(k) *C. jambhiri* lush.	Kachai Lemon, Jatti Khatti
		(l) *C. latipes*	
		(m) *C. depressa* Hayata.	
		(n) *C. volkarmeriana*	
		(o) *C. tiawanica*	
		(p) *C. limettoides*	Columbia, Soh Synteng
		(q) *C. limetta*	
		(r) *Citrus karna*	Khrana Khatta
		(s) *C. limonia*	Rangpur Lime
		(t) *Fortunella* sp.	Golden Orange, Golden Bean, Marumi, Meiwa, limau pagar
		(u) *Pocirus trifoliate* Raf	
		(v) Citrange	Coleman, Cunningham, Morton, Rusk
2.	Mango	(a) *Mangifera indica*	Bappakai, Chandrakaran, Kensington, Kitchner, Kurukkan, Muvandan, Mylepelian, Nekkare, Olour, Valaikolumban, Goa, Goa Kasargod, Olour, Nileswar Dwarf, Moreh (Khongnambi) Salem Peach, Prior, Starch, Cambodiana, Carabao, Corazont, Paho, Pahutan, Pico, Senora and Strawberry
3.	Almond	(a) *Prunus dulcis*, syn. *Prunus amygdalus*	Nonpareil, Mission, Price, Sonora, Carmel, johlyn, Jiml
4.	Langsat	(a) *Lansium domesticum*	Conception, Uttaradit, Paete.
5.	Jamun	(a) *Syzigium cuminii* Skeels	Konkan Bahadoli
6.	Mangosteen	(a) *Garcinia mangostana*	Jolo (variety from Philippines)

Growth hormones play a significant role in the appearance of polyembryonic seeds. Haccius (1953, 1955, 1957, 1960) demonstrated that in the undifferentiated *Eranthis hiemalis* embryos, differentiation was transiently arrested in the presence of 2,4-phenoxyacetic acid (2,4-D) at 500–1000 mg/l, while a high mitotic activity was preserved. After the inhibitory effect was relieved, differentiation was resumed, development of an adventive embryo was initiated and a twin was formed. The treatment of such embryos with citratephosphate buffer, pH 3.5–4.5, suppressed the meristematic activity of the embryo cells, but the suspensor cells remained viable and started active division to form normal or abnormal embryos. Further experiments with 2,4-D confirmed the formation of twins by splitting of the embryo in other plants as well: *Datura* (Sanders, 1950), *Triticum* (Ferguson *et al.*, 1979), *Zea* (Erdelska and Vidovencova, 1992).

Degree of Polyembryony

The degree of polyembryony is the frequency of multiple seedlings in a seed. It varies with species and varieties and the variation also occurs with location, environment and position of fruits on tree.

Citrus and Related Species

The degree of polyembryony in Citurs varies greatly. Variation is not only observed within the species and cultivars of the same species, but not within the tree of same cultivars, in trees from different localities, even on the same tree and different seasons. Some are strongly polyembryonic, others are weakly polyembryonic.

Strongly Polyembryonic

Majority of the cultivated species and their cultivars are strongly polyembryonic while a few are weaky polyembryonic. The list is as follows.

Table 6.4: Polyembryony in different *Citrus* Species

Species/Cultivar	Percentage of Polyembryony	References
Sweet orange	40-95	Webber, 1900
Sour orange	75-85	Torres, 1936
Grapefruit	60-95	Frost, 1926, Torres, 1936
Mandarins	10-100	Frost, 1926, Torres, 1936
Rough lemon	100	Toxopeus *et al.*1930
Acid lime	78	Sengupta *et al.*1970
C. pentinfera	90	Gurgel and Soubihe, 1951
C. pennivesiculata	88	Chandra and Sarkar, 1965
C. maderaspatana	80	Chandra and Sarkar, 1965
Trifoliate orange	5	Iwamasa *et al.*, 1967
Dancy tangerine	40	Iwamasa *et al.*, 1967
Sweet orange	17	Iwamasa *et al.*, 1967
Sikkim mandarin (*C. reticulate*)	55-67	Kishore *et al.*, 2008

The study on Sikkim mandarin (*C. reticulate*) the maximum number of seedlings/seed was observed to be 6 and the per cent occurrence of multiple seedlings decreases with the increase in the seedling number/seed. The per cent of one seedling/seed was 51.5 followed by two seedlings/seed (36.1). The per cent of 3, 4, 5 and 6 seedlings/seed was 7.5, 1.8, 0.6 and 0.1 respectively (Figures 6.5–6.7). It was also observed that the degree of polyembryony was positively correlated with the occurrence of three or more number of seedlings (Kishore *et al.*, 2008).

Some of the promising hybrids, being used as rootstocks seem to be practically 100 per cent polyembryonic. *Citrus jamberi, C. karnal.* Calamondin and grapefruit, mandarin and sweet orange cultivars are reported to be 100 per cent polyembryonic.

Weakly Polyembryonic

True lemon (C. *limon*): There is general agreement about the true lemons being weakly polyembryonic. It is reported that probably the nucellar embryos cease to develop at an earlier stage (Ueno *et al.*, 1967). While true lemons are weakly polyembryonic, some of the indigenous lemon like Galgal and Sadhapal are stongly polyembryonic (90-95 per cent) (Motilal, 1963).

Table 6.5: Polyembryonic Percentage and No. of Seedlings/Stone of Different Polyembryonic Mango Varieties Under Allahabad Condition

Polyembryonic Varieties	Polyembryonic Percentage (per cent)	No. of Seedling/Stone
Movandan	68.42	2.11
Goa	75.00	2.92
Nekkare	72.00	2.60
Bappakai	73.68	2.21
Moreh	74.00	2.65

Roshan *et al.*, 2009.

Table 6.6: Number of Seedlings/Stone in Polyembryonic Varieties of Mango

Polyembryonic Varieties	Average Number of Seedlings/Stone		
	1987	1988	Mean
Prior	2.06	2.12	2.09
Kensington	2.08	2.12	2.10
Vellaikulamban	1.70	1.75	1.72
Chandrakaran	2.16	2.04	2.10
Bappakai	1.49	1.58	1.53
Mylepelian	1.84	1.87	1.85
EC 95862	2.08	1.70	1.89
Kithner	2.16	1.87	2.02
Nakari	1.32	2.06	1.69
Muvandan	1.93	2.07	1.97
Olour	2.85	1.81	2.33
Starch	2.94	1.83	2.33
Peach	2.76	2.75	2.75
Kurukan	2.33	2.41	2.37

Singh and Reddy, 1990.

Figure 6.5: Citrus Seed with Two Seedlings

Figure 6.6: Citrus Seed with Three Seedlings

Figure 6.7: Citrus Seed with Four Seedlings

Figure 6.8: Mango Stone with Two Seedling

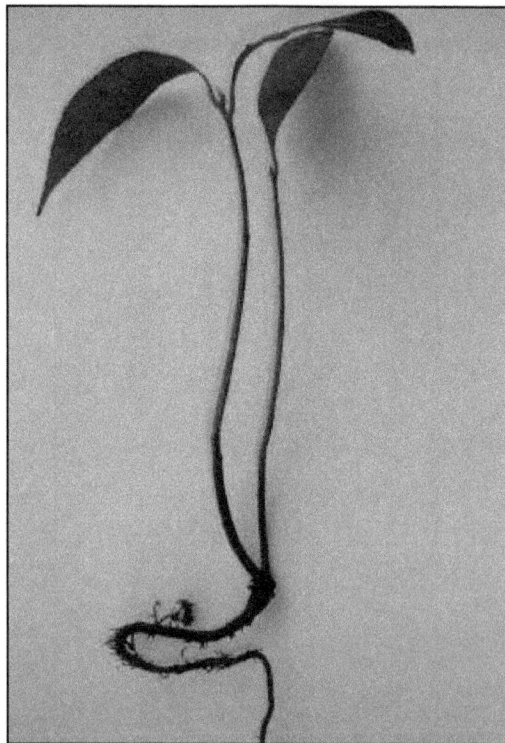

Figure 6.9: Mango Seedling having Common Root System

Mango

Important commercial varieties of mango grown in India are monoembryonic (Hayes, 1953). Naik (1941) and Singh (1960) also reported that a large number of mango varieties are monoembryonic, but consider a few exhibit polyembryony. Sen and Mallik (1940), from a survey conducted in West Coast of India reported that out of 400 varieties examined only 10 are polyembryonic, and even in these polyembryonic varieties there was a fair percentage of monoembryonic seed.

Identification

On germination, usually both sexual and nucellar seedlings will arise from the same seed. It is very difficult to distinguish and identify with certainty the nucellar seedlings in the nursery. However, it is of utmost importance to identify the seedlings to discard the unwanted ones. Further, the most of citrus species and other related genera contain both mono and polyembryonate seeds together in the lot. Identification and rejection of monoembryonate seeds prior to sowing lessens the burden of rejection of unwanted seedlings on germination, besides providing uniform seedlings to a larger extent.

Morphological characters of seeds like size, appearance, density etc. may be useful in identifying the type of seed. Singh (1965) reported no relation between size of seed and its embryonic nature. However, polyembryonic seeds are lighter probably due to air between the multiple embryos (Singh,1965) and have constricted surfaces. By adopting this technique one can discard the unwanted seed to get fairly uniform stock.

In the nursery, the nucellar seedlings are identified by their vigorous and more or less uniform growth. Generally sexual seedlings are crowded out by the nucellar seedlings. However, sexual seedlings if any, arise are weaker and vary from the seed parent. However, this is not a sure technique, as there is possibility of selection of a wrong seedling by rejection a right seedling. Further all the extra-embryos do not develop at the same time, seedlings at different stages of growth are noticed. Despite the above fact, the nucellar seedlings can be spotted by their same foliage and branching habit resembling the seed parent. It takes about one to five months after the date of sowing, to subject the seedlings for screening.

The trifoliate leaf character of *Poncirus trifoliata* is dominant (Marker gene) over the unifoliate character of citrus species. When such dominant parent is employed in crosses, the resultant seeds give rise to two types of seedling, one having the unifoliate leaves and the other having trifoliate leaves. This method can be used for producing genetically uniform seedlings for rootstock purposes and for estimation of percentage of nucellar embryony of any given variety. Though, theoretically speaking this technique is sound, in practice, certain limitations are experienced: (*i*) on large scale it is not possible (*ii*) purity of the pollen is not guaranteed; (*iii*) Absolute control over pollination is rather impossible (*iv*) it is not useful in cases where sexual incompatibility exist between *Poncirus trifoliata* and those *Citrus* species and cultivars; (*v*) Besides, the trifoliate characters, other morphological characters like thorns, wings articulation, pigment etc. at nursery stage can also be employed provided they are dominant.

Attempts were made to identify nucellar seedlings by means of colour tests. Furr and Reece (1946) for the first time tried to identify the nucellar seedlings, using Almen's Reagent and the extracts from the dried leaves. When the Alemn reagent is applied to the water extracts of dried leaf powders of different *Citrus* species, each shows a specific colour reaction. Nucellar seedlings develop the same colour as the seed parents, while sexual seedlings develop intermediate colour. However, this method is not useful when two species had similar colour reactions.

An intra-red analysis technique was tested as a means of distinguishing between nucellar and zygotic seedlings (Pieringer and Edwards, 1965). By employing this technique they could identify more zygotic seedlings than by the conventional method of identifying by morphological characteristics, in an unmarked population.

Many distinguishing characters for nucellar and zygotic seedlings at seedling stage have been proposed by various workers and some of these characters are proved to be a marker. Cytologically the nucellar cells destined to form adventive embryos can be distinguished from other cells of the nucellus by their dense cytoplasm and starchy contents (Wilms *et al.*, 1983). The zygotic embryo present at the micropyler region holds two original cotyledons of the seed. Zygotic embryo was usually larger than the other embryos and also took the most space in a seed (Figure 6.10), while, nuclear embryos were tiny, heart shaped and green and were crowded at the micropyler region and most of these could be easily separable (Das *et al.*, 2005). Nucellar embryos can be distinguished from the zygotic embryo by their lateral position in the embryo sac and irregular shape. Individual nucellar embryos had two cotyledons, and rudimentary radicals and plumules were present in between two cotyledons. Generally, the most vigorous and first germinating seedling of citrus is considered to be zygotic as it has the largest cotyledon and occupies most of the space of seed (Figure 6.11)

Similarly, in mango, the earlier concept was that the zygotic plantlets are being the weakest in polyembryonic mango seed because it probably degenerates due to competition with nucellar plantlets (Sachar and Chopra, 1957). The recent findings prove that zygotic plantlets are vigorous and positioned near micropyle and possess big cotyledons. The vigorocity of zygotic plants can be explained by a heterotic effect of cross between the female plant and unidentified male plant (Maria *et al.*, 2006).

Some identical twin seedlings showed a common primary root and in this case, seedlings germinate with common hypocotyls and primary root but the two epicotyles are distinctly separated from the cotyledonary node. In some other cases, during germination, the two hypocotyls and also the epicotyls of the identical twins are fused (Figures 6.12 and 6.13). Twins are seemed to develop by fission of the original zygotic embryo and the original two cotyledons of the zygotic embryo are also divided almost equally and shared among twins or triplets. The presence of twins further makes the identification process rather more complex.

The biochemical techniques are effective and expeditious tools to identify zygotic and nucellar seedlings at an early stage. Among biochemical techniques, isozyme markers are commonly used to distinguish nucellar (true-to-type) and zygotic (off-type) citrus seedlings (Moore and Castle, 1988; Anderson *et al.*, 1991). This technique is based on the presence or absence of protein bands in the sap of leaf, root or bark sample. Since the results may be influenced by the environment, as well as by the stage of plant and its organs, this method is relatively less reliable for seedling identification.

Molecular techniques by using molecular markers (RAPD, RFLP, SSR, ISSR, QTL, etc.) are the most recent, advanced, effective, expeditious and reliable tools to identify zygotic and nucellar seedlings at an early stage as results are not influenced by external factors. Das *et al.* (2007) identified the zygotic seedlings through RAPD markers and reported that zygotic seedlings (twin or triplets) usually had one or more extra band than those of the nucellar seedlings. They also observed variability in mandarins of north east by RAPD profiling. Maria *et al.* (2006) and Abirami *et al.* (2008) identified the nucellar seedlings of mango through RAPD and also determined the position of zygotic and nucellar embryos in an ovule. Carimi *et al.* (1998) used RFLP analysis of low and high- copy- number DNA to distinguish zygotic and nucellar seedlings and reported that in 82-88 per cent cases, zygotic embryos develop in early stage of ovule development. Nicasio *et al.* (2002) identified nucellar and zygotic seedlings with markers amplified with SSR (simple sequence repeat) and inter-simple sequence repeat (ISSR) primers.

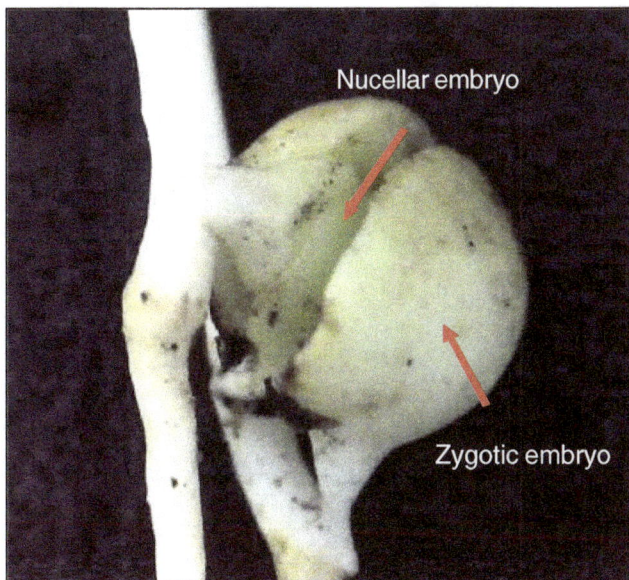

Figure 6.10: Zygotic and Nucelar Embryos of *Citrus reticulata*

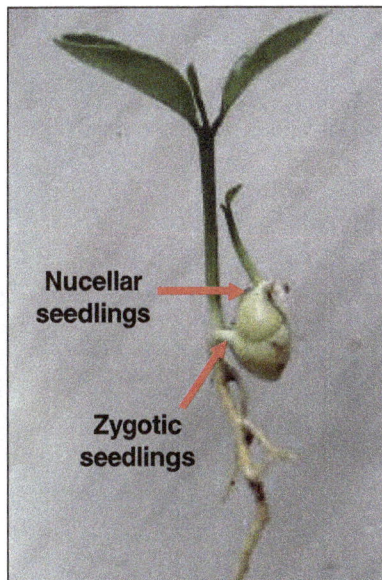

Figure 6.11: Zygotic and Nucelar Seedlings of *Citrus reticulata*

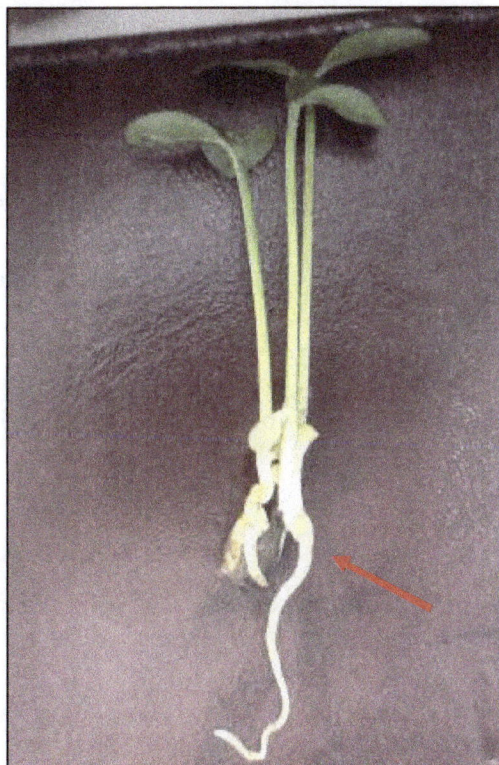

Figure 6.12: Zygotic Twins in *Citrus reticulata*

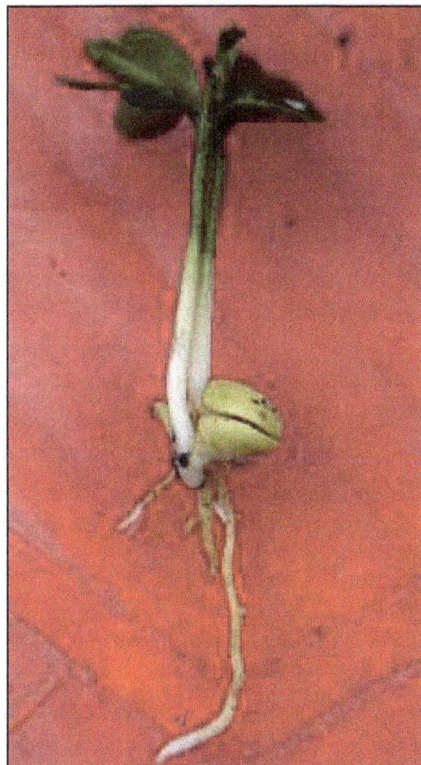

Figure 6.13: Zygotic Twin in *C. reticulate* having Equal Cotyledon Size

Figure 6.14: Mono (Left) and Poly Embryony (Right)

Inheritance of Muceller Embryony

Mango

Sturrock (1968) studied the genetics of mango polyembryony and came out with the following assumptions:

1. The polyembryony condition in mango is a recessive factor and is controlled probably by single pair of genes.

2. The condition becomes phenotypic only if the individual receives genes for the character from both parents and is homozygous for the character.

3. An individual that contains only one of the recessive genes would be heterozygous and would not display the character.

4. The involvement of more than one pair of genes would result in a much more complicated array of phenotypes in the progeny.

Citrus

Polyembryony in citrus *in vivo* is controlled by multiple recessive genes (Maheshwari and Rangaswamy, 1958). In monoembryonic species, these recessive genes may synthesize a potent inhibitor of embryogenesis.

Nucellar embryony is a genetic character, controlled by one or more number of genes, hence is inheritable (Parlevliet and Cameron, 1959). Parlevliet and Cameron (1959) studied the inheritance of nuycellar embryony in *Citrus*. They affected crosses between monoembyonic parent, which gave all monoembyonic offsprings indication that these monoembyonic varieties (Pummlo, Clementine mandarin, Temple mandarin, Wiling, Sukega) do not carry gene(s) which can produce polyembyrony by recombination. The crosses between monoembryonic and polyembronic parents produced both monoembryonic and polyembryonic offsprings, suggesting that one or more dominant or semi-dominant genes are present causing polyembryony. They further studied the offsprings obtained (Frost,1938) as a result of crosses between polyembryonic parents. Crosses between King and Willow Leaf mandarins (both polyembryonic) produced one monoembyronic hybrid (Wilking) when crosses with monoembryonic Pummelo and Clementine produced both polyembryonic and monoembryonic offsprings about 1:1 ratio. From these results,Parlevliet and Cameron (1959) postulated inheritance of polyembryony in citrus and came out with following findings.

☆ The polyembryony in citrus is generally controlled by a dominant gene having heterozygous allele (Pp) while homozygous recessive gene (pp) is present in monoembryonic citrus species.

☆ The hypothesis is supported by the findings obtained by crossing monoembryonic (pp) and polyembryonic (Pp) parents with the resultant ratio of 1:1.

☆ When polyembryonic parents were crossed, the ratio of polyembryonic and monoembryonic offspring was 3:1.

☆ A variable degree of polyembryony was recorded in polyembryonate offspring obtained by crossing monoembryonic and polyembryonic and between polyembryonic parents which implied the presence of minor genes affecting degree of polyembryony.

☆ The presence of modifier or duplicate genes should also be considered as in some crosses between monoembryonic and polyembryonic parents, and between polyembryonic parents, the progeny ratio varied greatly from 1:1 and 3:1 ratios.

Iwamaso *et al.* (1967) also obtained a mixture of mono and polyembryonate progeny from the crosses within polyembryonate parents and between both types, with a ratio approaching 1:1 in the latter case. These results lend support to the suggestion that a principle dominant gene controls the polyembryony, while a recessive gene controls mono-embryony in Citrus.

Horticultural Significance

Important relations of polyembryony to citrus propagation and breeding are reported (Webber, 1900). The main value of polyembryony in citrus lies in providing homogeneous seedlings, both for rootstock purpose and orchard tree. Swingle (1932) coined the word 'neopysis' for the rejuvenation processs exhibited by the nucellar seedlings over the old lines. The viruses are screened by a substance, possibly DNA which impregnates the embryo sac and adjacent tissue at the flowering time (Swingle, 1932). Hodgson (1961) enumerates the possible horticultural applications of polyembryony as follows:

1. True to type seedlings.

   Nucellar seedlings possess the genotype of mother plant only and are homogeneous. As such, they do not require propagation by usual vegetative means.

2. Genetically uniform rootstocks.

   The use of nucellar seedlings as stocks makes possible use of a stock variety in its most disease free and vigorous condition, because nucellar seedlings are virus-free and are vigorous than gametic seedlings. Gametic seedlings are heterozygous in nature. As such they manifest variable influence on the scion varieties in vigour, precocity, longevity, productivity etc. Therefore, they are not suitable for use as rootstocks. On the other hand, nucellar seedlings being genetically homozygous manifest uniform influence on the scion.

3. More vigorous seedling.

   Continuous vegetative propagation leads to decline in vigour in citrus. In such cases, it can be restored back by using nucellar seedlings due to existence of 'neophysis'.

4. Virus free seedlings and budwood.

   Nucellar seedlings are devoid of any known viruses. It is observed that the nucellar seedlings even from virus infected parent trees are entirely free fron infection. Besides providing virus free seedlings for planting and for rootstock purposes, the nucellar lines also provide virus-free budwood for propagation. Nucellar seedlings meet the requirements of budwood selection like freedom from virus, clonal purity and high yielding capacity with greater vigour.

5. The nucellar seedlings have a tap root and, therefore develop a better root system

6. Polyembryony helps in the large scale propagation of desired genotype.

7. Avoids the complications associated with sexual reproduction like pollinators and cross incompatibility.

Drawback of Polyembryony

The drawbacks in nucellar seedlings are the excessive thorniness and slowness in coming to bearing. In these aspects they resemble the ordinary sexual seedlings. Freedom from viruses, trueness to mother, higher vigour, larger production, homozygosity and the presence of tap root unlike cuttings and layers qualify them as orchard trees. The nucellar seedlings may grow for 6-8 years, before coming to bearing disqualifying them as orchard trees. Otherwise, the use of nucellar seedlings eliminates the question of stockscion incompatibility.

However, this drawback can be overcome to a large extent. Ringing such trees in early summer in Florida has been tried, with considerable success. The success of this operation depends on the age of the seedlings. It is effective on 3-4 years old seedling, but ineffective on 6-8 years old seedlings. The use of growth inhibitors like cycocel is also suggested for inducing flowering by suppressing the vegetative growth.

These undesirable qualities (thorniness and late bearing) can be avoided to some extent by utilising the budwood taken from the upper part of the old nucellar lines.

References

Abirami, K., Singh, S.K., Singh, Room, Mohapatra, T. and Kumar, Anand Raj, 2008. Genetic diversity studies on polyembryonic and monoembryonic mango genotypes using molecular markers. *Indian J. Hort.*, 65(3): 258–62.

Afzelius, K., 1928. Die Embryobildung bei *Nigritella nigra, Svensk. Bot. Tidskr.*, 22(1–2): 82–91.

Anderson, C.M., Castle, W.S. and Moore, G.A., 1991. Isozymic identification of zygotic seedlings in Swingle *Citrus paradisi* x *Poncirus trifoliata* nursery and field population. *Journal of American Society for Horticultural Sciences*, 116: 322–326.

Bedell, P.E., 1998. *Seed Science and Technology: Indian Forestry Species*. Allied Publishers Ltd., New Delhi.

Bhojwani, S.S. and Bhatnagar, S.P., 1999. Polyembryony. In: *The Embryology of Angiosperms*. Vikas Publishing House Pvt. Ltd., New Delhi, pp. 236–253.

Bouman, F. and Boesewinkel, F.D., 1969. On a Case of Polyembryony in *Pterocarya fraxinifolia* (Junglandaceae) and on Polyembryony in General. *Acta Bot. Neerl.*, 18(1): 50–57.

Camaron, J.W. and Soost, R.K., 1979. Sexual and nucellar embryony in F1 hybrid and advanced crosses in *Citrus* with *Poncirus. J. American Soc. Horti. Sci.*, 104(3): 408–410.

Campbell, C.W., 1967. Growing the Mangosteen in Southern Florida. *Proceedings of the Florida State Hortl. Soc.*, 79: 399–401.

Carimi, F., Tortorici, M.C., Pasquale, F.de. and Crescimanno, F.G., 1998. Somatic embryogenesis and plant regeneration from undeveloped ovules and stigma/style explants of sweet orange naval group [*Citrus sinensis* (L.) Osb.]. *Plant Cell, Tissue and Organ Culture*, 54(3): 183–189.

Chandra, D. and Sarkar, G., 1956. *Sci. and Cult.*, 31: 374–374.

Crete, P., 1944. Polyembryony in *Actinidia chinensis. Bulletin Society of Botany France*, 91: 89–92.

Cooper, D.C., 1943. Haploid–Diploid Twin Embryos in *Lilium* and *Nicotiana. American. J. Bot.*, 30(6): 408– 413.

Das, A., Mandal, Bidisha, Sarkar, J. and Chaudhuri, S., 2005. Variability in multiple embryo formation and seedling development of mandarin orange (*Citrus reticulata* Blanco) of north-eastern Himalayan region of India. *PGR Newsletter*, 143: 35–39.

Das, A., Mandal, Bidisha, Sarkar, J. and Chaudhuri, S., 2007. Occurrence of zygotic twins seedlings in mandarin orange plants of northeastern Himalayan Region. *Current Sci.*, 92(11): 1488–1489.

Dhillon, R.S., Kaundal, G.S. and Cheema, S.S., 1993. Nucellar embryony for propagating *Citrus. Indian Hort.*, 38: 44–45.

Ekdahl, J., 1941. Die Entwicklung von Embryo bei *Ulmus glabra* Huds, *Sven. Bot. Tidskr.*, 35(2): 143–156.

Erdelska, O. and Vidovencova, Z., 1992. Cleavage Polyembryony in Maize, Sex. *Plant Reprod.*, 5: 224–226.

Ernst, A., 1918. "Bustar dierungals Ursacheder Apogamic in Pflanzenreich. Eine Hypothese Zur experimentatieen verebumgs and Abslammungslehre, Java.

Esen, A. and Soost, K., 1977. Adventive embryogenesis in Citrus and its relation to pollination and fertilization. *American J. Bot.*, 64: 607–614.

Ferguson, J.D., 1979. McEwan, J.M., and Card, K.A., Hormonally induced polyembryos in wheat. *Physiol. Plant*, 45. 470–474.

Frost, H.B., 1926. Polyembryony, heterozygosis and chimeras in Citrus. *Hilgardia*, 1: 365–402.

Frost, H.B., 1938. Nucellar embryony and juvenile characters in clonal varieties of citrus. *J. Heredity*, 29: 423–432.

Furr, J.R. and Reece, P.C., 1946. Identification of hybrid and nucellar citrus seedlings by a modification of the rootstock color test. *Proc. American. Soc. Hort. Sci.*, 48: 141–146.

Furusato, K., 1953. Studies on polyembryony in Citrus. *Ann. Rept. Natl. Inst. Genet.* [Japan], 4: 56.

Guignard, L., 1901. La Double Fécondation dans le *Naias major, J. Botanique*, 15: 205–213.

Gurgel, J.T.A. and Soubihe, S.J., 1951. Ann. Esc. Sup. Agric. "Luiz de Queroz" Piracicaba, 8: 722–746.

Gómezand, P.M. and Gradziel, T.M. *Sexual Plant Reproduction*. Sexual polyembryony in almond, 16(3): 135–139.

Haccius, B., 1953. Histogenetische Untersuchungen an Wurzelhaube und kotyledonarscheide geophiler Keimpflanzen (*Podophyllum* und *Eranthis*). *Planta,* 41(5): 439–458.

Haccius, B., 1955. Experimentally induced twinning in plants. *Nature,* 176(4477): 355–356.

Haccius, B., 1957. Regenerationserscheinungen an pflanzlichen Embryonen nach Behandlung mit antimitotisch Wirksamen Substanzen, *Beitr. Biol. Pflanz.,* 34(1): 3–18.

Haccius, B., 1960. Experimentell induzierte Einkeimblattrigkeit bei *Eranthis hiemalis.* 2. Monokotylie durch Phenylborsaure. *Planta,* 54(5): 482–497.

Hayes, W.B., 1953. *Fruit Growing in India.* Kitabisthan, Allahabad.

Hodgson, R.W., 1961. *Ind. J. Hort.,* 18(4): 245–250.

Horn, C.L., 1943. The frequency of polyembryony in twenty varieties of mango. *Proc. American. Soc. Hort. Sci..,* 42: 318–320.

Iwamasa, M., Ueno, I. and Nishiura, M., 1967. Inheritance of nucellar embryony in citrus. *Bulletin Horticultural Research Station Japan Series,* B 7: 1–8.

Johri, B.M. and Tiagi, B., 1952. Floral morphology and seed formation in *Cuscuta reflexa* Roxb. *Phytomorphology,* 2(2–3): 162–180.

Juliano, J.B., 1937. Embryos of carabao mango, *Mangifera indica* L. *Philip. J. Agric.,* 25: 749–760.

Katyal, S.L., 1949. Studies on seed formation, embryo development and pollen germination in Morton Citrange. *Indian. J. Hort.,* 6(2): 21–23.

Kester, D.E. and Gradziel, T.M., 1996. Almond. In: *Fruit Breeding,* (Eds.) Janik J. and J.D. Moore. John Wiley and Sons, New York, pp. 1–97.

Kishore Kundan, Kumar, A., Rahman, H., Monika, N. and Pandey, Brijesh, 2008. Studies on degree of polyembryony in Sikkim Mandarin (*Citrus reticulate* Blanco). *Abs. 3rd Indian Horticulture Congress.* Nov. 6–9, OUAT, Bhubneshwar, p. 66.

Kobayashi, S., Ieda, I. and Nakantani, M., 1979. Studies on the nucellar embryogenesis in citrus. Formation of the primordial cells of the nucellar embryony on the ovule of the flower bud, and its meristematic activity. *J. Japanese Soc. Hortl. Sci.,* 48: 179–185.

Koltunow, A.M., 1993. Apomixis: Embryo sacs and embryos formed without meiosis or fertilization in ovules. *Plant Cell,* 5: 1425–1437.

Koltunow, A.M., Soltys, K., Nito, N. and Mcclure, S., 1995. Anther, ovule, seed and nucellar embryo development in *Citrus sinensis* cv. *Valencia. Canadian J. Bot.,* 73: 1567–1582.

Lebègue, A., 1948. Embryologie des Cruciféres: Polyembryonie chez l'Arabis lyallii S. Wats. *Bull. Soc. Bot. Fr.,* 95(7–9): 250–252.

Lebègue, A., 1949. Embryologie des Saxifragacées. Polyembryonie chez le *Bergenia delavayi* Engl., *Bull. Soc. Bot. Fr.,* 96(1–3): 38–39.

Lebegue, A., 1952. Polyembryony in Angiosperm. *Bulletin Society of Botany France,* 99: 329–369.

Maheshwari, P., Sachar, R.C. and Chopra, R.N., 1955. Embryological studies in mango. *Proceeding of 42nd Indian Science Congress,* Baroda, India, pp. 233.

Maheshwari, P. and Ranganswamy, N.S., 1958. Polyembryony and *in vitro* culture of embryos of *Citrus* and *Mangifera. Indian J. Hort.,* 15: 275–282.

Maria Cristina Rocha, Cordiero Alberto Carlos, Queiroz Pinto, Victor Hugo Vargas, Ramos Fabio Galop Falaro and Lilia Marta Santo Fraga, 2006. Identification of plantlets genetic origin in polyembryonic mango (*Mangifera indica* L.) Rosinha seeds using RAPD markers. *Revista Brasileira de Fruiticulture*, 28(3): 454–457.

Moore, G.A. and Castle, W.S., 1988. Morphological and isozymic analysis of open pollinated citrus-rootstock population. *J. Heredity*, 79: 59–63.

Naik, K.C., 1941. Studies in propagation of mango (*Mangifera indica* L.). *Indian J. Agric. Sci.*, 11: 736–738.

Nakatami, M., Ikeda, I. and Kobayashi, S., 1978. Bull Fruit Tree Res. Stn. (Ministry Agric. For) Ser E (Akitsu), 2: 25–35.

Narayanswamy, S. and Roy, S.K., 1960. Embryo sac development and polyembryony in *Sizygium cumini*. *Botanical Note*, 3: 273–284.

Naumova, T.N., 1992. *Apomixis in Angiosperms. Nucellar and Integumentary Embryony*, London: Boca Ration Ann. Arbor.

Nicasio Tusa, Loredana Abbate, Sergio Ferrante, Sergio Lucretti and Maria-Teresa Scaranol, 2002. Identification of zygotic and nucellar seedlings in citrus interploid crosses by means of isozymes, flow cytometry and ISSR–PCR. *Cellular and Molecular Biology Letters*, 7: 703–708.

Osawa, I., 1912. Cytological and experimental studies in citrus. *J. Col. Agric. Imp. Univ. Tokyo*, 4: 83–116.

Parlevliet, J.E. and Camaron, J.E., 1959. Evidence of the inheritance of nucellar embryony in citrus. *Proceedings American Society Horticultural Society*, 74: 252–260.

Peter, K.V., 2009. Polyembryony and its prevalence in horticultural crops. *Basics of Horticulture*. New India Publication Agency, pp. 11–22.

Renner, O., 1916. Zur Terminologie des pflanzlichen Generationswechsels. *Biol. Zentralbl.*, 36: 337–374.

Roshan, R.K., Nongallei Pebam and Singh, D.B., 2009. Evaluation of polyembrionic mango varieties at nursery stage under Allahabad condition. *Proceeding of International Conference on Horticulture*, pp. 446–448.

Roy, S.K., 1953. Embryology of *Eugenia jambos* L. *Curr. Sci.*, 22(8): 249–250.

Roy, S.K. and Sahai, R., 1962. The embryo sac and embryo of *Syzygium caryophyllifolium* D.C. *J. Ind. Bot. Soc.*, 41(1): 45–51.

Sanders, M.E., 1950. Development of self and hybrid *Datura* embryos in artificial culture. *American J. Bot.*, 37(1): 6–15.

Sarkar, K.R., 1961. *MSc. Thesis*, P.G. School, IARI, New Delhi.

Sachar, R.C., 1955. The embryology of *Argemone mexicana* L.: A reinvestigation. *Phytomorphology*, 5(2–3): 200–218.

Sachar, R.C. and Chopra, R.N., 1957. A study of endosperm and embryo in *Mangifera. Indian J. Agric. Sci.*, 27(1): 219–238.

Sen, P.K. and Mallik, P.C., 1940. The embryo of the Indian mango (*Mangifera indica* L.). *Indian J. Agric. Sci.*, 10: 750–760.

Sengupta, M.S., Roy, B.N. and Bose, T.K., 1970. *Allahabad Farmer*, 44(4): 237–239.

Sindhe, A.N.R., Swamy, B.G.L. and Govindappa, D.A., 1980. Synergid Embryo in *Penisetum squamulatum*. *Current Sci.*, 49: 914–915.

Singh, A.R. and Pandey, A.K., 1990. A note on extent of polyembryony in citrus species. *Advances in Hort. and Forestry*, 1: 99–101.

Singh, G. and Reddy, Y.T.N., 1990. A note on extent of polyembryony in mango. *Advances in Hort. and Forestry*, 1: 17–21.

Singh, L.B., 1960. *The Mango*. World Crop Series, Leonard Hill.

Shattuck, C.H., 1905. A morphological study of *Ulmus americana*. *Bot. Gaz.*, 40: 209–223.

Smirnov, A.G. and Grakhantseva, L.Sh., 1982. Some features of embryology of *Fritillaria meleagroides* (Liliaceae). *Bot. Zh.*, 67(4): 491–499.

Soares Filho, W.S., Lee, L.M. and Cunha Sobrinho, A.P., 1995. Influence of pollinators on polyembryony in citrus. *Acta Hort.*, 403: 256–265.

Solntseva, M.P., 1973. Types of embryo sac development and morophology of the embryo of *Rudbeckia laciniate* L. In: *Embriologiya pokrytosemennykh rastenii* (Embryology of Angioispermous Plants), Kovarskii, A.I. *et al.*, Eds., Chisinau: Shtiintsa, p. 47–57.

Solntseva, M.P., 1999. Problemy apogametii. *Bot. Zh.*, 84 (8): 1–23.

Strausburger, E., 1878. Javaische Zeitcher F. Nature wiss, 12: 647–670.

Sturrock, T.T., 1968. Genetics of mango polyembryony. *Proceedings of Florida State Horticulture Society* 81(1): 311–314.

Swamy, B.G.L., 1943. Gametogenesis and embryogeny of *Eulophia epidendraea* Fischer. *Proc. Natl. Inst. Sci., India*, 9: 59–65.

Swamy, B.G.L., 1946. The Embryology of *Zeuxine sulcata* Lindl. *New Phyto.*, 45(1): 132–136.

Swamy, B.G.L., 1948. Agamospermy in *Spiranthes cernua*. *Lloydia*, 11(3): 149–162.

Swingle, W.T., 1932. *Amer. J. Bot.*, 19: 839 (Abstract).

Traub, H.P., 1936. Artificial control of nucellar embryony in citrus. *Science (n.s.)*, 83: 165–166.

Torres, J.P., 1936. *Philip. J. Agric.*, 7: 37–58.

Toxopeus, H.J. 1930. Vereen, Landbouw Nederl–Ivdie Landbouw Tijdscher, 6: 39–405.

Toyama, T.K., 1974. Haploidy in peach. *Hort. Sci.*, 9: 187–188.

Webber, H.J., 1900. Complications in Citrus hybridization caused by polyembryony. *Science*, 11: 308.

Wlims, H.J., Van Went, J.L., Cresti, M. and Ciampolini, F., 1983. Adventitive embryogenesis in Citrus. *Caryologia*, 36: 65–78.

Yakovlev, M.S., 1967. Polyembryony in higher plants and principles of its classification. *Phytomorphology*, 17(1–4): 278–282.

2012, Nutri-Horticulture
Editor: Professor K.V. Peter
Published by: DAYA PUBLISHING HOUSE, NEW DELHI

Pages 123–142

Chapter 7

Computational Biology Applications

Aimy Sebastian and Vibin Ramakrishnan*

Institute of Bioinformatics and Applied Biotechnology,
Biotech Park, Electronic City Phase -1, Bangalore, INDIA
*E-mail: vibin@ibab.ac.in

Recent advancement in various fields of biotechnology has generated huge amount of data which are incomprehensible without generous support from various areas of bio-computing. Computational biology helps in the management of these data and also the extraction of relevant information that can potentially supplement experimental research. More over, various bio-computing tools generate huge amount of virtual data by simulating experiments beyond the resolution of lab equipments. Computational techniques thus become inevitable for further development of various disciplines in biology. Today, computational tools play a key role in the identification of genes from the sequenced genomes, analyze interaction between genes and gene expression analysis, analysis of protein -protein interaction, evolutionary studies, molecular modeling, protein structure prediction etc. Using computational biology techniques, it is possible to extract relevant information from the sequences of available model genomes. Different gene prediction and similarity search tools are available with which the genes in a given genome can be identified. From these gene sequences, it is possible to identify the proteins coded by them, their structure and function. Microarrays allow the expression analysis of these genes and Mass spectrometry helps in the identification of proteins. Clever use of these techniques can in principle contribute to the development of crops with high nutritional quality, insect resistance, drought resistance and even discover new varieties of plants which grow in poor quality soil.

In this chapter we attempt to provide a brief survey of bio-computing applications in some active areas of research that can potentially supplement the molecular biology efforts for crop improvement. We are also providing a fairly long list of biological databases, computational tools and their availability.

Biological Databases

Information storage and retrieval are the heart of any computational biology routine. Abundance of publicly accessible databases makes biological information available for the scientific community free of cost. These databases contain information from research areas such as genomics, proteomics, metabolomics, microarray gene expression, gene function, similarities of biological sequences and structures, phylogenetics etc.

Classification of Databases

Biological databases available online can be mainly classified into nucleotide sequence databases, protein sequence databases, protein structural databases, protein–protein interaction databases, metabolic pathway databases, microarray databases and specialized databases which are developed for some very specific applications.

Nucleotide Sequence Database

A nucleotide database contains DNA sequence information from different organisms. This information may be curated or non-curated. Some of the major nucleotide databases are:

☆ **DDBJ** - DNA Data Bank of Japan

☆ **EMBL** Nucleotide DB - European Molecular Biology Laboratory

☆ **UniProtKB** - Universal Protein Resource Knowledgebase

Protein Sequence Database

Some examples of protein sequence databases with collection of annotated protein sequences from different organisms are given below:

☆ **UniProt** - Universal Protein Resource (UniProt Consortium: EBI, Expasy, PIR)

☆ **PIR** - Protein Information Resource

☆ **Swiss-Prot** Protein Knowledgebase (Swiss Institute of Bioinformatics)

☆ **PROSITE** - Database of Protein Families and Domains

☆ **DIP** - Database of Interacting Proteins

☆ **Pfam** - Protein families database

Protein Structure Databases

These are repositories with 3-D structure of biological molecules such as proteins. Some of the well known structural databases are:

☆ Protein Data Bank(**PDB**) - Research Collaboratory for Structural Bioinformatics (RCSB)

☆ **CATH** - Protein Structure Classification

☆ **SCOP** - Structural Classification of Proteins

☆ **SWISS-MODEL** - Server and Repository for Protein Structure Models

☆ **ModBase** - Database of Comparative Protein Structure Models

Protein-Protein Interaction Databases

These databases contain both known and predicted protein interactions. The interactions include direct (physical) and indirect (functional) associations. Some of the protein-protein interaction databases are:

☆ **BioGRID** - A General Repository for Interaction Data sets

☆ **STRING** - STRING is a database of known and predicted protein-protein interactions.

☆ **DIP** - Database of Interacting Proteins

☆ **BIND** - Biomolecular Interaction Network Database

Metabolic Pathway Database

These databases represent pathway data at the level of individual interactions. Some of the important pathway databases are:

☆ **BioCyc** Database

☆ **KEGG** pathway Database

☆ **MANET** database

☆ **Reactome**

National Center for Biotechnology Information (NCBI)

The National Center for Biotechnology Information (**NCBI**) is a part of the United States National Library of Medicine (**NLM**), a branch of the National Institutes of Health. NCBI hosts various databases with nucleotide and protein sequences of different organisms, scientific literature and many more.

Literature databases maintained at NCBI have **PubMed** and **PubMed central** with research articles, **Books** with free online biomedical texts, **OMIM** (a catalog of human genes and genetic disorders and **Gene Reviews)**, peer-reviewed articles on genetic testing, diagnosis and management of inherited disorders.

GenBank is an annotated DNA sequence database maintained by NCBI. GenBank is part of the International Nucleotide Sequence Database Collaboration, which comprises of DNA DataBank of Japan (DDBJ), European Molecular Biology Laboratory (EMBL), and GenBank at NCBI. These three organizations exchange data on a daily basis. The Reference Sequence database[1](**RefSeq**) is a collection of comprehensive, integrated, non-redundant, well-annotated set of sequences, including genomic DNA, transcripts, and proteins. The NCBI maintains a **Nucleotide database** and a **Protein database** which provide a collection of sequences from several sources such as GenBank and RefSeq.

Some other important databases maintained at NCBI are

☆ **Genome database:** It provides views for a variety of genomes, complete chromosomes, sequence maps with contigs and integrated genetic and physical maps.

☆ **MMDB:** The Molecular Modeling Database (MMDB) contains 3D macromolecular structures, including proteins and polynucleotides.

☆ **3D Domains:** It has compact structural domains identified automatically in MMDB

☆ **Taxonomy:** The NCBI taxonomy database contains the names of all organisms that are represented in the genetic databases with at least one nucleotide or protein sequence.

☆ **GEO:** Gene Expression Ominibus is a microarray database maintained by NCBI

NCBI also provides a search engine **Entrez** for information search and retrieval from different databases hosted by them.

Advances in sequencing technologies made the genome sequence of many organisms available for scientific community. Databases like **Ensembl**[2] provide automatic annotation databases for human, mouse, other vertebrate and eukaryote genomes.

Some Resources Specific for Plants

Sequencing of many plant model genomes are completed and are also available in public databases. Some of the databases specific for plants are available in the following websites.

☆ http://www.plantgdb.org/

☆ http://www.gramene.org/

☆ http://www.arabidopsis.org/

☆ http://www.tigr.org/tdb/e2k1/ath1/

☆ http://www.cns.fr/externe/GenomeBrowser/Vitis/

☆ http://mips.helmholtz-muenchen.de/plant/index.jsp

☆ http://www.jcvi.org/potato/

Public databases for model organisms coupled with computational algorithms in data mining, present a new opportunity as well as a great challenge to researchers who intend to develop more focused tools for gene discovery and deployment.

Genome Sequencing and Annotation

Plant genome research is important for industries based on food, energy, and fiber. *Arabidopsis* is the first plant genome sequenced and is studied extensively[3,19,20]. It has several traits which make it a useful model for understanding the genetic, cellular, and molecular biology of flowering plants. *Arabidopsis* has one of the smallest genomes among plants. Because of the small size of the genome it is easy to sequence and map *Arabidopsis*. Since then, many other plants are sequenced over last few years.

Sequencing Techniques

Genome sequencing is identifying the order of DNA nucleotides, or bases, in a genome, precisely sequential order of Adenine(A)s, Cytosine(C)s, Guanine(G)s, and Thymine(T)s which make up an organism's DNA. Generally, for sequencing the DNA, first the chromosomes are broken into shorter fragments. Each fragment is then broken once again randomly into smaller pieces. These random fragments (or 'inserts') are then inserted into cloning vectors to amplify the DNA and they are organized into clone libraries (Figure 7.1). Then sequences on each fragment is read using different techniques. Current sequencing technologies can only read between 600 and 1000 base pairs of DNA. So the middle portion of the DNA fragments remains non-sequenced. Sequencing of both the ends lead to pairs of reads known as mate-pairs.[4-7]

Different sequencing strategies like whole genome shotgun sequencing and map-based sequencing (clone-by-clone method) are applied to sequence genomes. They differ in the methods they use in cutting the DNA into fragments, their assembly and whether they map the chromosome before sequencing. In map based sequencing, the genome has been broken down to relatively large fragments and clone them to make a library. Genome mapping techniques are employed to figure out the location of each clone in the genome. Then each clone is cut into smaller, overlapping pieces. Such small pieces are sequenced and overlaps are used to reconstruct the sequence of the whole clone and subsequently assembled into the whole genome based on the earlier constructed map. In the whole-genome shotgun method, a genome map is not created. Instead it breaks up the whole genome into very small pieces, clone them, sequence the pieces, and reassemble the pieces into the full genome sequence. Two important methods for sequencing genome are explained in the following two sections.

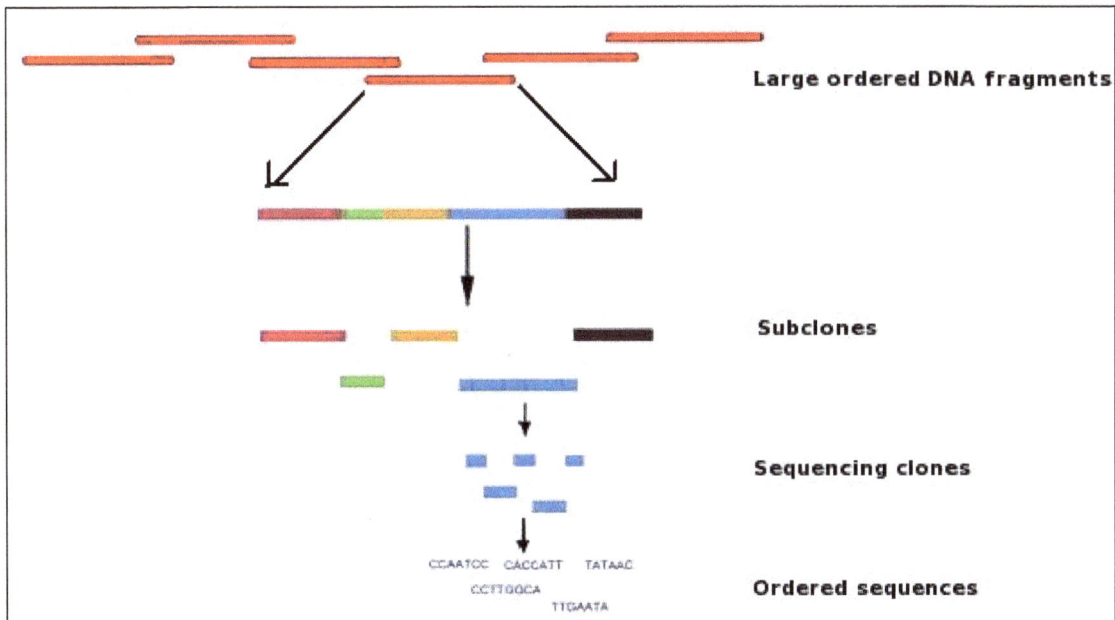

Figure 7.1: Steps Involved in DNA Sequencing

The Chain Termination Method or Sangers Method

Synthesis of a DNA strand over a template requires a primer, DNA polymerase and deoxy Nucleotide Phosphates (dNTPs). DNA polymerase enzyme extends the primer attached to the template strand by adding complementary nucleotides in 5'-3' direction resulting in a new double stranded DNA. In Sangers method[8], four different PCR reaction mixtures are prepared and in each mixture, along with template strand, fluorescent labeled (or radioactively labeled) primer, nucleotides and DNA polymerase. A small amount of dideoxynucleoside triphosphate (ddNTP) analogs to one of the four nucleotides (ATP, CTP, GTP or TTP) is also provided. These ddNTPs lack 3' hydroxyl group to which the next dNTP of the growing DNA chain is added. The addition of ddNTPs by DNA polymerase will cause the termination of growing chains at the site of coupling. Thus, chains terminated at all nucleotide positions are obtained after the PCR reaction (Figure 7.2). These chains with different lengths are separated using electrophoresis and sequence is read from the separated fragments.

The Chemical Degradation Method or Maxam-Gilbert Method

In this method[9], the sequence of a double-stranded DNA molecule is determined by treatment with chemicals that cut the molecule at specific nucleotide positions. The DNA fragment is labeled with ^{32}P at its 5' end. Then chemicals are used to break the DNA preferentially at each of the four nucleotide bases under conditions with which only one break per chain has been made. Thus four separate test tubes are prepared, one for each base (A, T, G, C). The fragments in the four reactions are arranged side by side in gel electrophoresis for size separation. Autoradiography of the gel yields a series of dark bands each corresponding to a radio-labelled DNA fragment, from which the sequence may be inferred.

After sequencing each short fragments of the genome, they are assembled[10-12] to make the entire genome based on the overlapping regions in the sequences (Figure 7.3). There are many computational

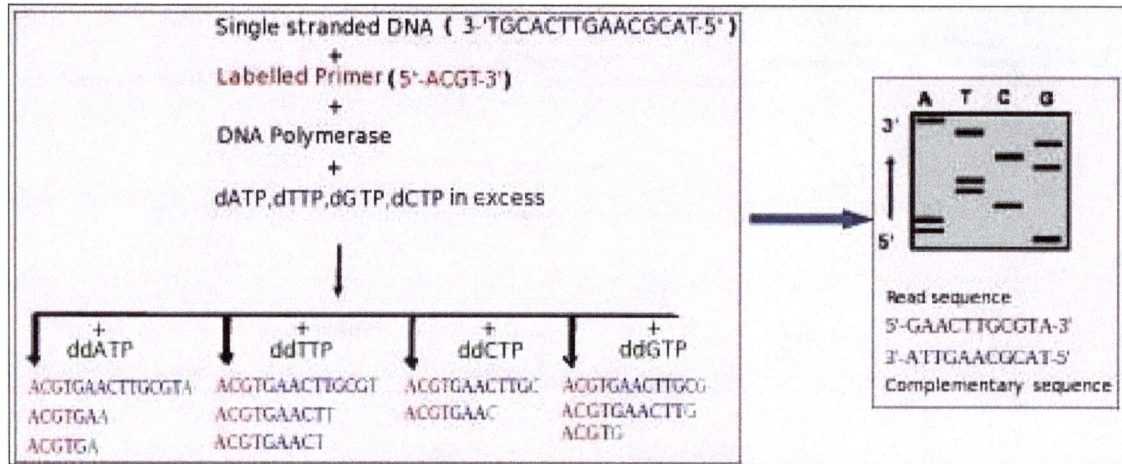

Figure 7.2: Sangers Method of Sequencing

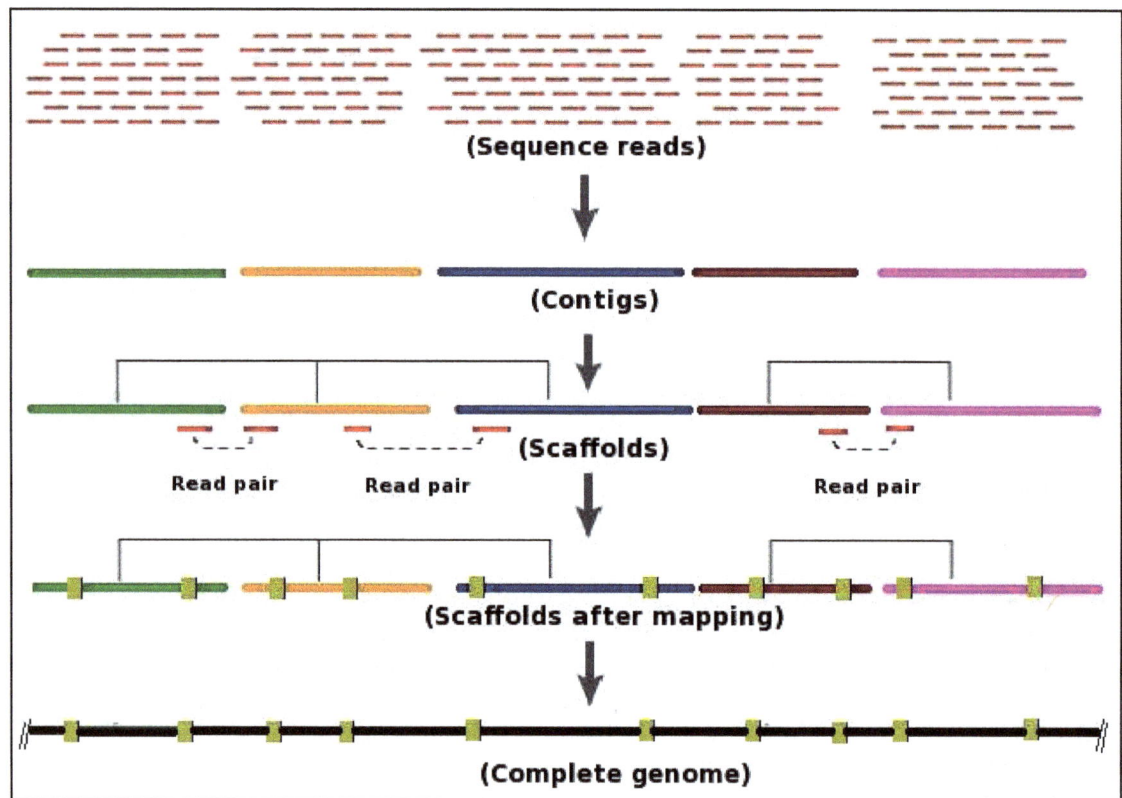

Figure 7.3: Long-range Sequence Assembly in Whole-genome Shotgun Sequencing.

All sequence reads generated are initially assembled into sequence contigs. Contigs are then organized into scaffolds based on the information in the read pair. Then scaffolds can be aligned relative to the source genome by the identification of already mapped, sequence-based landmarks in the sequence contigs.

tools like Phrap, Celera Assembler etc. to assist sequence assembly. Most of these programs perform a pairwise alignment of all sequences and merge the two sequences with maximum overlap. This process is performed until only one sequence is left. In map based sequencing the chromosomal maps generated are used to assemble the sequence fragments into the whole genome. In plants the genome assembly is a computationally hard problem because of the presence of repetitive content.

In the case of plants, the first model genomes sequenced are *Arabidopsis* and rice (*Oryza sativa* L.). Wheat, barley, maize, sorghum and millet seem to have genetic layout similar to rice; *i.e.*, the genes are almost in the same order, but the larger genomes have more junk DNA between genes. These findings suggest that improvement of other important food crops can benefit from information obtained from the model crop systems[15,16,19]. Some plant models in genomics are:

- ☆ *Arabidopsis thaliana*, which is the first plant genome sequenced
- ☆ Tomato which is a model plant for solanaceae genomics
- ☆ Rice, a model plant for cereal genomics
- ☆ Papaya, a tropical fruit tree
- ☆ Sugarcane, a tropical crop with a highly complex genome
- ☆ Cucumber, a member of cucurbit botanical family

Efforts in Sequencing Tomato and Potato Genome

The sequencing of Tomato genome as part of an International Project aims to sequence the gene-rich euchromatic portions of the twelve tomato chromosomes[13]. Tomato will act as a model plant for solanaceae genomics. The Solanaceae family includes the most variable of crop species in terms of their agricultural utility, as it includes the tuber-bearing potato, a number of fruit-bearing vegetables (*e.g.* tomato, eggplant, peppers, husk tomato), ornamental flowers, edible leaves, and medicinals. The Solanaceae are unique in that multiple crop species in this family are major contributors to fruit and vegetable consumption. The unique and ancient mode of Solanaceae evolution makes the family a unique subject to explore the basis of phenotypic diversity and adaptation to natural and agricultural environments.

The Potato Genome Sequencing Consortium (PGSC) has announced the release of the first draft sequence of the potato genome which is also a key member of the Solanaceae family. It is the world's third most important crop and the most important vegetable crop. Access to the potato genome sequence, the "genetic blueprint" of how a potato grows and reproduces, is anticipated to assist scientists to improve the yield, quality, nutritional value and disease resistance of different varieties of potato and also it will act as a model organism for other members of Solanaceae family. In addition to that, the potato genome sequence will permit potato breeders to reduce the time line currently needed to breed new varieties.

Genome Annotation

Genome sequence data available publicly in different databases are increasing day by day. This demands a high-throughput annotation scheme to extract biologically useful and timely information from the sequence data on a regular basis.

A major long-term goal of genomics is the potential ability to predict phenotype of an organism from its genotype. If we know the genome sequence of an organism, the functions of all its genes and the effects of specific alleles, then it may be possible to make strong inferences about the resulting phenotype. The term genome annotations refers to the process of finding particular gene and attach information that is relevant to its structure and function.

Sequence Alignment

The basic level genome annotation is performed based on sequence similarity using tools for sequence alignment and comparison. A sequence alignment is a method of arranging DNA, RNA or protein sequences to find the regions of similarity. It can be a pair-wise alignment in which one sequence is aligned with another or a multiple sequence alignment where many sequences are simultaneously aligned. *Global alignment* (Needleman-Wunsch algorithm)[22] is based on attempts to align entire lengths of the query and target sequences, where as a *local alignment* (Smith-Waterman algorithm)[21] searches for short matching segments (Figure 7.4).

BLAST and **FASTA** are the most widely used tools for sequence alignment[23-25]. **BLASTP** is a program which aligns one protein sequence against another and **BLASTN** aligns nucleotide sequences. **BLASTX** translates a nucleotide sequence and align it with the protein sequence. **TBLASTN** can be used to query a protein sequence against a translated nucleotide database. **TBLASTX** is used to query a translated nucleotide against its database. Multiple sequence alignment can be done using programs like **clustalW**[28]. Many other sequence alignment programs such as **LALIGN**, **Kalign**, **T-Coffee** are also been extensively used[26-28].

Structural Annotation

Structural annotation includes the identification of Open Reading Frames (ORFs), predicting genes, locating regulatory motifs etc[29-32]. ORF is a portion in the genome which contains the sequence that could potentially encode a protein. Tools like **ORF Finder** are widely used for the identification of ORFs. Gene finding in organism, especially prokaryotes start from searching for an open reading frame (ORF). Gene finding refers to the area of computational biology concerned with the identification of the stretches of sequence in the genome that are biologically functional. In eukaryotes, gene finding

```
gi|162458451|ref|NM_001111883.   CGCG------AGACAACTCACTGGTGAGAGACATAAGCCGAATGCAGCAGCGCAACTACG
gi|1843532|gb|U77952.1|MDU7795   ---ACTCAAAGGGTTA-CCTATAGTAAGAAATATCAGTGAGCTTCCACAGGATAACTATG
gi|22108|emb|X16308.1|           TGCG------AGATAACTCATTGGTGAGAGACATAAGCCAAATGCCGCAAAGCAGCTATG
gi|224117157|ref|XM_002331726.   TTCTATCAAAGGGTTG-CCACTTGTGAGGAATATCAGTGAGCTTCCACAGGATAACTATG
                                     *       *  * ** **   * ** **       * *   **      * *** *

gi|162458451|ref|NM_001111883.   GGAGGGAAGGATTCTCGCATATAACTGTCACAGGTGCTCTTGCTCACGGGACGAAGGAGG
gi|1843532|gb|U77952.1|MDU7795   GAAGGGGTGGTTTGGCGCACACAACTGTTGCTGGTTCGCTCTTGCATGGGTTGAAAGAGG
gi|22108|emb|X16308.1|           GGATTGAAGGATTGTCACATATAACAGTTGCTGGTGCGCTCAATCATGGGATGAAGGAGG
gi|224117157|ref|XM_002331726.   GAAGGGGAGGTTTGTCTCATATCACTCTTGCTGGTTCTGCCATGCATGGATTGAAAGAGG
                                  * *  *  ** **  * ** *  **  *  * *** *       ** **   *** ****

gi|162458451|ref|NM_001111883.   TGGAAGTGTGGCTACAAACATTTGGTCCAGGTCAAAGGACCCCAATCCACAGGCATTCTT
gi|1843532|gb|U77952.1|MDU7795   TCGAGGTTTGGCTACAAACATTTGCTCCAGGATCAGGCACACCAATACACAGGCATTCTT
gi|22108|emb|X16308.1|           TGGAAGTGTGGCTTCAGCAATAAGTCCAGGTCAAAGGACGCCGATCCACAGGCATTCCT
gi|224117157|ref|XM_002331726.   TAGAGGTATGGCTTCAAACATTTTCTCCAGGCTCGCGCACTCCAATCCACAGGCATTCTT
                                  * ** ** ***** ** *** *  ******    * ** ** * ********** *

gi|162458451|ref|NM_001111883.   GTGAAGAGGTTTTCATTGTCCTCAAGGGGAAAGGCACGCTCTTACTCGGGTCGAGCTCGC
gi|1843532|gb|U77952.1|MDU7795   GTGAAGAAGTTTTTGTTGTCCTAAAGGGAAGTGGAACTCTCTACCTTGCACCGAGCTCGC
gi|22108|emb|X16308.1|           GTGAAGTTTTCACTGTCCTCAAAGGGAAGGGTACGCTCTTGATGGGATCAAGCTCAC
gi|224117157|ref|XM_002331726.   GTGAAGAAATATTTGTTGTCCTCAAGGGAAGTGGAACTCTCTATCTTGCATCAAGTTCAC
                                  ******  * **  ****** ** ** ** * **   * *   * ** ** *
```

Figure 7.4: Multiple Sequence Alignment of Auxin Binding Protein (ABP)
in Four Plant Species (Generated using ClustalW)

is very different compared to prokaryotes as the eukaryotic genes are not continuous and are interrupted by intervening noncoding sequences called 'introns'. Gene finding or gene prediction can be done using programs like **GenScan**, **HMMgene** and **GeneMark**. Computational recognition of promoters and regulatory regions may reveal new genes in genomic DNA sequence and will improve gene prediction adapted to a coding region identification program. This will result in having better fundamental knowledge about regulation of eukaryotic gene expression and provide essential tools for manipulation of biologically important genes.

Functional Annotation

Functional annotation of genes details information pertaining to a gene's biological and biochemical functions, their interactions, expression and regulation. Tools such as **DAVID** and **Blast2GO (B2G)** can be used for functional annotation of the genes.

The Rice Genome Research Program generated huge amount of rice genome sequences which are annotated by searching the non-redundant protein database using **BLASTX**, searching the rice EST database using **BLASTN** software, and analyzing the sequence with **GenScan** to predict open reading frames and predicting exon/intron splice sites with Splice Predictor. These results are combined to make a final annotation of genes and elements, and their coordinates in a genome sequence[18].

Once genes are identified, the basis for traits such as disease resistance and stress tolerance is likely to be understood as it is in model organisms. A detailed study of these genes helps in understanding the biochemical signal transduction pathways of plant hormones or pathogen-related defense mechanisms. This allows more precise diagnosis in plant breeding programs and genetic modification by methods like inducing mutations. The sequences can also be used to detect genetic variation and to trace the evolutionary history of plants using phylogenetic analysis.

Proteomics Applications

Genes code for proteins which are the actual functional units of the body. Proteomics is the detailed and large scale study of proteins, mainly their structure and function.

Protein Structure Prediction

One important application of computational biology is the detection of protein structure and function from sequence[33]. Protein structure prediction includes different strategies applied for the prediction of three dimensional structure of the protein from its amino acid sequence. A protein function is by and large determined by the structure it assumes. The integral relationship between protein structure and its function facilitates function prediction with reasonable accuracy. Thus structure prediction becomes an inevitable area in the study of proteins.

Many methods are available which address the issue of protein structure prediction and in general they can be divided into four main areas; secondary structure prediction, homology modeling, Ab initio prediction and fold recognition or threading.

Secondary Structure Prediction

Secondary structure is the local structure, typically recognized by specific backbone torsion angles and main-chain hydrogen bond pairings (Figure 7.5). Secondary structure prediction algorithms connect amino acid sequences to secondary structure regions such as alpha-helices and beta-strands. Some of the most powerful methods for secondary structure predictions are based on artificial neural networks. There are many bioinformatics tools available that provide secondary structure predictions. Some of them are listed below.

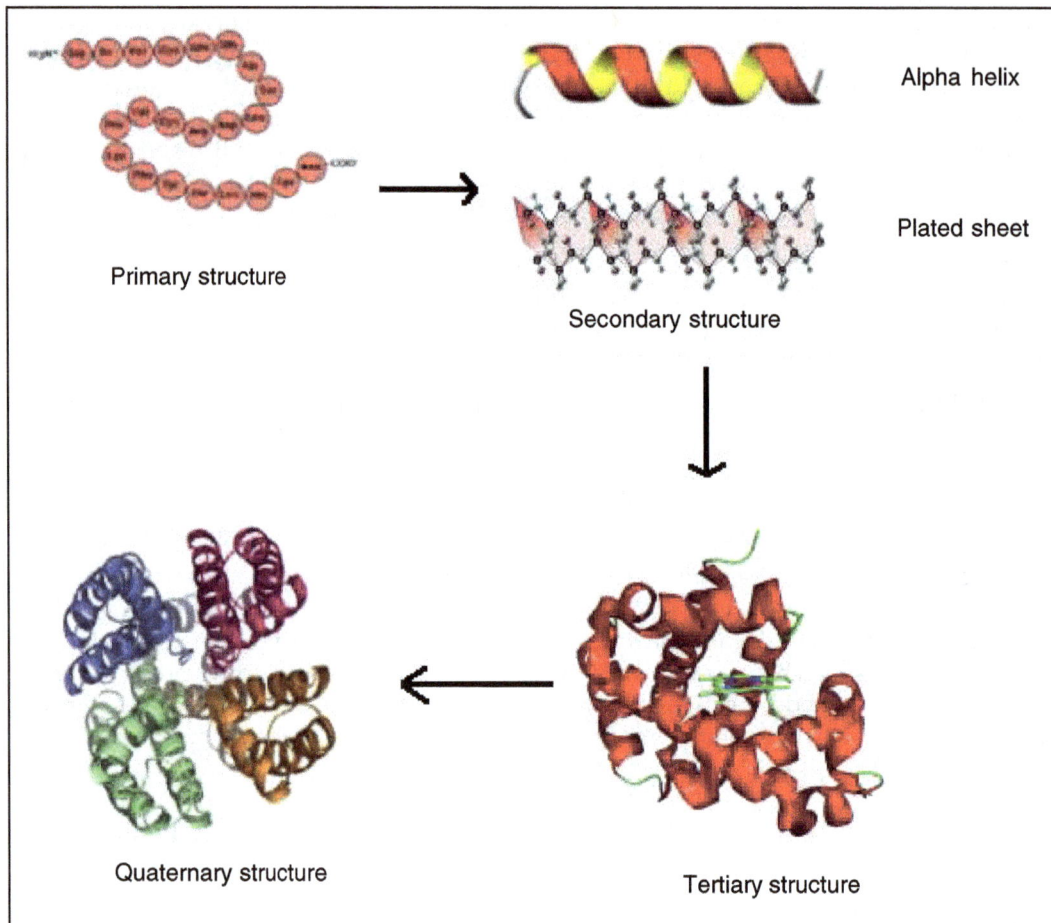

Figure 7.5: Protein Structural Hierarchy

Primary structure is the sequence of amino acid chain. Secondary structure generally consists of helices, sheets and turns formed as a result of repetitive hydrogen bonding interactions in prescribed patterns. Tertiary structure is a complete self folding functional unit formed as a result of unique association of different secondary structural elements. Quaternary structure is an association of multiple identical tertiary structure subunits performing a given function.

☆ **AGADIR** - An algorithm to predict the helical content of peptides

☆ **APSSP** - Advanced Protein Secondary Structure Prediction Server

☆ **Jpred** - A consensus method for protein secondary structure prediction at University of Dundee

☆ **PSIpred** - Various protein structure prediction methods at Brunel University

Homology Modeling

Homology modeling compares the query sequence with the sequence of another protein, which has a known structure. Main steps in homology modeling include.

☆ Sequentially align query sequence and homologue with the known structure

☆ Determining main chain segments in the sequence to represent regions containing insertions and deletions

☆ Replacing side chains of the amino acid residues that have been mutated in the query sequence

☆ Verify the model to detect and remove steric clashes between atoms that can make serious problems to structural integrity.

☆ Refinement of the model generated using energy minimization.

The quality of the model depends on sequence similarity and alignment algorithm. A homology modeling produces a good model if the sequence identity between the query sequence and the homologous sequence whose structure is known is greater than 40 per cent.

Some important tools available for homology modeling are the following:

☆ **SWISS-MODEL local** - An automated knowledge-based protein modeling server

☆ **3Djigsaw** - Three-dimensional models for proteins based on homologues of known structure

☆ **CPH models** - Automated neural-network based protein modeling server

☆ **ESyPred3D** - Automated homology modeling program using neural networks

☆ **Geno3d** - Automatic modeling of protein three-dimensional structure

Ab Initio Prediction

Ab initio- or de novo- protein modeling methods build three-dimensional protein models based on physical principles. This method of structure prediction involves modeling of all energetic contributions involved in the process of folding and then in finding the structure with the lowest free energy. These procedures tend to require vast computational resources and hence have only been carried out for tiny proteins. To predict protein structure de novo for larger proteins will require better algorithms and larger computational resources like powerful supercomputers or distributed computing. Tools like *HMMSTR/Rosetta* help in the prediction of protein structure from sequence.

Fold Recognition

Protein fold recognition, also known as protein threading predicts protein structures by using statistical knowledge of the relationship between the structure and the sequence[45]. Threading methods compare a target sequence against a library of structural templates and compared based on a scoring function. The scores are then ranked and it is considered that the sequence adopts the fold with the best score. Programs like *GenTHREADER* and *Threader* are available for fold recognition.

Protein Function Prediction

Protein function prediction from the available sequence and structure data is very challenging. The simplest method is clustering proteins into families relying on sequence similarity measures. It is assumed that members of the same family may possess similar or identical functions. Proteins are usually assigned to any particular family based on the presence of family-specific patterns, domains, or structural elements. Homology search tools such as **BLAST**, **FASTA** and multiple sequence alignment tools like **clustalw** are widely used to perform this operation. There are more sophisticated approaches, which detect domains using domain databases (**PFAM**, **SMART**, or **Super family**), which may also use the order of domains as a fingerprint for the protein, and classify proteins into families. At present classification of proteins into families based on structural similarities is very limited as the total number of structures available in **PDB**[34] is very small[36].

Evolutionary Studies in Plants

The term phylogeny refers to the evolution or historical development of a plant or animal species, or even a human tribe or similar group. The evolutionary relationship between different organisms or sequences are represented in a phylogenetic tree[38]. A dendrogram is the broad term used for this kind of tree representation. A phylogram is a tree structure where the branch lengths are proportional to the predicted or hypothetical evolutionary time between organisms or sequences (Figure 7.6). Cladograms are branched diagrams, similar in appearance to family trees which illustrate patterns of relationships where the branch lengths are normalized. Bioinformaticians produce cladograms representing relationships between sequences, either DNA sequences or amino acid sequences.

For generating the phylogenetic tree of related sequences, a multiple sequence alignment is performed initially. Simple tree generation algorithms work by calculating the genetic distance from the multiple sequence alignment result[39,40]. Complex algorithms are heuristic in nature and they use different optimization techniques. Programs such as **PHYLIP**, **DendroUPGMA**, **PhyML** etc. can be used for the generation of the phylogenetic tree.

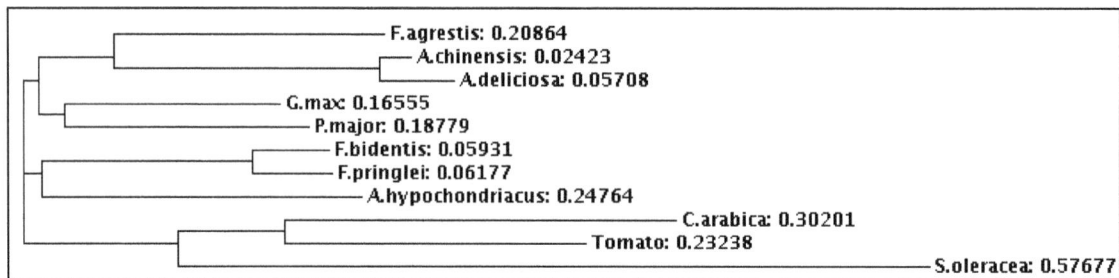

Figure 7.6: A Phylogram Representing the Evolutionary Relationship Between the Ribulose 1,5 Bisphosphate Carboxylase, Small Subunit Type I (rbcS1) mRNA in Different Plant Species

Microarray Data Analysis

Microarrays have become an important technology for the global analysis of gene expression in humans, animals, plants and microbes. Implemented in the context of a well-designed experiment, cDNA and oligonucleotide arrays can provide high-throughput, simultaneous analysis of transcript abundance for hundreds or thousands of genes.

A microarray is a small glass slide or a membrane on to which specific sequences representing genes are attached to specific locations called *spots*. The spot size will be less than 200 microns. Those sequences, specific to different genes, immobilized on the array are called *probes*. Each spot corresponds to a specific probe with known identity. Researchers use location of the probes in the array to identify a particular gene. The sequences used in a microarray can be a genomic DNA, cDNA or oligonucleotide. An oligonucleotide is a fragment of single stranded DNA with 5 to 50 nucleotides in length. The sequences in the sample to be analyzed are called *target*.

The principle behind microarray technology is DNA hybridization. Hybridization is the process of combining two single stranded complimentary molecules into one molecule. The target sequences are fluorescent labeled and are allowed to hybridize with the probe. Both the probe and the target sequences will hybridize if they are complementary to each other. In the case of a gene expression analysis the amount of mRNA hybridized to each spot on the array indicates the expression level of various genes. By quantifying the rate of hybridization we can identify the rate of expression of each gene. The rate of hybridization is proportional to the fluorescence emitted. So by measuring the fluorescent intensities we can quantify the rate of hybridization that in turn leads to the identification of differentially expressed genes[46] (Figure 7.7).

Though many softwares available for microarray analysis **Bioconductor, TM4** and **GeneSpring** are the ones that are frequently used.

Applications of Microarrays include

1. *Gene discovery*: DNA Microarray technology helps in the identification of new genes, know about their functioning and expression levels under various conditions.

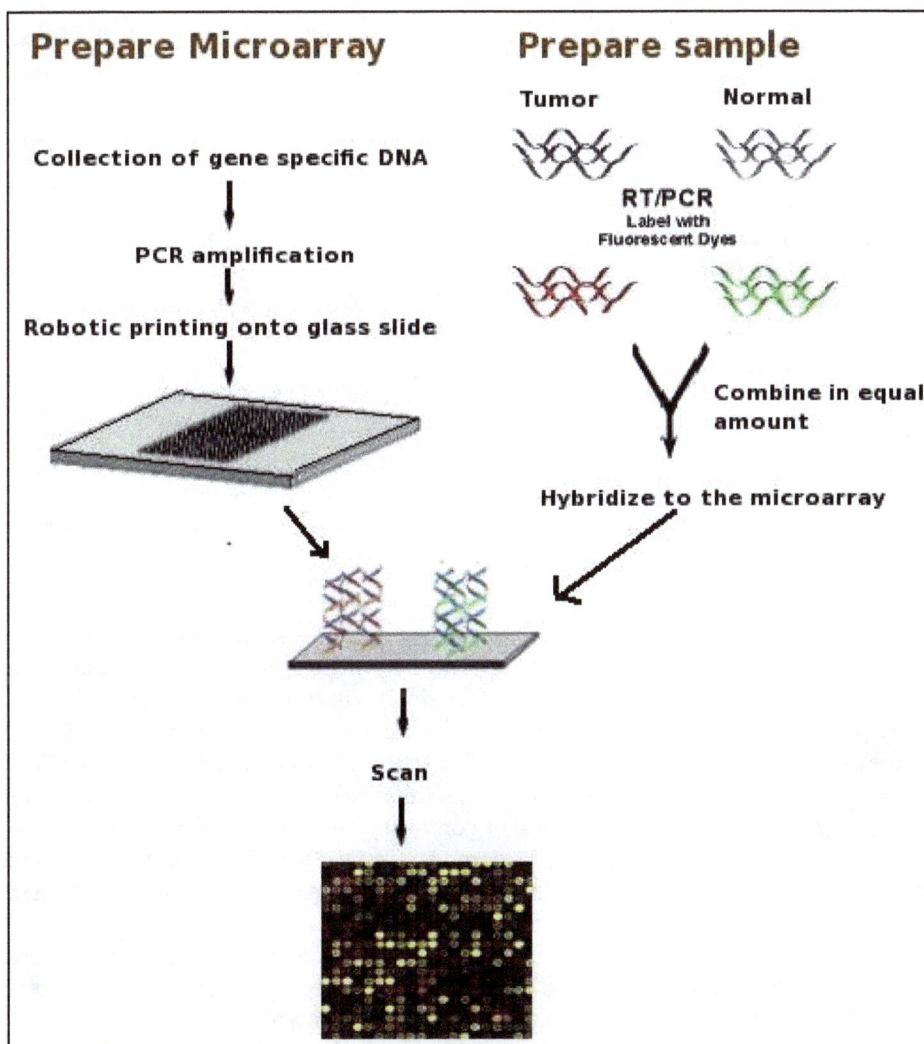

Figure 7.7: Overview of Microarray Technology

2. *Disease diagnosis*: DNA Microarray technology helps researchers learn more about different diseases such as heart diseases, infectious diseases and of cancer.

3. *Gene expression analysis*: Microarray is widely used to study the differential expression of genes.

4. *Toxicological research*: Microarray technology provides a robust platform to investigate the impact of toxins on the cells and their effect in their respective progeny. Toxicogenomics establishes correlation between responses to toxicants and the changes in the genetic profiles of the cells exposed to such toxicants.

5. *Mutation analysis*: Here researchers use genomic DNA. The genes might differ from each other by as less as a single nucleotide base. A single base difference between two sequences is known as Single Nucleotide Polymorphism (SNP) and detecting them is known as SNP detection.

In the case of plants, microarray has been used for studying the plant physiology[47–51]. Microarrays have already been used for the study of gene expression of cotton mutants with altered fibre characteristics, understanding gene expression pattern during elongation of maize roots at low soil water potential, identification and characterization of genes responding to temperature stress in Chinese cabbage, analysis of transcript abundance in barley under conditions of water deficit, profiling expression of aluminium-tolerance genes in wheat etc. Seki *et al.*(2002) used microarrays containing 1300 full-length cDNAs to identify 44 drought-inducible genes in Arabidopsis following two hour dehydration. Ozturk *et al.* (2002) recently used cDNA microarrays to identify large-scale changes in transcript abundance for barley (*Hordeum vulgare* L.) exposed to rapid (*i.e.* 6–10 h) drought shock treatments.

Mass Spectrometric Analysis

Mass spectrometry (MS) is an analytical technique frequently used to assess the composition of a sample or molecule. It is also used for elucidating the chemical structures of molecules, such as peptides and other chemical compounds. The MS principle consists of ionizing chemical compounds to generate charged molecules or molecular fragments and measurement of their mass-to-charge ratios.

Mass Spectrometry is critical in proteomics and functional genomics[52]. Protein sequencing and identification is often performed using Tandem mass spectrometry, also known as MS/MS. Each protein has a mass spectrum with unique characteristics. For performing a mass spectrometry experiment of a specific protein, the protein is separated by electrophoresis or chromatography and then digested with a site-specific protease such as trypsin. It will result in a set of peptides. After mass spectrometric analysis, these peptides will generate a set of fragment peaks from which the residual sequence of the peptide may be inferred. Tools like, **SEQUEST, Mascot, X!TANDEM** and **OMSSA** are used to identify peptides from databases, which are similar to the protein under study. For example **SEQUEST** (http://fields.scripps.edu/sequest/) converts the amino acid sequences in a protein database to fragmentation patterns and it is compared against the MS/MS spectrum generated on the target peptide. The program initially identifies amino acid sequences in the database which match the measured mass of the peptide. In the following step, it compares fragment ions against the MS/MS spectrum and generates a preliminary score for each amino acid sequence. A cross correlation analysis is then performed on 500 top scoring peptides by correlating theoretical, reconstructed spectra against the experimental spectrum and the results are generated accordingly.

Mass spectrometry is widely used in crop improvement as well. MacLeod *et al.* (1998) have used mass spectrometry for quantifying selenium content in plants, soils and sludges. Peltier *et al.* studied thylakoid proteins using mass spectrometry. Chang *et al.* (2000) used mass spectrometric techniques to study the patterns of protein synthesis and tolerance of anoxia in root tips of maize seedlings acclimated to a low-oxygen environment. The proteins of the 30S and 50S ribosomal subunits in spinach chloroplasts were identified by a combination of 2-DE, chromatography, MS, and Edman sequencing[53-57]. Mass Spectrometric data can also be made useful in genome annotation.

Computational Primer Design

Primers are essential components of polymerase chain reaction (*PCR*), DNA sequencing as well as modern microarray systems which utilize appropriate probes. Success of all these procedures greatly depends on the proper design of the primer.

Primer design involves various parameters such as string-based alignment scores, specificity, melting temperature, primer length and GC content[58,59]. A primer specificity is an important factor that should be taken care of while designing a primer. Primer specificity is very much dependent on primer length. In general, oligonucleotide primers of length 18-24 are seen to be highly specific. For a PCR reaction two primers are added and they should be designed in such a way that their melting temperature should be the same so that the reaction can proceed in an efficient way. While designing the primer, care should be taken so that the designed primers are not with an intra-primer homology beyond 3 bases. If the primers are self complementary they may form partially double stranded structure. Also primers should not be complementary to each other. Another important point to be taken care of is the GC content. Ideally a primer should have 45-55 per cent GC. It should also be taken care that the primer does not have a poly G or poly C stretch that may cause non specific binding.

Even though computational programs cannot replace an experienced researcher, they can assist in efficient design of primers. There are many programs available for primer designing. Some of them are **Primer 3**[60], **NetPrimer** and **Primo Pro**[61].

Glossary of Computational Tools

Bioinformatics Tools	*Website*
Sequence Alignment(Pairwise allignment)	
BLAST	http://blast.ncbi.nlm.nih.gov/Blast.cgi
FASTA	http://www.ebi.ac.uk/Tools/fasta/
LALIGN	http://www.ch.embnet.org/software/LALIGN_form.html
Sequence Alignment(Multiple sequence alignment)	
clustalW2	http://www.ebi.ac.uk/Tools/clustalw2/index.html
T-Coffee	http://www.tcoffee.org/
Kalign	http://www.ebi.ac.uk/Tools/kalign/index.html
ORF finding	
ORF Finder	https://www.dna20.com/toolbox/ORFFinder.html

Bioinformatics Tools	Website
Gene Prediction	
GeneMark	http://exon.gatech.edu/GeneMark/
HMMgene	http://www.cbs.dtu.dk/services/HMMgene/
GENSCAN	http://genes.mit.edu/GENSCAN.html
Functional Annotation	
DAVID	http://david.abcc.ncifcrf.gov/
Secondary Structure Prediction	
Agadir	http://agadir.crg.es/
APSSP	http://imtech.res.in/raghava/apssp/
PSIpred	http://bioinf.cs.ucl.ac.uk/psipred/
Jpred	http://www.compbio.dundee.ac.uk/www-jpred/
Homology Modeling	
SWISS-MODEL	http://swissmodel.expasy.org/workspace/
3Djigsaw	http://bmm.cancerresearchuk.org/~3djigsaw/
CPH models	http://www.cbs.dtu.dk/services/CPHmodels/
ESyPred3D	http://www.fundp.ac.be/sciences/biologie/urbm/bioinfo/esypred/
Geno3d	http://geno3d-pbil.ibcp.fr/cgi-bin/geno3d_automat.pl?page=/GENO3D/geno3d_home.html
Ab – inito structure prediction	
HMMSTR/Rosetta	http://www.bioinfo.rpi.edu/~bystrc/hmmstr/server.php
Threading	
GenTHREADER	http://bioinf.cs.ucl.ac.uk/psipred/
Threader	http://bioinf.cs.ucl.ac.uk/threader/
Phylogenetic Analysis	
PHYLIP	http://evolution.genetics.washington.edu/phylip.html
DendroUPGMA	http://genomes.urv.cat/UPGMA/
PhyML	http://atgc.lirmm.fr/phyml/
Microarray Analysis	
Bioconductor	http://www.bioconductor.org/
TM4	http://www.tm4.org/
GeneSpring	http://www.chem.agilent.com/en-US/products/software/lifesciencesinformatics/pages/gp35082.aspx

Bioinformatics Tools	Website
Mass Spectrometry data analysis	
MASCOT	http://www.matrixscience.com/
X!TANDEM	http://www.thegpm.org/TANDEM/
OMSSA	http://pubchem.ncbi.nlm.nih.gov/omssa
Primer Design	
Primer 3	http://frodo.wi.mit.edu/primer3/
Primo Pro	http://www.changbioscience.com/primo/primo.html
NetPrimer	http://www.premierbiosoft.com/netprimer/

References

1. Pruit, K.D. and Maglott, D.R., 2001. RefSeq and LocusLink: NCBI gene–centered resources. *Nucleic Acids Res.*, 29: 137–140.

2. Hubbard, T. *et al.*, 2002. The Ensembl genome database project. *Nucleic Acids Research*, 30: 138–141.

3. Marra, M. *et al.*, 1999. A map for sequence analysis of the Arabidopsis thaliana genome. *Nature Genet.*, 22: 265–270.

4. Staden, R.A., 1979. Strategy of DNA sequencing employing computer programs. *Nucleic Acids Research.*, 6(7): 2601–2610.

5. Church, G.M., 2006. Genomes for all. *Sci. Am.*, 294(1): 46–54.

6. Weber, J.L. and Myers, E.W., 1997. Human whole-genome shotgun sequencing. *Genome Res.*, 7: 401–409.

7. Anderson, S., 1981. Shotgun DNA sequencing using cloned DNase I-generated fragments. *Nucleic Acids Res.*, 9: 3015–3027.

8. Sanger, F. and Coulson, A.R., 1975. A rapid method for determining sequences in DNA by primed synthesis with DNA polymerase. *J. Mol. Biol.*, 94: 441–448.

9. Maxam, A.M. and Gilbert, W.A., 1977. New method for sequencing DNA. *Proc. Natl Acad. Sci., USA*, 74: 560–564.

10. Myers, E.W., *et al.*, 2002. On the sequencing and assembly of the human genome. *Proc. Natl. Acad. Sci., USA*, 99(7): 4145–4146.

11. Pop, M., Salzberg, S.L. and Shumway, M., 2002. Genome sequence assembly: Algorithms and issues. *IEEE Computer*, 35(7): 47–54.

12. Myers, E.W., 1995. Toward simplifying and accurately formulating fragment assembly. *J. Comp. Bio.*, 2(2): 275–290.

13. Mueller, L., 2009. A snapshot of the emerging tomato genome sequence. *Plant Genome*, 2009.

14. Meldrum, D., 2000. Automation for genomics, part two: sequencers, microarrays, and future trends. *Genome Res.*, 10: 1288–1303.

15. Paterson, A.H. *et al.*, 2000. Comparative genomics of plant chromosomes. *Plant Cell*, 12: 1523–1540.

16. Sasaki, T. and Sederoff, R.R., 2003. Genome studies and molecular genetics: The rice genome and comparative genomics of higher plants. *Current Opinion in Plant Biology*, 6(2): 97–100.

17. Hubbard, T. and Birney, E., 2000. Open annotation offers a democratic solution to genome sequencing. *Nature*, 403: 825.

18. Sasaki, T. and Burr, B., 2000. International rice genome sequencing project: The effort to completely sequence the rice genome. *Current Opinion in Plant Biology*, 3(2): 138–142.

19. Rensink, W.A. and Buell, C.R., 2004. Arabidopsis to rice: Applying knowledge from a weed to enhance our understanding of a crop species. *Plant Physiol.*, 135(2): 622–629.

20. Meinke, D.W., Cherry, J.M., Dean, C., Rounsley, S.D. and Koornneef, M., 1998. *Arabidopsis thaliana*: A model plant for genome analysis. *Science*, 282(5389): 662–682.

21. Smith, T.F. and Waterman, M.S., 1981. Identification of common molecular subsequences. *Journal of Molecular Biology*, 147: 195–197.

22. Needleman, S.B. and Wunsch, C.D., 1970. A general method applicable to the search for similarities in the amino acid sequence of two proteins. *J. Mol. Biol.*, 48(3): 443–453.

23. Altschul, S.F., Gish, W., Miller, W., Myers, E.W. and Lipman, D.J., 1990. Basic local alignment search tool. *J. Mol. Biol.*, 215: 403–410.

24. Lipman, 1985. Rapid and sensitive protein similarity searches. *Science*, 227(4693): 1435–1441.

25. Pearson, 1988. Improved tools for biological sequence comparison. *Proceedings of the National Academy of Sciences of the United States of America*, 85(8): 2444–2448.

26. Thompson, J.D., Plewniak, F. and Poch, O.A., 1999. Comprehensive comparison of multiple sequence alignment programs. *Nucleic Acids Res.*, 27: 2682–2690.

27. Notredame, C., Higgins, D.G. and Heringa, J.T., 2000. Coffee: A novel method for fast and accurate multiple sequence alignment. J. Mol. Biol., 302(1): 205–217.

28. Thompson, J.D., Higgins, D.G. and Gibson, T.J., 1994. "CLUSTAL W: Improving the sensitivity of progressive multiple sequence alignment through sequence weighting, position-specific gap penalties and weight matrix choice". *Nucleic Acids Res.*, 22: 4673–4680.

29. Guigo, R., Agarwal, P., Abril, J.F., Burset, M. and Fickett, J.W., 2000. An assessment of gene prediction accuracy in large DNA sequences. *Genome Res.*, 10: 1631–1642.

30. Solovyev, V. and Salamov, A., 1997. The gene-finder computer tools for analysis of human and model organisms genome sequences. *ISMB*, 5: 294–302.

31. Kulp, D., Haussler, D., Reese, M.G. and Eeckman, F.H.A., 1996. Generalized hidden Markov model for the recognition of human genes in DNA. *ISMB*, 4: 134–142.

32. Gelfand, M.S., Mironov, A.A. and Pevzner, P.A., 1996. Gene recognition via spliced sequence alignment. *Proc. Natl. Acad. Sci., USA*, 93: 9061–9066.

33. Baker, D. and Sali, A., 2001. Protein structure prediction and structural genomics. *Science*, 294(5540): 93–96.

34. Berman, H.M. *et al.*, 2000. The protein data bank. *Nucleic Acids Res.*, 28(1): 235–242.

35. Zhang, Y. and Skolnick, J., 2005. The protein structure prediction problem could be solved using the current PDB library. *Proc. Natl. Acad. Sci., USA*, 102: 1029–1034.

36. Veeramachaneni, V. and Makowski, W., 2004. Visualizing sequence similarity of protein families. *Genome Res.*, 14: 1160–1169.

37. Bateman, A. *et al.*, 2000. The Pfam protein families database. *Nucleic Acids Res.*, 28: 263–266.

38. Sidow, A. and Bowman, B.H., 1991. Molecular phylogeny. *Curr. Opin. Genet. Dev.*, 1: 451–456.

39. Koonin, E.V., Aravind, L. and Kondrashov, A.S., 2000.The impact of comparative genomics on our understanding of evolution. *Cell*, 101: 573–576.

40. Phillips, A., Janies, D. and Wheeler, W. 2000. Multiple sequence alignment in phylogenetic analysis. *Molecular Phylogenetics and Evolution*, 16(3): 317–330.

41. Rubin, G.M. *et al.*, 2000. Comparative genomics of the eukaryotes. *Science*, 287: 2204–2215.

42. Green, P. *et al.*, 1993. Ancient conserved regions in new gene sequences and the protein databases. *Science*, 259: 1711–1716.

43. Marti-Renom, M.A., Stuart, A.C., Fiser, A., Sanchez, R., Melo, F. and Sali, A., 2000. Comparative protein structure modeling of genes and genomes. *Annu Rev Biophys Biomol Struct.*, 29: 291–325.

44. Mount, D.M., 2004. *Bioinformatics: Sequence and Genome Analysis*. Cold Spring Harbor Laboratory Press.

45. Jones, D.T., Taylor, W.R. and Thornton, J.M., 1992. A new approach to protein fold recognition. *Nature*, 358: 86–89.

46. Schena, M., Shalon, D., Davis, R.W. and Brown, P.O., 1995. Quantitative monitoring of gene expression patterns with a complementary DNA microarray. *Science*, 270(5235): 467–470.

47. Seki, M., Narusaka, M., Ishida, J., Nanjo, T., Fujita, M., Oono, Y., Kamiya, A., Nakajima, M., Enju, A., and Sakurai, T. *et al.*, 2002. Monitoring the expression profiles of 7000 *Arabidopsis* genes under drought, cold and high-salinity stresses using a full-length cDNA microarray. *Plant J.*, 31: 279–292.

48. Ozturk, Z.N. *et al.*, 2002. Monitoring large-scale changes in transcript abundance in drought- and salt-stressed barley. *Plant Molecular Biology*, 48: 551–573.

49. Schenk, P.M., Kazan, K., Wilson, I., Anderson, J.P., Richmond, T., Somerville, S.C. and Manners, J.M., 2000.Coordinated plant defense responses in *Arabidopsis* revealed by microarray analysis. *Proc. Natl. Acad. Sci., USA*, 97: 11655–11660.

50. Girke, T., Todd, J., Ruuska, S., White, J., Benning, C. and Ohlrogge, J., 2000. Microarray analysis of developing *Arabidopsis* seeds. *Plant Physiol.*, 124: 1570–1581.

51. Richmond, T. and Somerville, S., 2000. Chasing the dream: Plant EST microarrays. *Curr. Opin. Plant Biol.*, 3: 108–116.

52. Andersen, J.S. and Mann, M., 2000. Functional genomics by mass spectrometry. *FEBS Lett.*, 480: 25–31.

53. Chang, W.W., Huang, L., Shen, M., Webster, C., Burlingame, A.L. and Roberts, J.K., 2000. Patterns of protein synthesis and tolerance of anoxia in root tips of maize seedlings acclimated to a low-oxygen environment, and identification of proteins by mass spectrometry. *Plant Physiol.*, 122: 295–318.

54. Kuster, B., Mortensen, P. and Mann, M., 1999. Identifying proteins in genome databases using mass spectrometry. In: *Proceedings of the 47ᵗʰ ASMS Conference on Mass Spectrometry and Allied Topics*, Dallas, TX, 1897–1989.

55. MacLeod, J.A., Gupta, U.C., Milburn, P. and Sanderson, J.B., 1998. Selenium concentration in plant material, drainage and surface water as influenced by Se applied to barley foliage in a barley-red clover-potato rotation. *Canadian Journal of Soil Science*, 78: 685–688.

56. Peltier, J.B., Friso, G., Kalume, D.E., Roepstorff, P., Nilsson, F., Adamska, I. and van Wijk, K.J., 2000. Proteomics of the chloroplast: Systematic identification and targeting analysis of lumenal and peripheral thylakoid proteins. *Plant Cell*, 12: 319–342.

57. van Wijk, K.J., 2000. Proteomics of the chloroplast: experimentation and prediction. *Trends Plant Sci.*, 5: 420–425.

58. Haas, S., Vingron, M., Poustka, A. and Wiemann, S., 1998. Primer design for large scale sequencing. *Nucleic Acids Res.*, 26: 3006–3012.

59. Robertson, J.M. and Walsh-Weller, J., 1998. An introduction to PCR primer design and optimization of amplification reactions. *Methods Mol. Biol.*, 98: 121–154.

60. Rozen, S. and Skaletsky, H., 2000. Primer3 on the WWW for general users and for biologist programmers. *Methods Mol. Biol.*, 132: 365–386.

61. Li, P., Kupfer, K.C., Davies, C.J., Burbee, D., Evans, G.A. and Garner, H.R., 1997. PRIMO: A primer design program that applies base quality statistics for automated large-scale DNA sequencing. *Genomics*, 40: 476–485.

2012, Nutri-Horticulture
Editor: Professor K.V. Peter
Published by: DAYA PUBLISHING HOUSE, NEW DELHI

Pages 143–153

Chapter 8

Breeding Methods of Self Pollinated Crops

T. Vanaja

College of Agriculture, Kerala Agricultural University,
P.O. Padannakad, Kerala, INDIA
E-mail: vtaliyil@yahoo.com

Importance of Mode of Reproduction

The mode of reproduction, which in turn is the system of pollination and fertilization, becomes important in classical or traditional breeding, because it determines a crop's genetic composition. Genetic composition is the deciding factor to develop suitable breeding and selection methods and is also essential for artificial manipulation of crops to breed improved types (Dabholkar,2006). Genetic variability is identified, increased and exploited in conventional plant breeding through sexual reproduction. Knowledge of reproductive parts of a plant and reproductive process is very important as far as a breeder is concerned.

Flower Parts

A flower consists of four floral organs namely petals, sepals, stamens and pistil. Usually petals are large and brightly coloured; sepals are small and inconspicuous, and both do not take part directly in reproduction. It is the stamens and pistil which are directly involved in reproduction. Stamen consists of a filament which supports an anther containing numerous pollen grains. Pistil consists of basal ovary in which seeds are developed after fertilization, a style and stigma. The number of ovules within an ovary vary from one as in rice to several in tomato.

Types of Flowers

Flowers may be complete flowers or incomplete flowers. Flowers having all four floral parts are termed as complete flowers (*e.g.*: Cow pea, Tomato) and those lacking one or two parts are known as incomplete flowers (*e.g*: Rice, Black pepper). Based on the presence of reproductive flower parts, flowers are again classified as perfect flowers and imperfect flowers. Perfect flowers are bisexual having stamens and pistil in the same flower (*e.g.*: Rice, Black pepper, Cow pea etc), while either of these essential organs absent in unisexual flowers. Unisexual flowers are either staminate or pistillate flowers.Crops in which staminate and pistillate flowers are seen in the same plant are known as monoecious (*e.g.* corn and cassava). Crops in which staminate and pistillate flowers are seen in different plants are known as dioecious (e.g:hemp and papaya).

Pollination and Fertilization

Anther has four cavities which contain microspore mother cells. Each microspore mother cell undergoes two successive nuclear divisions during meiosis and produces a tetrad of four microspores. Each microspore develops into a pollen grain. Transfer of pollen grains from stamen to stigma is pollination; the means of transfer vary in different crops. The stigma captures the pollen for which it has either branched stigmatic surface or sticky surface. Mature pollen grains germinate on the stigma and the pollen tube formed passes through the style of the flower and enters through micropyle. By the division of generative nucleus of pollen, two male gametes or sperms are formed which move along with the growth of pollen tube and enters into the embryo sac. Within the ovule, each megaspore mother cell undergoes the same type of meiotic division as in microspore mother cell to get egg. After the two sperms are ejected into the embryo sac, one sperm fuses with the egg cell to form zygote, and the other fuses with the secondary nucleus formed by the fusion of polar nuclei to form endosperm.

Self Pollinated Crops

Self pollinated crops are crops of self fertilizing species, which are basically homozygous identical genotypes (autogamous). Self pollination is the transfer of pollen from anther to stigma of the same flower or to the flower in the same plant or within the same clone. The structure of flower itself will give a strong indication of mode of pollination. In self pollinated crops, the flower will be of cleistogamous type (Hayes, 1942). At the time of anthesis, the position of anthers and pistil and the time of their maturity will give an idea about mode of fertilization. In a bisexual flower, if stigma is enclosed by stamina column, the mode of pollination will be self.

Self fertilizing method of reproduction is considered so effective because of decreasing genetic variation and fixation of highly adapted genotypes; high rate fixing of most of the loci,and in mixed stands of self and cross pollinating crops, the self fertilizing plants can donate pollen to both plant types unlike the cross fertilizing plants. Choice of parents and the identification of the best plants in segregating generations with definite goals are the critical steps in the improvement of self fertilizing crops.

Common Self Pollinated Crops

Barley	Jute	Potato	Vetch
Bean	Lentil	Rice	Wheat
Black pepper	Millet	Sesame	
Chick pea	Mungbean	Soyabean	
Cow pea	Oat	Tobacco	
Crambe	Peanut	Tomato	
Linseed	Pea		

In these crops, the extent of cross pollination varies from none to 4 or 5 per cent. A breeder working with self pollinated crops should be aware of this fact. In order to determine the percentage of cross pollination, two different cultivars pure for different forms of an easily recognizable and inheritable trait are planted. Plants of cultivar having recessive character of the trait are planted surrounded by plants of cultivar having dominant trait. The seeds collected from plants recessive for the trait are grown and plants having dominant trait are noted and percentage of natural cross pollination estimated (Sleper and Poehlman, 2006).

Mechanisms Favouring Self Pollination

Floral mechanisms which favour self pollination and prevent cross pollination are:

1. Pollen grains shed before flower opens
2. Flowers do not open at all
3. Shortly after the anthers open, stigma grows through the stamina column
4. Stamens and stigma are enclosed in a floral part after flower opening

Genetic Significance of Pollination Method

In self pollinated crops, the plants will be homozygous because, upon self pollination loci with identical genes will remain homozygous and loci with contrasting genes will segregate and produce homozygous and heterozygous progeny in equal number. With each successive self fertilization, heterozygosity is reduced by 50 per cent. After successive generations of self pollination, the proportion of heterozygous loci becomes very low. Even though a 100 per cent theoretical homozygosity is not possible, plants selected from a mixed population or progenies obtained after hybridization and selection will take five to eight generations of selfing to attain practical state of homozygosity so that the progenies will be uniform in appearance and performance. Breeding procedures in self pollinated species depend upon the genetic makeup of the population. A mixed population of self pollinated crops consists of plants with different homozygous genotypes. If single plant progenies are raised separately they will be uniform even though the populations differ from each other. In a self pollinated population, a heterozygous progeny may appear either due to cross pollination between two different homozygous plants or due to mutation.

Improvement Methods in Self Pollinated Crops

Improvement methods in crop plants include, selection from existing variability for a particular trait, release of genetic variation or create new gene combination through hybridization which in turn include inter varietal hybridization (crossing within the germplasm pool) and interspecific hybridization (crossing with foreign genotypes), followed by segregation; creation of new genetic variability through mutagenesis and polyploidization; introduction of alien DNA (transgenes). The choice of breeding method depends on the objective of the breeding programme, inheritance of the trait, and reproductive system of the crop species. The major groups of methods of breeding are described below

Introduction

In olden days, plant breeders objective was to increase yield by selecting superior genotypes growing elsewhere in the country by introduction or making selection from existing land races. In this method, crop species and varieties are obtained from other countries. Many crops and varieties growing in one country are introductions. Soybean is a famous introduction crop to the agriculture of United States of America.

Selection

Techniques of Selfing

In controlled selfing, two methods are adopted. One method is space isolation in which plants are kept apart at enough space to prevent chance of cross pollination. The distance of isolation varies depending upon crop, weather condition and natural barriers to the spread of pollen. The other method to ensure self pollination is covering with paper or cloth bags.

Pure Line Selection

According to Johannsen pure lines are descendants of a single, homozygous, self fertilized organism. The seeds from selected plants are not added together unlike in mass selection, instead separately sown as lines or offspring families to perform offspring tests. This is done to study the breeding behaviour of the selected plants.

Progenies of individual plants of self pollinated crops breed true to type. Two main sources of selection are available for production of new varieties.

1. Introduction of improved or relatively unimproved strains or varieties cultivated in a wide range of climatic and soil conditions, both domestic and exotic.

2. Well adapted local cultivars which contain varying biotypes. Origin of these may be hybrids or pure lines which altered through mechanical mixtures, natural crossing, or mutations.

The rice varieties series Ptb 1 to Ptb 34 from Regional Agricultural Research Station, Pattambi, Kerala, India are pure line selections.

Mass Selection

Mass selection is applied to a limited extend in self fertilizing plants and is an effective method for the improvement of land races. As this method of selection depends mainly on phenotype and performance of plants, it will only be effective for highly heritable traits. This method is used to improve the overall population. The seeds from selected plants are bulked for the next generation. The disadvantage of this method is that, if there is strong genotype x environment interaction for the trait of interest, selection based on phenotype will lead to low improvement.

Hybridization and Selection

When it is realized that the existing variability was exhausted, the breeders turn their attention to hybridize between different genotypes to get new recombination of different desirable genes scattered in a single genotype.

The detailed outline of the steps to be followed in a hybridization programme are detailed below:

Objectives of Crossing

Objective of crossing will be combining different traits in different varieties or species. Sometimes the recombination of genetic factors leads to formation of new desirable characters not present in parents. Occasionally, transgressive segregation occurs in quantitative characters like yield, earliness, height of plant, and resistance to lodging. Selection of parents that is already relatively satisfactory for these characters will enhance the probability of obtaining desired end result (Hayes, 1942).

Selection of Parents

Initially suitable parents having desirable traits will be selected. Selection of type of breeding depends upon availability of traits in genotypes. If the trait of interest is present in the same species, inter-varietal hybridization, and if the trait is present in different species, inter-specific hybridization will be carried out. Another point a breeder has to concentrate while selecting parent is that, those genotypes frequently used in hybridization programme should be avoided to prevent the narrowing down of genetic base. Always we have to concentrate to widen the genetic base of progenies by selecting proper parents for hybridization. If more varieties having desirable traits are available as parents, breeder can consider the combining ability, ability to transfer the trait to the progenies. Bauer and Leon (2008) reported that using breeding values in parental selection in self pollinated crops seems to be superior to conventional selection strategies, where selection is often based on several traits which are correlated among each other.

Purification of Parents

Even in case of strict self pollinating crops, there is chance of small percentage of cross pollination (2-5). Hence after selecting parents, they will be subjected to one generation selfing to eliminate the possibility of effect of crossing happened in the past. These kinds of pure breeding parents are subjected to hybridization.

Demarking Male and Female Parents

Demarking male and female parents from the selected parents is also very important. The parent in which the trait of interest is present, is an otherwise inferior one for other desirable traits, has to be taken as male parent (donor parent). The variety which is good for all desirable characters except the character we are looking for is available, is taken as female parent (recurrent parent). If both parents are important for one or another character, it is better to take the one with recessive character as female parent and the plant with dominant character as the male parent so that selfs can be discarded when F_1 is grown.

Making Crosses Between Pure Breeding Parents

If so many good recurrent parents are available we can go for line x tester crossing (one donor crossing with a series of recurrent parent).If a good recurrent parent is not available, all are having the trait we are looking, but agronomically not good, we can go for diallel crossing (all possible combination of crossing). Hybridization can be done either in the field or green house. Green house is better because it protects from foreign pollen and attack from birds. Moreover, crosses in green house can be done during any time of the year. If the parents differ in their duration, different planting time should be adopted to get synchronized flowering. Sowing of seeds can be done either in rows or in pots with sufficient space within row and between rows.

Handling the Hybrid Materials

As the F_1 seeds are precious materials, care should be taken while harvesting and sowing of the F_1 seeds. F_1 seeds are sown either in pots or in field with sufficient spacing.

Selection from Segregating Generation

If pedigree method is adopted, the F_2 and succeeding generations are grown in spaced planted progeny rows of 25 -50 plants. Superior recombinant lines having desirable traits will be selected in the succeeding generations. The parents should be grown after 10- 30 progeny rows so that frequent comparison can be done while selecting progenies.The best available standard also can be planted. Selection for special traits like disease resistance, plant height, heading time, colour of glumes, type of awn, lodging habit etc. is carried out in F_3 generation onwards. For disease resistance trait, the screening should be begun F_2 generation onwards. In the case of biotic resistance breeding programmes, separate nurseries are grown for each generation for pest and disease screening. Lines from duplicate nursery grown under normal conditions are selected discarding those lines susceptible in the disease nursery. Depending upon the breeding objective, quality test can be started at any time possible. Special characters may be studied by growing the generations from F_3 onwards under special environment that will bring desirable differentiations.

Techniques of Crossing

Depending on crop and prevailing environmental conditions, crossing different strains requires special crossing techniques. Understanding structure of flower is very important. Important features of techniques of crossing summarized by Hayes and Garber (1921) are as follows:

1. Carefully study the structure of flower

2. Determine which flowers produce the larger, healthier seeds and which set seeds the most freely.

3. Learn the normal time and method of blooming of flowers, and length of time the pistil remains receptive and the pollen grains capable of functioning.

4. Procure the necessary instruments and see that these are of efficient kind for the work to be undertaken.

5. Be careful not to injure the flowering parts any more than is necessary. Do not remove the surrounding flower parts *e.g.*, petals, glumes etc., unless necessary.

6. A few crosses carefully made are of much greater value than many pollinations carelessly executed.

General Procedure of Artificial Hybridization

1. *Emasculation*: Removal of anthers from female parent.

2. *Pollination*: Transferring pollen grains collected from matured flowers of selected male parent.

3. *Bagging*: After pollination, the flowers will be covered with cloth bag or butter paper bag to prevent contamination with foreign pollen. The bag can be removed after setting period or can be kept till harvest. Direct comparison of progenies of both male and female parents with that of F_1 and segregating generations of crosses are desirable when genetic studies are to be carried out.

In addition to increase of yield, a breeder also concentrates to incorporate tolerance/resistance to biotic and abiotic stresses, and quality/improved nutritional traits. These can be achieved either through intervarietal hybridization, if the trait is present in another variety which is otherwise less yielding, or through interspecific hybridization, if the trait is present in different species. For example, saline tolerant rice cultivars were developed through intervarietal hybridization between high yielding saline susceptible varieties and saline tolerant low yielding land races (Vanaja *et al.*,2009). Interspecific hybrids of black pepper having partial resistance to *Phytophthora* foot rot disease were developed by hybridizing high yielding, *Phytophthora* susceptible black pepper variety with a wild species of black pepper resistant to *Phytophthora* foot rot (Vanaja *et al.*, 2008). While breeding for yield, and other biotic/abiotic stress resistance, quality of the produce should be given due importance before release to commercial cultivation.

Different Methods of Breeding Self Pollinated Crops through the Method of Hybridization

1. Pedigree selection
2. Bulk population
3. Single seed descent
4. Back cross
5. Multiple cross
6. Dihaploidy

Pedigree Selection

F_1 hybrids obtained by crossing selected parents are subjected to natural selfing to produce segregating F_2 generations. The size of segregating generation depends how complex the trait is. Large size is essential for complex traits. The offsprings of selected progeny in the F_2 generation are planted in separate lines. Single plants are taken as source of new offspring lines, until genetic uniformity is reached. Seed yield of several plants per line is combined and mixed to have more seed for testing. In the early phase of selection, usually traits with high heritability which are quick and easy to measure are concentrated. In the later phase of selection, when plots are the units of assessment, traits like yield with lower heritability are assessed and taken as basis for selection. This method is very labour intensive. The early elimination of inferior populations and subsequent concentration of selection efforts within superior populations are assumed to result in increased efficiency. The success of this method depends on accurate evaluation of heterogeneous populations.

Bulk Population

This method is simple and cheap and involves less work than pedigree selection in the earlier generations. This type of selection is applicable with crops which are usually planted at high planting densities, *e.g.* small grain crops. It is necessary to plant large populations to ensure that the better segregates are selected. Segregating generations are subjected to another single plant selection step.

Single Seed Selection

In this method, the entire F_2-generation is transformed into a generation of homozygous plants (true-to-type breeding). This method is hence used to decrease the time that is passing with genotypes not yet being homozygous, but does not eliminate weak plants early such as in other methods. There is no selection of superior plants in the F_2 generation. Record keeping is not necessary in early generations and modification is possible.

Backcrossing

It is a selection method for the upgrading of genotypes. This is a type of repeated crossing and selection, in which F_1 hybrid, obtained by crossing an otherwise desirable variety(recurrent parent) with a donor parent (wild or exotic)having the particular desirable trait(*e.g.* pest/disease resistance) which is lacking in the recurrent parent, is crossed with the recurrent parent (Hayes, 1942). Grow 50 -200 back crossed seeds in spaced progeny rows. After each backcross, those hybrid plants which 'still' harbour the desired genes are identified and selected and are backcrossed again with the recurrent parent. In some cases, depending the nature of trait, the progenies of the selected plant need to be studied before proceeding to next back cross. After 2 -6 back crossings and selection, the population is handled like that of pedigree. This technique is easy when the traits of interest are easily inherited, dominant and easily identified in hybrid plants and also free from linkage with genes of undesirable characteristics. One advantage of the back crossing method is that extensive testing, for other than the desired trait, is not necessary. Marker assisted backcrossing is routinely applied in breeding programs for gene introgression.

Multiple Crosses

In this method suggested by Harlan and Martini, compound crossing is done. This method of crossing is advocated when desirable traits are spread in different cultivars and these cultivars have several undesirable traits also. In the first phase, individual crosses will be done followed by double crossing with different F_1s and finally double hybrids are subjected to one more crossing. In multiple crossing, large segregating population needs to be grown.

Doubled Haploid

In this method, haploids are produced by chromosome elimination in wide crosses, ovule culture or by anther and micro spore culture, and chromosome numbers of the haploid plants are doubled with use of colchicine. Compared with chromosome elimination and ovule culture, anther and micro spore culture are mostly used because of their ability to produce haploid plants with much larger quantity. Genetic improvement cycles can be shortened in this method in comparison to pedigree or bulk methods. Like the single seed selection method, early generations are not subjected to selection, but most of the lines are eliminated during field evaluation trials. As in single seed selection method, the genetic material from F_1 is transformed as fast as possible into a bulk of homozygous plants. Doubled haploid technology enhances 'forward breeding' by allowing hybrids to be bred with new traits without locking up the germplasm, and by without negative side effects. This method is very labor intensive and the most expensive of the procedures which increase the number of generations per year. The plants must be genetically stable for the success of this method.

Methods Employed in Selfing and Crossing of a Few Important Self Pollinated Crops

Rice

The different methods of emasculation in rice are summarized below.

1. Immersion of heads of rice in water at 40 -44°C for 10 minutes destroys the viability of pollen without injuring other floral organs as suggested by Jodon (Hayes, 1942). Treatment at 0 - 6°C, also gives result but less effective. Large mouthed thermous jug containing hot water was used and treatment done in the morning prior to normal anthesis. According to Jodon, emasculation by hot or cold water eliminated injury to the glumes, which germinated well, was obtained when florets so emasculated were pollinated.

2. Another method of emasculation is clipping away the 6 anthers using a small forceps through a slanting opening made by clipping the uppermost part of lemma. This process is done in the evening or morning before the anthers shed their pollen.

3. Wet cloth method: It is done during morning before the normal blooming time of heads. The head which blooms on that particular day will be selected and held it in a wet towel in our hand and blow using our mouth, through the tip of the head for a few minutes. Due to the combined action of temperature of the blowing air and presence of water in the towel, a hygroscopic pressure develops which leads to force opening of glumes before the normal anthesis time. After removing the anthers from the opened florets counting 6 numbers, those florets which do not open immediately below the emasculated florets were clipped off to

distinguish between pollinated and non-pollinated seeds after maturing the grains. From my experience of rice breeding, I feel this method is the better one for the best result.

The florets are pollinated after breaking mature anthers collected from male parent within the emasculated florets. Those pollinated panicles are covered with butter paper bag with the help of a supporting stake for a period of 2-3 days.

Wheat, Oats and Barley

Hayes and Garber (1921) reviewed that the period from 5 pm to 7 am was the right time for emasculation and pollination. According to Sumenson, chilling wheat plants for periods of 15 – 24 hours, at -2.7 to 2.2°C, resulted in a marked reduction of self fertilized florets through killing of pollen (Hayes 1942). Different varieties showed variation in tolerance to chilling. From the heads of plants of female parents, all but 8- 15 florets on a spike are removed before anthesis. Anthers are removed from the remaining flowers using a small forceps before anther dehiscence, and the head is enclosed in a paper bag. About two days after, matured anthers are collected from selected male plants, crushed and applied on the stigma of the emasculated florets. Paper bags are allowed to remain on the pollinated spike till harvest.

Potato

It is wise to enclose the flower clusters in small cloth bags in selfing the potato for careful genetic experiments (Hayes, 1942).Otherwise bagging is unnecessary. Emasculation is done by removing the anthers with a small forceps or scraping off the anthers with a small knife. Pollination is effected by tapping a flower of the male parent gently,so that pollen is spilled on to the thumbnail and then applied to the stigma of the emasculated flower.

Rye

For self pollination in Rye, several heads of a plant are enclosed in a parchment bag before blooming. An eyelet can be placed on the top of the bag and tie to a stake for support. Bags are kept until harvest. For effecting control crosses, the same technique applied in wheat, Oats and Barley may be used.

Mutation Breeding

In mutation breeding, seeds, pollen, or vegetative propogules are treated with either physical mutagen (X- or gamma rays) or chemical mutagen (EMS - Ethyl Methyl Sulphonate).Applications of mutation methods become more complex when mutations from recessive to dominant trait or traits under polygenic control would be the target. Rutger pointed out that mutation breeding in rice, a self fertilizing crop, resulted in more cultivars than in any other crops. In rice, Rutger also mentioned that chemical mutagens produced higher mutation frequencies but strongly advocated the use of ionizing radiation because of practical problems encountered with chemical mutagens (Van Harten, 1998).

Analysis and Exploitation of Variation

The heritable genetic variations like additive, and additive x additive interaction are fixable, but dominant, and dominant x dominant interactions are non fixable. A survey of estimates of different components of variations (Roy, 2000) has shown that the additive genetic component predominated in autogamous species. In self fertilizing crops, the interactions heterozygote x heterozygote, and heterozygote x homozygote are less important than homozygous x homozygous interaction. The fixable type of variation can be exploited under two situations.

1. When this type of variation is already present in the population: In a population of pure breeding genotypes of self fertilizing crops based on progeny performance test,the better lines are selected.

2. When variation is created in the population: In this case, variability is created by inter varietal/interspecific/inter generic hybridization, mutation or polyploidy. Dominance and other epistatic type of variation occurring along with additive variation in the early generation segregating population are non fixable, and are not utlilized and their presence impedes selection. Pure line breeding makes use of additive, and additive x additive type of variations.

Breeding for Biotic Stress Resistance

Even though breeding for biotic stress resistance is similar to other traits, the breeder has to deal with two series of heritable factors: (1) the heritable characters of the host plant and (2) heritable differences in the organism. The early generations of selection or segregating progenies are grown in controlled epidemic conditions, so that the resistant strain can be selected and others discarded.

Breeding for Abiotic Stress Resistance

Of all the parameters which limit crop production, abiotic environmental stresses contribute most significantly to reduction in potential yield (Flowers and Yeo, 1995). Abiotic stresses arise from extremes of climate. The stress factors especially drought and salinity negatively effect plant growth and development. Progress in the development of salt/water stress tolerant varieties is slow because, crop responses to salt/water stress are made up of a number of complex and interrelated, morphological, physiological and biochemical process;Water/salt stress phenomenon and the mechanism of Water/salt stress resistance interact; Stress tolerance and high yields are incompatible. Yield in the absence of salt or drought itself is a complex trait influenced by many component traits. Further, under these stress conditions, these traits will interact with the stress environment and the resistance mechanisms which make breeding for yield under these abiotic stresses really complicated.

Breeding Self-Pollinated Crops through Marker Assisted Selection

In traditional plant breeding, improvement methods include development of segregating germplasm and the selection of those genotypes with the best performance for traits of interest through direct trait evaluation. The efficiency of selection can be improved by Marker-assisted selection (MAS). In MAS, breeders select for traits of interest through genetic markers linked to genes controlling the trait or through backcrossing for the recurrent parent genome (Diers, 2004). For some traits, MAS could potentially increase the efficiency of selection compared to phenotypic selection.

Participatory Plant Breeding (PPB)

In order to improve the suitability of the varieties produced to specific local farming situations, the new approach recommends farmer participatory varietal selection for breeding programmes (Bennet and Khush, 2003). Participatory Plant Breeding is the latest strategy in the area of plant breeding to integrate end users based participatory approach. Participatory Plant Breeding is based on a set of methods that involve close farmer –researcher collaboration to bring about plant genetic improvement within a crop. It is expected to produce more benefits than the traditional global breeding model in situations where a highly centralized approach inappropriate (Morris and Bellon, 2004). Participatory plant breeding methods designed to incorporate the perspective of farmers- usually by inviting farmers to participate in variety evaluation activities, realizing the fact that modern varieties developed for favorable production conditions have not always diffused readily into marginal environments. By involving farmers in the genetic improvement process, plant breeding programmes will be able to produce better varieties which will be adopted more widely and generate greater benefits on aggregate. PPB provides a means of assessing so-called 'subjective traits'. In food crops these include taste, aroma, appearance, texture and other characteristics which determine the suitability of a particular variety for culinary use. These traits are difficult to measure quantitatively because they are a function of human perceptions.

A recent achievement in salinity resistance in rice, incorporating participatory plant breeding (Vanaja *et al.*, 2009) is briefly explained. In order to get saline resistant high yielding varieties, inter-varietal hybridization was carried out between saline susceptible rice varieties, and saline tolerant traditional cultivars. As there was no reliable laboratory based selection criteria for salinity resistance, all filial generations were raised directly in the problem area ensuring farmers' participation for selection of promising genotypes suitable to the saline flooded areas.

References

Bauer, A.M. and Leon, J., 2008. Multiple trait breeding values for parental selection in self-pollinated crops. *Theoretical and Applied Genetics*, 116(2): 235–242.

Bennet, J. and Khush, G.S., 2003. Enhancing salt tolerance in crops through molecular breeding: A new strategy. In: *Crop Production in Saline Environments*, pp. 11–65.

Dabholkar, A.R., 2006. Reproductive systems in crop plants. In: *General Plant Breeding*. Concept Publishing Company, New Delhi, India.

Diers, B.W., 2004. Breeding self pollinated crops through marker assisted selection. In: *Encyclopedia of Plant and Crop Science*. Affiliation to University of Illinoise, Urbana, Illinoise, USA.

Flowers, T.J. and Yeo, A.R., 1995. Breeding for salinity resistance in crop plants: Where next? *Aust. J. Plant Physiol.*, 22: 876–884.

Hayes, H.K., 1942. A classification of methods of breeding sexually propagated plants. In: *Methods of Plant Breeding*. McGraw-Hill Book Company Inc., pp. 56 –57.

Hayes, K. and Garber, J., 1921. *Breeding Crop Plants*. McGraw Hill Book Company, Inc., New York, p. 328.

Morris, M.L. and Bellon, M.R., 2004. Participatory plant breeding research: Opportunities and challenges for the international crop improvement system. *Euphytica*, 136: 21–35.

Roy, D., 2000. *Plant Breeding: Analysis and Exploitation of Variation*. Alpha Science International Limited, Pangboum, UK, p. 701.

Sleper, D.A. and Poehlman, J.M., 2006. *Breeding Field Crops*, 5th edn. Blackwell Publishing, p. 417.

Vanaja, T., Neema, V.P., Mammootty, K.P. and Rajeshkumar, R., 2008. Development of a promising interspecific hybrid in black pepper (*Piper nigrum* L.) for *Phytophthora* foot rot resistance. *Euphytica*, 161(3): 437–445.

Vanaja, T., Neema, V.P., Mammootty, K.P., Rajeshkumar, R., Balakrishnan, P.C., Jayaprakash Naik and Raji, P., 2009. Development of first non-lodging and high yielding rice cultures for saline *Kaipad* paddy tracts of Kerala, India. *Current Science*, 96(8): 1024–1028.

Van Harten, A.M., 1998. *Mutation Breeding: Theory and Practical Applications*. The press syndicate of the University of Cambridge, p. 353.

2012, Nutri-Horticulture
Editor: Professor K.V. Peter
Published by: DAYA PUBLISHING HOUSE, NEW DELHI

Pages 155–169

Chapter 9

Biochemistry and Physiology of Latex Production in Rubber (*Hevea brasiliensis*)

N. Usha Nair, Molly Thomas and S. Sreelatha

Rubber Research Institute of India,
Kottayam, Kerala, INDIA
E-mail: usha@rubberboard.org.in

Hevea brasiliensis, the most important commercial source of natural rubber is a forest tree indigenous to the tropical rain forests of Central and South America and belongs to family *Euphorbiaceae*. Natural rubber (*cis*-1, 4-polyisoprene) is a vital raw material used in the manufacture of a wide range of industrial and domestic products in the medical, automobile and defense industries. There are over 2000 species of plants from about 300 genera as well as two fungal genera (Archer *et al.*, 1963a; Backhaus, 1985) known to contain rubber, in their latex but most of them make short chain rubber molecules which are poor in quality for commercial use.

The physiology of *Hevea brasiliensis* is unique. Biosynthesis of latex, the economic product, is confined to the latex vessels which occur exclusively in the phloem region. The latex vessels are derived from the cambium and are arranged as concentric rings in the bark. Between the vessels in each ring, there are anastomoses which allow withdrawal of latex from a large area of bark by means of a single tapping. Latex biosynthesis depends on the number, diameter and anatomical characters of latex vessel system and physiological and biochemical factors. The capacity of the latex vessels to synthesize and regenerate latex drained during each tapping is critical and is accomplished in the interval between two successive tapping. Figure 9.1 represents the schematic three-dimensional image of latex vessel distribution in the bark of *Hevea*.

Composition of Latex

Latex is a specialized form of cytoplasm containing a suspension of rubber and non-rubber particles in an aqueous serum (Frey- Wyssling, 1929; Southorn, 1961; Archer *et al.*, 1969). Latex contains all the sub cellular organelles of non-photosynthetic cells like vacuoles, plastids, mitochondria, nuclei, endoplasmic reticulum and ploysomes (d'Auzac and Jacob, 1989; de Fay *et al.*, 1989). Besides rubber

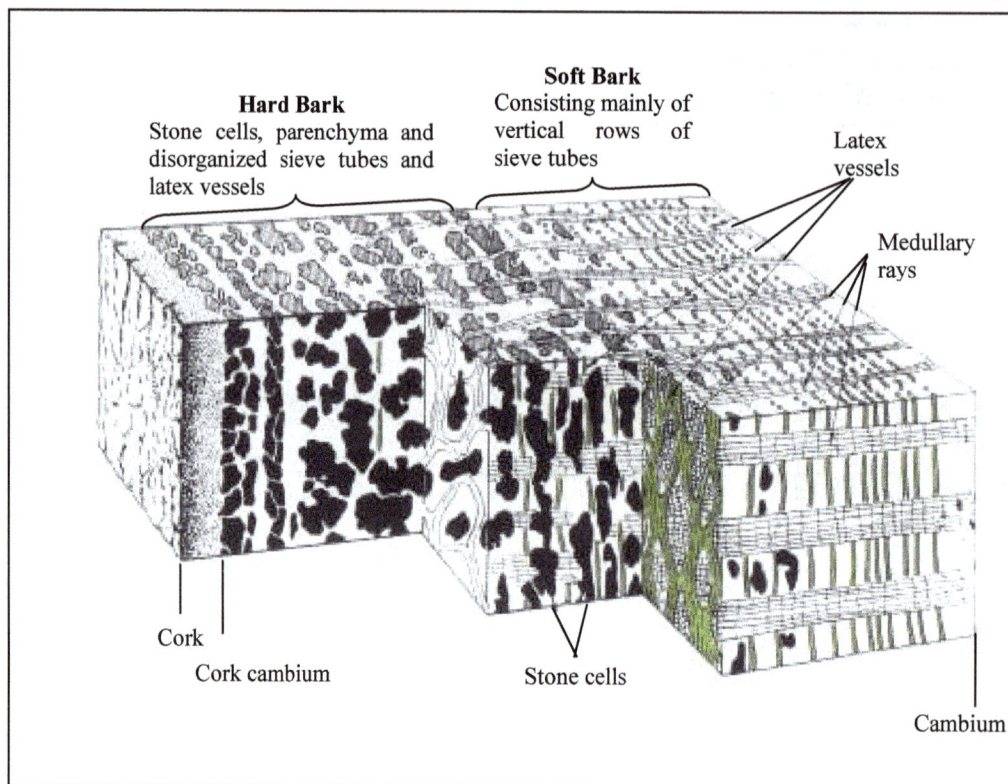

Figure 9.1: The Three-dimensional Diagram of *Hevea brasiliensis* bark (x10).
Adapted from Gomez (1982).

and water, fresh latex contains carbohydrates, proteins, lipids and inorganic salts (Archer *et al.*, 1963b).

Using ultra centrifugation at 59,000g for one hour, latex can be separated into a white upper layer of rubber particles, an orange or yellow layer containing Frey-Wyssling particles, an aqueous serum named C - serum and a bottom fraction containing greyish yellow gelatinous sediments (Cook and Sekhar, 1953). The serum contains most of the soluble substances including amino acids, proteins, carbohydrates, organic acids, inorganic salts and nucleotide materials (Archer *et al.*, 1969). The bottom fraction consists mainly of lutoid particles and also includes varying amounts of Frey-Wyssling particles, mitochondria and other particulate components of plant cell having a density greater than that of serum.

Rubber particles make up to 30 to 45 per cent of the volume of latex and are found as spherical or pear shaped particles (Dickenson, 1969) with sizes ranging 0.02 to 3.00 μm with the majority in the region of 0.1 μm (Southern and Yip, 1968; Tata and Yip, 1968). Rubber particles are strongly protected in suspension by a film of adsorbed protein and phospholipids (Archer, 1964). This protein-phospholipid layer imparts a net negative charge to the rubber particle contributing to colloidal stability (Bowler, 1953). The lutoid particles amounting to 10 to 20 per cent of the volume are sub-cellular membrane bound bodies ranging in size from 2 to 5 μm (Southorn and Yip, 1968). The membrane

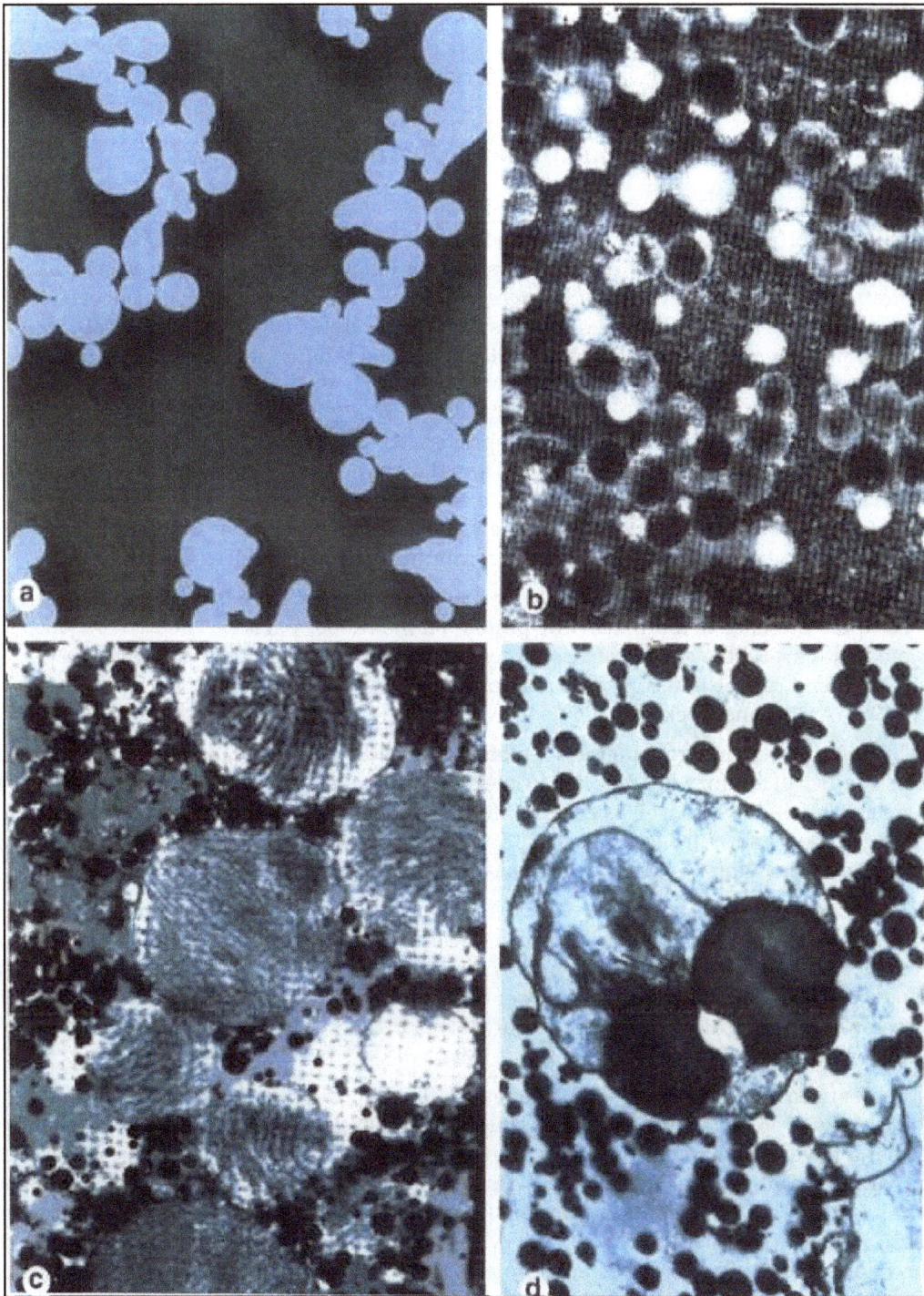

Figure 9.2: Rubber and Non-rubber Particles in Latex
(*a*) Rubber particles; (*b,c*) Lutoid particles; (*d*) Frey-Wyssling complex.

encloses a fluid serum known as B-serum composed of proteins, ions, hydrolases and defense proteins (Southorn and Yip, 1968). Frey-Wyssling particles, enclosed in a typical double membrane, are spherical and yellow coloured, ranges in size from 4 to 6 μm and constitute one to three per cent of volume (Dickenson, 1964).

Organic Non-rubber Constituents of Latex

Carbohydrates

Quebrachitol (methyl-inositol), sucrose and glucose are the major soluble carbohydrates in latex (Low, 1978). Quebrachitol is the most concentrated single component in the serum phase. The concentration of quebrachitol varies with clones and ranges from one to three per cent of whole latex. It is a major contributor to the osmotic pressure of the cytosol (d'Auzac and Jacob, 1989). The concentration of sucrose in latex also varies with clones and is influenced by yield stimulation.

Proteins

The total protein content of fresh latex is approximately one per cent of which about 20 per cent is adsorbed on rubber particles, an equal quantity found in the bottom fraction and the remainder in the serum phase (Archer *et al.*, 1963b). The existence of a lipo-protein envelope on the surface of rubber particles was established by Bowler (1953). The adsorbed proteins and phospholipids on the rubber particle impart a net negative charge thereby contributing to the colloidal stability of latex.

The isoelectric point of rubber particles in fresh latex varies from 4.0 to 4.6 and the variation is ascribed to the presence of more than one protein on the rubber particle. The two enzymes found associated with the rubber surface are isopentenyl pyrophosphate polymerase (Archer *et al.*, 1963a; Lynen, 1967) and rubber transferase (Archer *et al.*, 1963b; 1966; McMullen and McSweeney, 1966; Lynen, 1967; Archer and Cockbain, 1969).

The major protein in serum phase is α-globulin (Archer *et al.*, 1969; Hahn *et al.*, 1971). C - Serum contains nearly half the enzymes detected in latex. Enzymes of the glycolytic pathway (Bealing, 1969; d'Auzac and Jacob, 1969) as well as enzymes for isoprenoid pathway (Witisuwannakul *et al.*, 1990 and Suvachitanont and Witisuwannakul, 1995) are found in C - serum. Twenty seven enzymes were separated by electrophoresis of which, 17 existed in multiple forms (Jacob *et al.*, 1978).

Proteins in the bottom fraction are the soluble proteins in lutoid serum (B - serum) and in the lutoid membrane. Hevein is the major protein in B - serum which accounts for about 70 per cent of the water soluble proteins in the bottom fraction (Archer *et al.*, 1969). It is a low molecular weight anionic protein (about 5000 Da), having about five per cent sulphur (Archer, 1960; Tata, 1975; 1976), with a single chain of 43 amino aids, rich in cysteine and glycine. It is involved in the coagulation of latex by bringing together rubber particles (Gidrol *et al.*, 1994).

Lutoids from young latex vessels contain a protein deposited in the form of bundles of micro fibrils each having probably a double helical structure (Archer *et al.*, 1963b). This microfibrillar protein, which has a lower isoelectric point than hevein, comprises about 40 per cent of the total protein of lutoid particles in young latex vessels but is absent in older vessels. Two basic proteins, a major one identical with hevamine A, a cationic protein (Archer, 1976) and a minor one have lysozyme and chitinase activities (Tata, 1980; Tata *et al.*, 1983). The basic proteins of bottom fraction are involved in the mechanism of latex flow. Some defense proteins like chitinase, β- 1, 3 glucanase and lysozymes are accumulated in this fraction.

Lipids

Lipids of fresh latex consist of fats, waxes, sterols, sterol esters and phospholipids. Lipids associated with rubber and non-rubber particles in latex play a vital role in the stability and colloidal behavior of latex. There is distinct clonal variation in the amount of lipids extractable from rubber cream and bottom fraction (Nair *et al.*, 1993). Total lipids constitute about 1.6 per cent of latex, of which 54 per cent constitute neutral lipids, 33 per cent glycolipids and 14 per cent phospholipids (Hasma and Subramaniam, 1986). Triglycerides and sterols are the main components of the neutral lipids of rubber particles. Lutoid stability as indicated by bursting index, is negatively correlated with the phospholipid content (Sherif and Sethuraj, 1978). The constituents of the phospholipids are mainly phosphatidylethanolamine (PE), phosphatidylcholine (PC) and phosphatidylinositol (PI).

Nucleic Acids and Polysomes

Hevea latex contains both ribosomal RNA and soluble RNA, DNA and messenger RNA (Tupy, 1969a). Polyribosomes were also discovered in the serum phase of latex (Coupe and d' Auzac, 1972) functional with regard to protein biosynthesis and contributed to the regeneration of the contents of the laticiferous cells between two tappings.

Other Constituents

Most of the classic amino acids are reported in latex. The major proportion of free amino acids is located in the cytosol. The predominant amino acids in the cytosol are glutamic acid, alanine and aspartic acid. Nucleotides in latex are important as co-factors and are intermediates in biosynthetic processes. Low molecular weight thiols, which are the main reducing agents in latex, include glutathione and cysteine. Ascorbic acid is also a very important reducing agent in latex. These two reducing components are involved in the redox potential of latex. Malic acid and citric acid make up 90 per cent of organic acids in latex. Total concentration of inorganic ions in fresh latex is about 0.5 per cent, the major ions being potassium, magnesium, copper, iron, sodium, calcium and phosphate (Archer *et al.*, 1963b).

Latex Metabolism and its Role in Natural Rubber Production

Latex harvesting (tapping) results in a loss of cell constituents from the laticifer. Latex vessels are filled with viscous latex under hydrostatic pressure usually between 10 and 15 atmosphere in the early morning. During latex harvesting when the vessels are cut, the pressure at the location of the cut is released and latex exudes. This expulsion results in the displacement of latex along the length of latex vessels owing to the strong forces of cohesion existing in the liquid phase. A fall in pressure in the latex vessels follows and consequently water from the surrounding tissues enters the latex vessels owing to gradient in water potential. This dilution makes the latex less viscous resulting in an enhanced flow rate. The turgor pressure gradually recovers as the latex flow slows down and reaches to normal when the flow ceases.

Latex Vessel Plugging

Latex contains destabilizing factors located in the lutoid particles. The dilution reaction following tapping alters the osmotic concentration of the latex. The lutoids are damaged by mechanical and shearing forces. Lutoid damage leads to release of lutoid serum which contains Hevein, a lectin like protein. Hevein induces coagulation of latex by forming bonds between rubber particles by fixing to the N-acetylglucosamine (Glc NAc) groups borne by a 22 kDa receptor protein localized in the surface of the rubber particle (Gidrol *et al.*, 1994). Chitinase in the lutoid helps to prevent this by releasing

GlcNAc moieties from the 22 kDa receptor to block the hevein binding site. Recently, a *Hevea* latex lectin (HLL) binding protein was isolated from C-serum. This protein has an anti coagulating role, important for maintaining the colloidal stability by preventing the coagulation of latex (Witissuwannakul, *et al.*, 2008).

Latex harvesting process (tapping) results in a loss of cell constituents from the latex vessels. Regeneration of the lost constituents takes place in the lacticiferous tissue between two successive intervals of tapping and increases with latex production. Regeneration mechanism can become a limiting factor for production if the interval between two tappings is short. Regeneration of latex requires an intense metabolic activity involving reconstitution of all the sub cellular elements with their enzymatic functions. This requires reactions which can provide energy or reducing capacity required for the anabolic process leading to isoprene synthesis.

Rubber Biosynthesis

The biochemical composition of latex very clearly indicates that the biological function of the laticiferous tissue is the synthesis of rubber which forms 35-45 per cent of fresh weight and over 90 per cent of dry weight of latex. Rubber from *Hevea brasiliensis* is a high molecular weight polymer composed of isoprene units linked together to form a polymer and the geometric configuration of double bonds is exclusively *cis* (Bunn, 1942; Golub *et al.*, 1962). Rubber biosynthesis was studied extensively especially in *Hevea brasiliensis* by (Archer *et al.*, 1963a; Archer and Cockbain, 1969; Archer and Audly, 1987; Light and Dennis 1989) and to a small extent in other natural rubber producing species such as *Parthenium argentatum* (guayule) (Cornish and Backhaus, 1990; Backhaus *et al.*, 1991) and *Ficus elastica* (Siler and Cornish, 1994).

The synthesis of *cis* poly isoprene can be divided schematically into two distinct phases. The first phase is the conversion of sugars into acetate or acetyl CoA the precursor (s) of the polymer. It simultaneously provides energy in the form of ATP and reducing power in the form of NADPH. The second is a chain of reactions in which *cis*-polyisporene is synthesized from acetate (or acetyl CoA).

Glycolysis is the major catabolic process of sugar in latex which produces acetate or acetyl CoA molecules, which are the precursors of *cis* polyisoprene and also of ATP and NADPH which are required in polyisoprene synthesis. The sequence of reactions occurs in the cytoplasm and certain stages control its activity. It is reported that invertase is a key enzyme in glycolysis and that it controls its intensity. Its activity is extremely pH dependent with an optimum of about 7.3 to 7.5. The enzyme involved in the synthesis of sucrose viz; pyrophosphate and sucrose synthetase can also effectively slow down glycolytic activity. Stimulation of *Hevea* by ethrel increases the pH of latex and the catabolism of sugars. This phenomenon is one of the factors for the increased production obtained with stimulation.

Rubber biosynthesis is the major synthetic process in latex since over 90 per cent of dry matter in latex is *cis*-polyisoprene. The individual steps in the synthesis of rubber from sucrose are well established (Lynen, 1969). Biosynthesis of rubber can be divided into three stages (1) generation of acetyl CoA (2) conversion of acetyl CoA to isopentenyl diphosphate via mevalonic acid (3) polymerization of IDP to rubber.

Generation of Acetyl CoA

Sucrose in latex is the primary source of acetate and acetyl CoA essential for the biosynthesis of rubber. Acetate forms the basic precursor of rubber biosynthesis in all rubber plants. Using acetate labeled with ^{14}C, it was proved that latex used as a medium converted it to labeled rubber, indicating presence of the entire enzyme system necessary for the conversion of acetate to rubber. The pool of acetyl CoA in latex is conceivably generated from sucrose and *Hevea* latex contains all the enzymes required for the metabolism of sucrose to pyruvate. Other possible sources of acetyl CoA are via, β- oxidation of fatty acids or via the metabolism of amino acids especially leucine.

Conversion of Acetyl CoA to Isopentenyl Diphosphate

Two molecules of acetyl CoA condense to form aceto acetyl CoA. Aceto acetyl CoA condenses with another molecule of acetyl CoA resulting in β-hydroxy-β- methyl glutaryl CoA (HMG-CoA). Mevalonic acid is derived from HMG-CoA in a nicotinamide adenine dinucleotide phosphate linked reduction (NADPH) which also occurs in *Hevea* latex. The activity of the enzyme HMG-CoA reductase in latex is surprisingly low and may be a limiting factor in rubber biosynthesis (Lynen, 1969). Three genes *hmg*1, *hmg*2 and *hmg*3 were discovered for the rubber tree of which *hmg*1 gene was considered responsible for rubber biosynthesis (Kush *et al.,* 1990). Mevalonic acid synthesis takes place on the endoplasmic reticulum (Archer, 1980).

Conversion of mevalonate to isopentenyl diphosphate (IDP) was confirmed to proceed via 5-diphospho - mevalonate in *Hevea* latex. The enzymes required for these conversions, mevalonate kinase and phoshpho mevalonate kinase are present in latex. All the other enzymes required for the conversion of HMG-CoA to IPP are also found in latex serum.

Polymerization to Rubber

The mechanism of polymerization of IPP is elucidated in relation to terpene biosynthesis (Lynen *et al.,* 1959). Two steps are involved in the process: (1) the first step in the biosynthesis of all isoprenoid from IDP is the isomerization of IDP to dimethylallyl diphosphate (DMADP) mediated by the enzyme isopentenyl diphosphate isomerase (IDP isomerase). It was suggested that the requirement for IDP isomarase in *Hevea* latex is very small (Archer and Audley, 1987). (2) condensation of DMADP with IDP by *cis* - polyprenyl transferase, to give a molecule each of pyrophosphate and gernyl pyrophosphate (C 10). This C-10 molecule has allelic structure and repeats the condensation with another molecule of IDP. The propagation repeated several times result in the formation of natural rubber with high molecular weight. It is the *cis*-1, 4 prenyl transferase or rubber transferase which is firmly associated with the rubber particles responsible for elongating the rubber particle. The stereo-specificity of rubber transferase enzyme in latex ensures a *cis* configuration for each double bond.

Siler and Cornish (1995) suggested that the production of initiator molecule is a very essential step in rubber biosynthesis. Recent evidence suggests that rubber initiation can proceed from the trans components geranyl diphosphate (C 10), farnesyl diphosphate (C 15) or geranyl geranyl diphosphate (C 20) implying that one of the terminals of the rubber molecule can be trans (Archer *et al.,* 1982). Trans prenyl transferase is the enzyme that catalyses the trans addition of IDP to DMADP to form short trans prenyl initiator molecules up to C 20.

The conversion of IDP takes place at the surface of existing rubber particles and this is predominantly a chain extension process on already existing rubber chains which carry allyl pyrophosphate end groups (Archer and Audley, 1967). Rubber transferase (*cis*-1-4-prenyl transferase) forms an integral part of the outer layer surrounding the rubber particle and acts as a conduit for the growing chain. The termination of rubber biosynthesis which is the release of rubber from the enzyme has not been studied in detail. In *Hevea*, the chain termination process might be quite consistent as the event yields poly isoprene with molecular weight distribution of about 100 to 1000 kDa (Subramaniam, 1980).

CO$_2$ and H$_2$O
↓
Intermediates
↓
Sucrose
↓
Glucose   +   Fructose   HMG-CoA
↓
G6P
↓
F6P
↓
FDP   Acetoacetyl-CoA   MVA
↓   ↓
GAP ⇄ DHAP   MVAP
↓   ↓
1,3- PGA   MVAPP
↓   ↓
3-PGA   IPP
↓
2-PGA   Acetyl-CoA
↓
PEP
↓
CO$_2$+ Acetaldehyde ← Pyruvate   Acetate   Rubber
↓↑   ↑↓   (from sucrose?)
Ethanol   Lactate

[+ATP]   [-ATP]

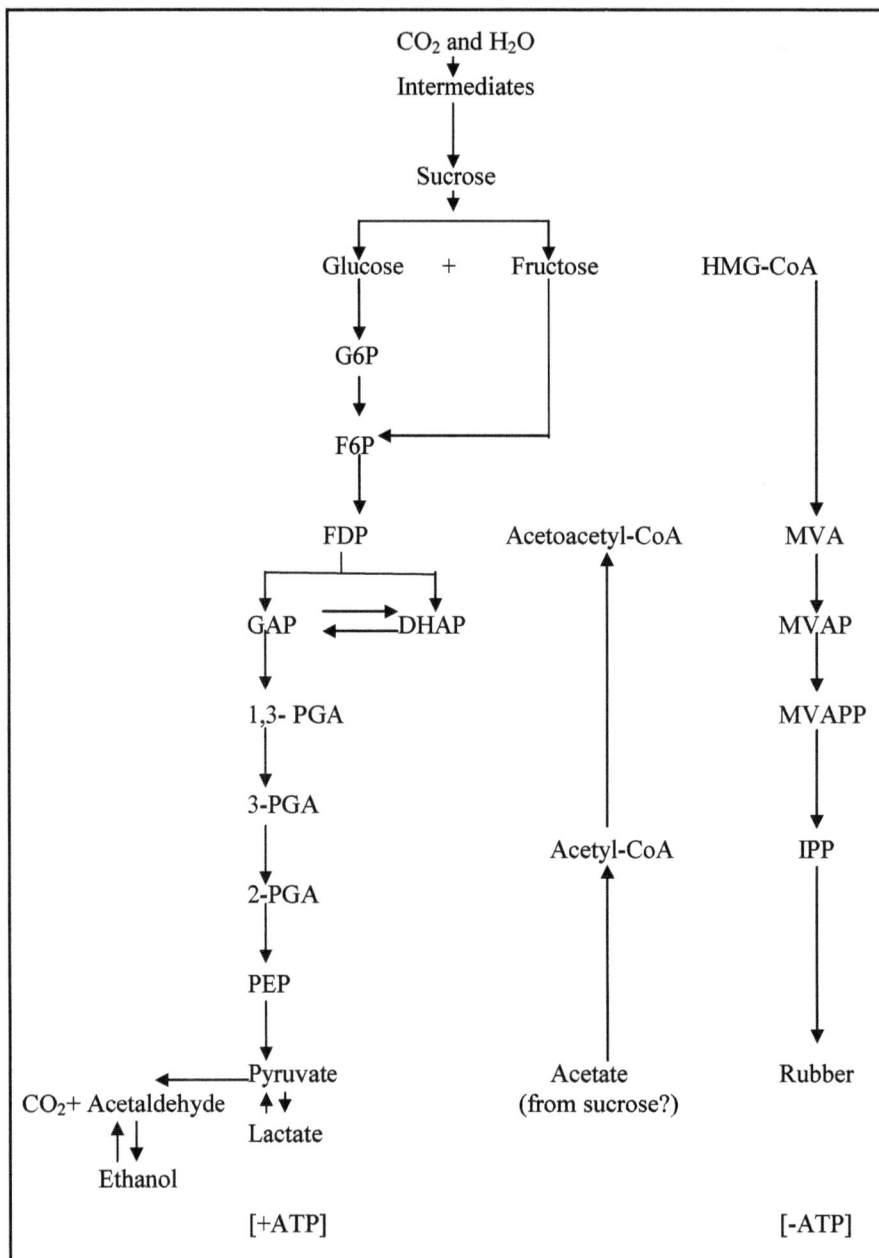

Figure 9.1: Pathway of Rubber Biosynthesis*

G6P: Glucose-6-phosphate; PEP: Phosphoenol pyruvate; F6P: Fructose-6-phosphatre; ATP: Adenosine triphosphate; FDP: Fructose-1, 6-diphosphate; Acetyl-CoA: Acetyl-coenzyme A; GAP: Glyceraldehyde-3-phosphare Acetoacetyl-CoA- aceotoacetyl-coenzyme A; DHAP: Dihydroxyacetone phosphate; HMG-CoA: β-hydroxy-β–methylglutaryl-Coenzyme A; 1,3-PGA: 1 - 3-phosphoglyceric acid; MVA: Mevalonic acid; 3-PGA: 3-phosphoglyceric acid; MVAP: Mevalonic acid 5-phosphate; 2-PGA: 2-phosphoglyceric acid; MVAPP: Mevalonic acid 5-pyrophosphate; IPP: isopentenyl pyrophosphate

*Adapted from Moir, 1969.

Factors Influencing Latex Production

Important physiological requirements of a rubber tree with high productivity are high net assimilation rate for growth and high ratio of partition of assimilates into rubber (Templeton, 1969). Therefore, any high yielding tree must have capacity for a high rate of girthing to produce a large trunk and thus ensuring high potential yield. Rate of girth increment on tapping is comparatively less in clones of higher productivity than those of lower productivity, indicating that rubber formation and growth process compete with each other. Formation of rubber is a high energy requiring process which could lead to depression of growth after tapping (Bonner, 1967; Chua, 1967). A peculiar situation in Hevea is that the yield of latex from a tree is determined not only by the inherent factors and environment but also by the exploitation methods (Sethuraj, 1985). The yield of rubber from a tree on every tapping is determined by the volume of latex and the percentage of rubber it contains (DRC).

Latex Yield Limiting Factors

Two primary factors limiting production are latex flow which governs the quantity of latex collected at tapping and latex regeneration between two successive tapping. Important biochemical parameters connected with latex production are total solid content (TSC), bursting index, sucrose, pH, inorganic phosphorus (Pi), magnesium and thiols (RSH).

Total Solid Content

Rubber constitutes over 90 per cent of the TSC of latex and under certain conditions, TSC reflects the biosynthetic activity. A decrease in rubber content indicates inadequate regeneration which can become a limiting factor in production. On the other hand, a high TSC can limit production by making flow difficult (Van Gils, 1951).

Bursting Index

Bursting index is a measure of lutoid stability. A high bursting index of lutoids indicates a high destabilization of latex and an early plugging (Jacob *et al.*, 1989).

Sucrose

Sucrose is the initial molecule involved in rubber synthesis. A high sucrose content in latex may indicate good loading of latex vessels and an active metabolism. It may also indicate low metabolic utilization of sucrose and low productivity. Excessive exploitation can cause a decrease in sugar content (Jacob *et al.*, 1989).

pH

The extent of latex regeneration between tapping depends on the metabolic activity of laticifers. Some of the enzymes which play key role in latex metabolism are sensitive to variations in pH, which thus is a metabolic regulator.

Inorganic Phosphorus (Pi)

Inorganic phosphorus reflects the energy metabolism and is necessary for the production of nucleic acid required in rubber biosynthesis. Direct correlation between Pi content of latex and production was established in many clones (Subronto, 1978; Eschbach *et al.*, 1984; Thomas *et al.*, 2000).

Magnesium

Magnesium is an activator as well as inhibitor of many enzymes involved in latex metabolism and therefore, has a complex link with production (Jacob *et al.*, 1989).

Thiols (RSH)

Glutathione is the major fraction of the RSH and most reactive. It is found mainly in the latex cytosol. Thiols trap toxic forms of oxygen thereby protecting the membranes of latex organelles. They are also activators of key enzymes in latex (d'Auzac *et al.*, 1982), such as invertase and pyruvate kinase. The regeneration of reduced glutathione from its oxidized form (GSSG) is also important in maintaining the stability of latex.

ATP

Biochemical energy availability is a major factor in latex flow and regeneration (Jacob *et al.*, 1997). The increases in total adenine nucleotide content are directly linked to the intensity of the metabolism and thereby yield. Direct correlation between ATP content and production was established in many clones (Sreelatha *et al.*, 2004).

Latex Diagnosis

Certain biochemical parameters in latex exhibit a high degree of correlation with production and therefore could be used to assess the production capacity of clones and describe clones effectively despite differences of environment and age (Eschbach *et al.*, 1983; Jacob *et al.*, 1986; Prevot *et al.*, 1986). The four parameters which are directly or indirectly related to latex production are sucrose, thiols, inorganic phosphorus and drc. Characterization of clones based on these parameters and laticifer functioning is well established (Esbach *et al.*, 1984; Serres *et al.*, 1988; Gohet *et al.*, 2003; Usha Nair *et al.*, 2003; Nair, 2003; Nair and Kavitha, 2006). These physiological parameters help to characterize clones as having active, slow or intermediate metabolism. An active metabolism is generally associated with high Pi, low sugar and RSH contents and a high DRC. It is also possible to define the optimum exploitation system for a clone based on threshold levels of latex diagnosis parameters (Gohet *et al.*, 2003; Nair *et al.*, 2004).

References

Archer, B.L 1980. Polyisoprene. (In) *Encyclopedia of Plant Physiology*: 8. Secondary Plant Products (Eds. E.A. Bell and B.V. Charlwood). Springer-Verlag, New York, pp. 309–328.

Archer, B.L., 1960. The proteins of *Hevea brasiliensis* latex: 4. Isolation and characterization of crystalline hevein. *Biochemical Journal*, 75(2): 236–270.

Archer, B.L., 1964. Site and mechanism of rubber biosynthesis from mevalonate. *Proceedings of the Natural Rubber Producers' Research Association Jubilee Conference*, 1964, Cambridge, London, pp. 101–112.

Archer, B.L., 1976. Hevamine – a crystalline basic protein from *Hevea brasiliensis* latex. *Phytochemistry*, 15: 297–300.

Archer, B.L. and Adley, B.G., 1967. Biosynthesis of rubber. *Advances in Enzymology*, 29: 221–257.

Archer, B.L. and Audley, B.G., 1987. New aspects of rubber biosynthesis. *Botanical Journal of the Linnean Society*, 94: 181–196.

Archer, B.L. and Cockbain, *e.g.*, 1969. Rubber transferase from *Hevea brasiliensis* latex. *Methods in Enzymology*, 15: 476–480.

Archer, B.L., Audley, B.G. and Bealing, F.J., 1982. Biosynthesis of rubber in *Hevea brasiliensis*. *Plastics and Rubber International*, 7: 109–111.

Archer, B.L., Audley, B.G., Cockbain, *e.g.* and McSweeney, G.P., 1963a. The biosynthesis of rubber: Incorporation of mevalonate and isopentenyl pyrophosphate into rubber by *Hevea brasiliensis* latex fractions. *Biochemical Journal*, 89: 565–574.

Archer, B.L., Audley, B.G., McSweeney, G.P. and Hong, T.C., 1969. Studies on composition of latex serum and bottom fraction particles. *Journal of the Rubber Research Institute of Malaya*, 21(4): 560–569.

Archer, B.L., Barnad, D., Cockbain, E.G., Cornforth, J.W., Cornoforth, R.H. and Popjak, G., 1966. The stereochemistry of rubber biosynthesis. *Proceedings of the Royal Society, London*, Series B, 163: 519–523.

Archer, B.L., Barnard, D., Cockbain, E.G., Dickenson, P.B. and McMullen, A.I., 1963b. Structure composition and biochemistry of *Hevea* latex. In: *The Chemistry and Physics of Rubber Like Substances* (Ed. Bateman,L.) Maclaren and Sons Ltd., London, pp. 41–72.

Backhaus, R.A., 1985. Rubber formation in plants: A mini review. *Israel Journal of Botany*, 34: 283–293.

Backhaus, R.A. Cornish, K., Chen,S.F., Huang,D.S., and Bess, V.H., 1991. Purification and characteristics of an abundant rubber particle protein from guayule. *Phytochemistry*, 30 (8): 2493–2497.

Bealing, F.J., 1969. Carbohydrate metabolism in *Hevea* latex: Availability and utilization of substrates. *Journal of the Rubber Research Institute of Malaya*, 21(4): 445–455.

Bonner, J., 1967. Rubber biogenesis. (In): *Biogenesis of Natural Compounds* (Ed. Bernfeld,P.). Pergamon Press Inc., New York.

Bowler, W.W., 1953. Electrophoretic mobility of fresh *Hevea* latex. *Industrial Engineering Chemistry*, 45: 1790.

Bunn, C.W., 1942. Molecular structure and rubber like elasticity: 1. The crystal structures of Gutta percha, rubber and polychloroprene. *Proceedings of Royal Society*, 180 (80): 40.

Chua, S.E., 1967. Physiological changes in *Hevea* trees under intensive tapping. *Journal of the Rubber Research Institute of Malaya*, 20(2): 100–105.

Cook, A.S. and Sekhar, B.C., 1953. Fractions from *Hevea brasiliensis* latex centrifuged at 59,000g. *Journal of the Rubber Research Institute of Malaya*, 14: 163–168.

Cornish, K. and Backhaus, R.A., 1990. Rubber transferase activity in rubber particles of guayule. *Phtochemistry*, 29(12): 3809–3813.

Coupe,M and d'Auzac, J., 1972. Demonstration of functional polysomes in the latex of *Hevea brasiliensis* (kunth) Mull.Arg. Comptes rendues de l Academic des Science de Paris,274,series D.1031.

d'Auzac, J. and Jacob, J.L., 1969. Regulation of glycolysis in the latex of *Hevea brasiliensis*. *Journal of the Rubber Research Institute of Malaya*, 21(4): 417–444.

d'Auzac, J. and Jacob, J.L., 1989. The composition of latex from *Hevea brasiliensis* as a laticiferous cytoplasm. *In: Physiology of Rubber Tree Latex* (Eds. J.d'Auzac, J.L. Jacob and H. Chrestin). CRC Press, Florida, pp. 60–88.

d'Auzac, J. Cretin, H., Marin, B. and Lioret, C., 1982. A plant vacuolar system: The lutoids from *Hevea brasiliensis* latex. *Physiologie Vegetale*, 20: 311–331.

Fay, E. de., Hebant, H. and Jacob, J.L., 1989. Anatomical organization of the laticifers system in the bark of *Hevea brasiliensis*. In: Physiology of Rubber Tree Latex. pp: 15–29.

Dickenson, P.B., 1964. The ultra structure of the latex vessel of *Hevea brasiliensis*. *Proceedings of the Natural Rubber Producers' Research Association Jubilee Conference*, 1964, Cambridge, London, pp. 52–66.

Dickenson, P.B., 1969. Electron microscopical studies of latex vessel system of *Hevea brasiliensis*. *Journal of the Rubber Research Institute of Malaya*, 21(4): 543–559.

Eschbach, J.M., Roussel, D., van de Sype, H., Jacob, J.L. and d'Auzac, J., 1984. Relationship between yield and clonal physiological characteristics of latex from *Hevea brasiliensis*. *Physiologie Vegetale*, 22(3): 295–304.

Eschbach, J.M., Van de Sype, H., Roussel, D. and Jacob, J.L., 1983. The study of several physiological parameters of latex and their relationships with production mechanisms. *International Rubber Research Development Board Symposium*, 1983, Beijing, China.

Frey–Wyssling, A., 1929. Microscopic investigations on the occurrence of resins in *Hevea* latex. Archf. Rubber cult. Ned–Indie., 13: 371

Gidrol, X., Chrestin, H., Tan, H.L. and Kush, A., 1994. Hevein a lectin–like protein from *Hevea brasiliensis* (rubber tree) is involved in the coagulation of latex. *Journal of Biological Chemistry*, 269: 9278–9283.

Gohet. E., Chantuma, P., Lacrotte. R., Obouayeba, S., Dian. K., Clemant–Demange, A., Kurnia, D. and Eschbach, J.M., 2003. Latex clonal typology of *Hevea brasiliensis*: physiological modeling of yield potential and clonal response to ethephon stimulation. Proceedings of the International Workshop on Exploitation Technology, India, pp., 199–217.

Golub, M.A., Fuqua, S.A. and Bhacca, N.S., 1962. High resolution nuclear magnetic resonance spectra of various polyisoprenes. *Journal of American Chemical Society*, 84(2): 4981.

Gomez, J.B., 1982. Anatomy of *Hevea* and its influence on latex production. Malaysian Rubber Research and Development Board, Monograph No. 7, Kuala Lumpur. P. 21.

Hahn, A.M., Hermann, J., Marianti., Mariano, A and Walugeno, K., 1971. Studies on α–globulin: *Hevea* and destabilizing protein of *Hevea brasiliensis* latex. *Communications of the Research Institute of Estate Crops*, 1.

Hasma, H. and Subramaniam, A., 1986. Composition of lipids in latex of *Hevea brasiliensis* Clone RRIM 501. *Journal of Natural Rubber Research*, 1(1): 30–40.

Jacob, J.L., Eschboch, J.M., Prevot, J.C., Roussel, D., Lacrotte, R., Chrestin, H. and d'Auzac, J., 1986. Physiological basis for latex diagnosis of the functioning of the lactiferous system in rubber trees. *Proceedings, International Rubber Conference*, 1985, Kuala Lumpur, Malaysia, 3: 43–65.

Jacob, J.L., Nouvel, A. and Prevot, J.C., 1978. Electrophorese et mise en evidence d'activities enzymatiques dans le latex d'*Hevea brasiliensis, Revue Generale des Caoutchoucs et Plastiques*, 582: 87.

Jacob, J.L., Prevot, J.C., Lacrotte, R., Gohet, E., Clement, A., Gallois, R., Joet, T., Pujade Renaud, V and d'Auzac, J., 1997. The biological mechanism controlling *Hevea brasiliensis* rubber yield. IRRDB Symposium on Natural rubber V2. Physiology and Exploitation and Crop protection and planting methods.,14–15 October, Ho chi Minh city,Vietnam, pp: 1–11.

Jacob, J.L., Prevot, J.C., Roussel, D., Lacrotte, R., Serres, E., d'Auzac, J., Eschbach, J.M. and Omont, H., 1989. Yield–limiting factors, latex physiological parameters, latex diagnosis and clonal typology. In: *Physiology of Rubber Tree Latex* (Eds. J. d'Auzac, J.L. Jacob and H. Chrestin). CRC Press Inc., Florida, pp. 345–382.

Kush, A., Goyvaerts, E., Cheye, M.L. and Chua, N.H., 1990. Laticifer–specific gene expression in *Hevea brasiliensis* (rubber tree). *Proceedings of the National Academy of Science*, USA, 87: 1787–1790.

Light, D.L. and Dennis, M. S., 1989. Purification of a prenyltransferase that elongates *cis*.polyisoprene from the latex of *Hevea brasiliensis*. *Journal of Biological Chemistry*, 264: 18589–18597.

Low, F.C., 1978. Distribution and concentration of major soluble carbohydrates in *Hevea* latex, the effects of ethephon stimulation and the possible role of these carbohydrates in latex flow. *Journal of the Rubber Research Institute of Malaysia*, 26(1): 21–32.

Lynen, F., 1969. Biochemical problems of rubber synthesis. *Journal of the Rubber Research Institute of Malaysia*, 21(4): 389–406.

Lynen, F., 1967. Biosynthetic pathways from acetate to natural products. *Pure and Applied Chemistry*, 14(1): 137.

McMullen, A.I. and McSweeney, G.P., 1966. The biosynthesis of rubber: Incorporation of isopentenyl pyrophosphate into purified rubber particles by a soluble latex serum enzyme. *Biochemical Journal*, 101: 42.

Moir, G.F.J., 1969. Latex metabolism. *Journal of the Rubber Research Institute of Malaya*, 21(4): 407–416.

Nair, N.U., Thomas, M., Sreelatha, S., Simon, S.P., Vijayakumar, K.R. and George, P.J., 1993. Clonal variation in lipid composition in the latex of *Hevea brasiliensis* and its implication in latex production. *Indian Journal of Natural Rubber Research*, 6 (1&2): 143–145.

Nair, N.U., 2003. Latex diagnosis for optimizing production in clone RRII 105. International workshop on Exploitation Technology, 15–18, December 2003, pp., 191–198.

Nair, N.U., R.B. Nair., Thomas, M. and Gopalakrishnan J., 2004. Latex diagnosis in relation to exploitation systems in clone RRII 105. *Journal of Rubber Research*, 7(2): 127–137.

Nair, N.U. and Kavitha K. Mydin 2006. Physiological criteria for evaluating production of newly evolved hybrid clones. International Natural Rubber Conference, 13–14 November, Ho Chi Minh City, Vietnam.

Prevot, J.C., Jacob, J.L., Lacrotte, R., Vidal, A., Serres, E., Eschbach, J.M. and Gigault, J., 1986. Physiological parameters of latex from *Hevea brasiliensis*: Their use in the study of the laticiferous system, typology of functioning production mechanisms, effect of stimulation. *Proceedings of the IRRDB Rubber Physiology and exploitation Meeting*, 1986, Hainan, China, pp. 136–157.

Serres, E., Clement, A., Provet J.C., Jacob, J.L., Commere, J., Lacrotte, R. and Eschbach, J.M., 1988. Clonal typology of laticiferous producing vessels in *Hevea brasiliensis*. CR. IRRDB Exploitation, Physiol. Amel. *Hevea* (Jacob J.L. and Prevot, J.C. eds). Pp. 321–246.

Sethuraj, M.R., 1985. Physiology of growth and yield in *Hevea brasiliensis*. *Proceedings, International Rubber Conference*, 1985, Kuala Lumpur, Malysia, 3: 3–19.

Sherif, P.M. and Sethuraj, M.R., 1978. Role of lipids and proteins in the mechanism of latex vessel plugging in *Hevea brasiliensis*. *Physilogia Plantarum*, 42: 351–353.

Siler, D.J. and Cornish. K., 1994. Identification of *Parthenium argentatum* rubber particles protein immunoprecipitated by an antibody that specifically inhibit rubber transferase activity. *Phytochemistry*, 36(3): 623–627.

Siler, D.J. and Cornish. K., 1995. Biochemical studies of *cis*–1,4–polyisoprene (natural rubber) biosynthesis in *Parthenium argentatum*, a new source of hypo allergenic latex. Plant physiology, 42, Annual meeting. American Society of Plant Physiologists, July 29th to August 2nd.

Southorn, W.A., 1961. Microscopy of *Hevea* latex. *Proceedings of the Natural Rubber Research Conference*, 1960, Kuala Lumpur, p. 766.

Southorn, W.A. and Yip, E., 1968. Latex flow studies: 3. Electostatic considerations in the colloidal stability of fresh *Hevea* latex. *Journal of the Rubber Research Institute of Malaya*, 20(4): 201–215.

Sreelatha,S., Simon, P.S. and Jacob, J., 2004. On the possibility of using ATP concentration in latex as an indicator of high yield in *Hevea brasiliensis*. *Journal of Rubber Research*, 7(1): 71–78.

Subramaniam, A., 1980. Molecular weight and molecular weight distribution of natural rubber. *Rubber Research Institute of Malaysia Technology Bulletin*,4.

Subronto, 1978. Correlation studies of latex flow characters and latex minerals content. Proceedings, International Rubber *Research and Development Board Symposium*, 1978. Kuala Lumpur, Malaysia (Preprint: Part 1, Session 1).

Suvachitanont, W and Witisuwannakul, R., 1995. 3–hydroxy–3–methylglutaryl CoA synthase in *Hevea brasiliensis*. *Phytochemistry*, 40: 757–761.

Tata, S.J., 1975. A study of the proteins in the heavy fraction of *Hevea brasiliensis* latex and their possible role in the destabilization of rubber particles. Thesis for a degree of Masters of Science, University of Malaya, Kuala Lumpur, Malaysia.

Tata, S.J., 1976. Hevein. Its isolation, purification and some structural aspects. *Proceedings International Rubber Conference, 1975, Kuala Lumpur, Rubber Research Institute of Malaysia*. p. 499.

Tata, S.J., 1980. Distribution of proteins between the fraction of *Hevea* latex separated by ultracentrifugation. *Journal of the Rubber Research Institute of Malaysia*, 28(2): 77–85.

Tata, S.J. and Yip, E., 1968. A protein fraction from B serum with strong destabilizing activity on latex. Rubber Research Institute of Malaya, Kula Lumpur (Research Archive No. 59).

Tata, S.J., Beintema, J.J. and Balabakaran, S., 1983. The lysozyme of *Hevea brasiliensis* latex: Isolation, purification, enzyme kinetics and a partial amino–acid sequence. *Journal of the Rubber Research Institute of Malaysia*, 31(1): 35–48.

Templeton, J.K., 1969. Partition of assimilates. *Journal of the Rubber Research Institute of Malaya*, 21(3): 259–263.

Thomas, M., Thomas, K.U., Sobhana, P., Nair, R.B. and Jacob, J., 2000. Path coefficient analysis of yield and yield attributes in the *Hevea brasiliensis* clone RRII 105. *Indian Journal of Natural Rubber Research*, 13(1&2): 103–107.

Tupy 1969a. Nucleic acids in latex and production of rubber in *Hevea brasiliensis*. *Journal of the Rubber Research Institute of Malaya*, 21: 468.

Van Gils, G.E., 1951. Studies on the viscosity of latex: 1. Influence *of dry rubber content. Archives of Rubber Cultivation*, 28: 61.

Witisuwannakul, R., Pasitkul, P., Jewtragoon, P and Witisuwannakul, D., 2008. *Hevea* latex lectin binding protein in C–serum as an anti latex coagulating factor and its role in a proposed new modal for latex coagulation. *Phytochemistry*, 69: 656–662.

Witisuwannakul, R., Witisuwannakul, D and Dumkong, S., 1990. *Hevea* Calmodulin: regulation of the activity of latex 3–hydroxy–3–methylglutaryl CoA reductase. *Phytochemistry*, 29: 1755–1758.

2012, Nutri-Horticulture   *Pages 171–194*
Editor: Professor K.V. Peter
Published by: DAYA PUBLISHING HOUSE, NEW DELHI

Chapter 10

Water Requirement of Horticultural Crops

E.J. Joseph

Scientist, Centre for Water Resource Development and Management,
Kozhikode, Kerala
E-mail: jej@cwrdm.org

Crop plants are normally grown in the soil. Plant roots take up water and transfer it through the root tissues to the conducting vessels of the plant, which in turn pass it on to the mesophyll tissues of the leaves. After moving through these tissues in the liquid phase along a potential gradient, water reaches the evaporation sites which are primarily the walls of the sub-stomatal cavities. The escape of water from the sub-stomatal cavities into the atmosphere is a process of molecular diffusion of water vapour. The leaves maintain their continuity of structure with the stem, which has conducting tissues called xylem and phloem. The stem maintains its continuity with the root, which eventually is in contact with the sources of water, the soil. Thus, water moves from soil to atmosphere in a soil-plant-atmosphere continuum.

The knowledge of the basic soil-plant-water relationships and crop water requirements are essential for scientific water management and high water use efficiency in crop production. The planning of cropping pattern and water management, based on characteristics of the resources, *viz.,* water, soil, crop and climate will go a long way in improving the productivity of crops.

Water

Water is essential for every living organism. It constitutes about 80 to 90 per cent of most plant cells and tissues in which there is active metabolism. Water is a reactant or reagent in many physiological processes including photosynthesis and hydrolysis of starch to sugar. The unique properties of water, including its dipolar character, hydrogen bonding and latent heat, make it biologically an important compound. The other significant properties of water, which need mention, are its characteristics as a solvent and its cohesive and adhesive properties. Growth of plants is controlled by the rates of cell division and cell enlargement and by the supply of organic and inorganic compounds required for the synthesis of new protoplasm and cell walls. The permeability of cell walls and membranes to water results in a continuous liquid phase extending throughout the plant in which the translocation of solutes occurs. Water is needed to maintain sufficient turgidity for growth of cells and to maintain the form and position of leaves and new shoots.

Soil

Soil serves as the storage reservoir for water to be used by plants. When ample water is present in the root zone, plants can obtain their daily water requirements for proper growth and development. As the plants continue to use water, the water available in the root zone decreases, and unless water is added through irrigation or precipitation, the plants cease to grow and finally wilt. The quantity of water to be applied in each irrigation and frequency of irrigation are dependent on the properties of soil and the crop to be irrigated.

Soil is a three-phase system comprising of the solid phase, made of mineral and organic matter, the liquid phase called the soil moisture, and the gaseous phase called the soil air. The main component of the solid phase is the soil particles, the size and shape of which give rise to pore spaces of different geometry. The most important soil properties influencing irrigation are its infiltration characteristics and water holding capacity. Other soil properties like soil texture, soil structure, capillary conductivity, soil profile conditions and depth of water table are also important.

The relative proportion of sand, silt and clay determines the soil texture and it influences considerably the other phases (water and air) contained in the spaces in the soil matrix. Loam soils contain almost equal amounts of sand, silt and clay and are considered as the most favourable for plant growth, because they hold more available water and cations than sand and are better aerated and easier to work than clay. The arrangement of individual soil particles with respect to each other into a pattern is called soil structure. Soil structure has a pronounced effect on such soil properties like erodibility, porosity, hydraulic conductivity, infiltration, and water holding capacity. The measurable properties of soil structure are porosity, aggregation, cohesiveness and permeability. For optimum crop growth, the soil structure should be such that the infiltration capacity is large, the percolation capacity is medium and aeration is sufficient. A massive compact soil restricts aeration and root spread.

Since the pore spaces are as important as the solid particles, soil structure is also sometimes defined as the arrangement of small, medium and large pores into a structural pattern. The pore spaces are filled with water and air in varying proportions, depending on the amount of moisture present. Large pores induce aeration and infiltration, medium-sized pores facilitate capillary conductivity, and small pores induce greater water holding capacity. Two major classes of soil pores recognized are capillary pores and non-capillary pores. Capillary pores contain the water which remains after the completion of free drainage of the soil. Non-capillary pores, or aeration pores are filled with air after the drainage of the soil to field capacity. The large non-capillary porosity of sandy soils results in better drainage and aeration, but low water holding capacity. The large proportion of small capillary pores in clayey soils results in better water holding capacity, but poor aeration and drainage.

Soil-water Relationships

Water is retained by a soil particle in the form of a thin film around it, and is contained in the numerous small pores of the soil matrix with forces, such as surface tension capillarity, cohesion and adhesion. Water is held in the soil in the following forms: 1) *Hygroscopic water:* It is the water held tightly to the surface of soil particles by adsorption forces with a tension of 31 atmospheres or more, 2) *Capillary water:* It is the water held by the forces of surface tension ranging from one-third atmospheres to 31 atmospheres and forms continuous film around soil particles and in capillary spaces, 3) *Gravitational water:* It is the water held at a tension below one-third atmosphere that moves freely in response to gravity and drains out of the soil. Between 31 atmospheres and 15 atmospheres, capillary adjustment is sluggish. Capillary water is the major source of water used by plants and it functions physically and chemically as the soil solution. The major factors influencing the quantity of capillary water in soils are the soil structure, soil texture and organic matter. The finer the texture of the mineral soil particles, the greater is its capillary capacity. Granular soil structure and high organic matter content in soil increases its capillary capacity. Capillary water is capable of movement upwards, downwards or laterally, but the rate of movement becomes less as the capillary water zone moves farther from the water table below. Therefore, deep water table (more than 2 to 3 metres from root zone) is not much of direct use to the crops.

When water is added to a soil either by precipitation or irrigation, it is distributed around the soil particles where it is held by adhesive and cohesive forces. Water infiltrates into the soil and gradually fills the soil pores completely and continues to move in the soil mass. The principal factors governing the infiltration of water into the soil are the initial moisture content of the soil, condition of soil surface, hydraulic conductivity of soil profile, soil texture, porosity, degree of swelling of soil colloids and organic matter, vegetative cover, duration of rainfall or irrigation and viscosity of water. Infiltration rates are generally lower in soils of heavy texture (clayey soil) than in soils of light texture (sandy soil). The downward movement of water ceases after about 48 to 72 hours of the rainfall or irrigation and the water retained in the soil at this stage is the upper limit of the available soil moisture for plants and is called *field capacity* (FC). As evaporation and transpiration continue, soil moisture content reaches a level when plants are unable to extract it. The moisture content at this stage is termed *permanent wilting point* (PWP). The range of soil moisture between field capacity and permanent wilting point is called *available water content* (AWC) for plant growth. The available water content increases with the fineness of texture and content of organic matter. The available water content for different textural groups of soil is given in Table 10.1.

Table 10.1: Available Water Content for Different Soil Textural Groups

Soil Texture	Field Capacity (Per cent)	Permanent Wilting Point (per cent)	Bulk Density (g cc⁻¹)	Available Water Content (mm per metre depth of soil profile)
Sandy	5 to 10	2 to 6	1.5 to 1.8	50 to 100
Sandy loam	10 to 18	4 to 10	1.4 to 1.6	90 to 160
Loam	18 to 25	8 to 14	1.3 to 1.5	140 to 220
Clay loam	24 to 32	11 to 16	1.3 to 1.4	170 to 250
Clay	32 to 40	15 to 22	1.2 to 1.4	200 to 280

Movement of water in unsaturated soils involves both liquid and vapour phases. Water vapour movement goes on within the soil and also between soil and atmosphere. Diffusion of water vapour is caused by a vapour pressure gradient as the driving force. The vapour pressure of soil moisture increases with the increase in soil moisture content and temperature, and it decreases with the decrease in soluble salt content. In a coarse-textured soil, pores become free of liquid water at relatively low tensions and when the soil dries out there is little moisture left for vapour transfer. But, a fine-textured soil retains substantial quantities of moisture even at high tensions. Maximum water vapour movement in soils occurs just before the permanent wilting point is reached.

Measurement of Soil Moisture

The measurement of soil moisture is important because of the significance of soil moisture in scheduling of irrigation and estimating the quantity of water to be applied per irrigation. The measurement of changes in soil moisture with time is essential for the estimation of *evapotranspiration*. Moisture content in soil is expressed either by the quantity of water (on weight basis or volume basis) in a given quantity of soil, or by the stress or tension under which the water is held by the soil. Soil moisture tension is a measure of the tenacity with which water is retained in the soil and shows the force per unit area that must be exerted to remove water from the soil. The tenacity is generally measured in terms of the potential energy of the free water in the soil. It is expressed in atmospheres, the average air pressure at sea level.

There are several methods of soil moisture measurement, and the choice depends on how precisely and quickly the information is needed. In the universal *gravimetric method*, soil samples are taken from the field using an auger and the loss in weight of the sample when dried in an oven at 105° to 110°C to a constant weight is determined. A *tensiometer* measures the force with which the surrounding soil is pulling the water held by it in a porous porcelain cup. A *resistance block*, with embedded electrodes, and *potentiometer* determines soil moisture indirectly by measuring the resistance to the flow of electricity through the soil matrix. A *neutron moisture meter* determines soil moisture content by measuring the number of hydrogen nuclei that are present in a unit volume of soil, since the number of hydrogen nuclei is a direct function of the number of water molecules present in the same volume of soil.

Plant-water Relationships

The roots, stem and leaves of a plant provide a continuum for water movement from soil to atmosphere. The three main phases of plant-water relationships are: water absorption, water conduction and translocation, and water loss or transpiration. The water potential gradient from soil to atmosphere could generate the driving force for water movement from the soil through the plant to the atmosphere. The rate of water movement is everywhere proportional to the potential gradient and inversely proportional to the resistance to flow. The approximate magnitudes of water potential in the soil-plant-atmosphere system are given in Table 10.2.

Water absorption by roots depends on the supply of water at the root surface. As the soil dries out from a saturated state, the rate of water movement in the soil decreases rapidly. Absorption is regulated by the size and distribution of roots, and several soil factors such as soil temperature, soil water potential, concentration of soil solution, aeration etc. The rate of water absorption is also controlled by the rate of transpiration. The movement of water through the root and conducting elements of the leaf xylem to the leaves is initiated and largely controlled by the transpiration from the leaves in response to the water potential gradient extending from the soil water, through the plant to the atmosphere. Transpiration, though an energy-controlled process, is modified by the soil, plant and atmospheric factors, which govern the potential gradients in the various parts of the water path to the leaf surface.

Table 10.2: Approximate Magnitudes of Water Potential in the Soil-plant-atmosphere System

Components	Water Potential (Bar)
Soil	-0.1 to -20
Leaf	-5 to -50
Atmosphere	-100 to -2000

Plaut and Moreshet, 1973.

The factors influencing the water relations of plants, and thus their growth and yield responses, may be grouped into the following:

1. *Soil factors* like soil moisture content, soil texture, soil structure, bulk density, salinity, soil fertility, soil temperature, soil aeration, and drainage,

2. *Plant factors* like type of crop, density and depth of rooting, rate of root growth, aerodynamic roughness of the crop, resistance to flow of water and vapour in different plant parts, drought tolerance and varietal effects,

3. *Weather factors* like sunshine, atmospheric temperature, relative humidity, wind and rainfall, and

4. *Miscellaneous factors* like soil volume, crop spacing, and crop and soil management.

The movement of water from soil to the atmosphere through the plants is a continuous process during which several tasks are performed. Plant nutrients, which are absorbed by roots in soil solution move upwards through the vascular system, along with water in ionic form, to the leaves. Out of the huge quantities of water taken up by plants, major portion is lost as transpiration. When transpiration of water occurs through the stomata, carbon dioxide diffuses into the leaf tissues, where it aids in photosynthesis. Evaporation mainly takes place from the soil surface beneath the crop. Since both evaporation and transpiration occur simultaneously from the plants and the surrounding environment, it is difficult to isolate them.

Evaporation

Evaporation is the process whereby liquid water is converted to water vapour (vaporization) and removed from the evaporating surface. Water evaporates from a variety of surfaces, such as lakes, rivers, pavements, soils and wet vegetation. The essential requirements in evaporation process are: 1) the source of heat to vapourize the liquid water, and 2) the presence of a gradient of concentration of water vapour between the evaporating surface and the surrounding air. The source of energy for evaporation may be solar energy, the air blowing over the surface, or the underlying surface itself. Evaporation from the land surface is mainly influenced by the degree of wetness of the surface, temperature of air and soil, atmospheric humidity and wind velocity. The density of vegetation is an important factor controlling evaporation from land surface. Evaporation rate reduces rapidly with the increase in depth below the surface. The mulch materials over soil surface restrict air movement, maintain a high vapour pressure near the soil surface, and shield the soil from solar energy, all of which reduce evaporation.

Transpiration

Transpiration consists of the vaporization of liquid water contained in plant tissues and the vapour removal to the atmosphere. Crops predominately lose their water through stomata, which are small openings on the plant leaf through which gases and water vapour pass. The water, together with plant nutrients, is taken up by the roots and transported through the plant. The vaporization occurs within the leaf, namely in the intercellular spaces, and the vapour exchange with the atmosphere is controlled by the stomatal aperture. Nearly all water taken up are lost by transpiration and only a tiny fraction is used within the plant.

Transpiration is the dominant factor in plant-water relations because evaporation of water produces the energy gradient which causes the movement of water into and through the plants. The rate of transpiration depends on the supply of energy to vaporize water, the vapour pressure or concentration gradient in the atmosphere which constitutes the driving force, and the resistance to diffusion in the vapour pathway.

The major factors which influence transpiration are the climate, the soil and the plant factors. The important climatic factors which control transpiration are light intensity, atmospheric vapour pressure, temperature and wind. The soil factors are those governing the water supply to the roots, and the plant factors include the extent and efficiency of root systems in moisture absorption, the leaf area, leaf arrangement and structure and stomatal behaviour.

Evapotranspiration or Consumptive Use

The combination of two separate processes, whereby water is lost on the one hand from the soil surface by evaporation and on the other hand from the crop by transpiration, is referred to as evapotranspiration (ET) or consumptive use (CU). Evaporation and transpiration occur simultaneously and there is no easy way of distinguishing between the two processes. Apart from the water availability in the topsoil, the evaporation from a cropped soil is mainly determined by the fraction of the solar radiation reaching the soil surface. This fraction decreases over the growing period as the crop develops and the crop canopy shades more and more of the ground area. When the crop is small, water is predominately lost by soil evaporation, but once the crop is well developed and completely covers the soil, transpiration becomes the main process of water loss. At sowing, nearly 100 per cent of ET comes from evaporation, while at full crop cover; more than 90 per cent of ET comes from transpiration.

Reference Crop Evapotranspiration

The evapotranspiration rate from a reference surface, not short of water, is called the reference crop evapotranspiration or reference evapotranspiration and is denoted as ET_0. The reference crop evapotranspiration is defined as the rate of evapotranspiration from a hypothetical grass reference crop with an assumed crop height of 0.12 m, a fixed surface resistance of 70 s m^{-1} and an albedo of 0.23, closely resembling the evapotranspiration from an extensive surface of green grass of uniform height, actively growing, well-watered and completely shading the ground. Due to the ambiguities in definitions, the use of other earlier denominations such as potential ET is nowadays not encouraged. The concept of the reference evapotranspiration was introduced to study the evaporative demand of the atmosphere independently of crop type, crop development and management practices. As water is abundantly available at the reference evapotranspiring surface, soil factors do not affect ET. Relating ET to a specific surface provides a reference to which ET from other surfaces can be related. It obviates the need to define a separate ET level for each crop and stage of growth. ET_0 values measured or calculated at different locations or in different seasons are comparable as they refer to the ET from the same reference surface. The only factors affecting ET_0 are climatic parameters. Consequently, ET_0 is a climatic parameter and can be computed from weather data. ET_0 expresses the evaporating power of the atmosphere at a specific location and time of the year and does not consider the crop characteristics and soil factors.

Factors Affecting Evapotranspiration

Weather parameters, crop characteristics, management and environmental aspects are factors affecting evaporation and transpiration.

Weather Parameters

The principal weather parameters affecting evapotranspiration are radiation, air temperature, humidity and wind speed. Several procedures have been developed to assess the evaporation rate from these parameters. The evaporation power of the atmosphere is expressed by ET_0.

Crop Factors

The crop type, variety and development stage should be considered when assessing the evapotranspiration from crops grown in large, well-managed fields. Differences in resistance to transpiration, crop height, crop roughness, reflection, ground cover and crop rooting characteristics result in different ET levels in different types of crops under identical environmental conditions. Crop evapotranspiration (ETc) refers to the evaporating and transpiring demand from crops, grown in large fields under optimum soil water, excellent management and environmental conditions, and achieve full production under the given climatic conditions.

Soil Cover

When soil is covered with different mulch materials, crop evapotranspiration will be reduced.

Determination of Evapotranspiration

The principal methods for direct measurement of evapotranspiration are: (1) lysimeter experiments, (2) field experiments, (3) soil moisture depletion studies, (4) water balance method and (5) climatological method.

Lysimeter Experiments

By isolating the crop root zone from its surrounding environment and controlling the processes which are difficult to measure, the different components in the soil water balance equation can be determined with greater accuracy. This is done in lysimeters where the crop grows in isolated tanks filled with either disturbed or undisturbed soil. In precision weighing lysimeters, where the water loss is directly measured by the change of mass, evapotranspiration can be obtained with an accuracy of a few hundredths of a millimetre, and small time periods such as an hour can be considered. In non-weighing lysimeters, the evapotranspiration for a given time period is determined by deducting the drainage water collected at the bottom of the lysimeters from the total water input. A requirement of lysimeters is that the vegetation both inside and immediately outside of the lysimeter be perfectly matched (same height and leaf area index). This requirement has historically not been closely adhered to in many of the lysimeter studies, which resulted in erroneous and unrepresentative ETc and crop coefficient (Kc) values in certain cases. As lysimeters are difficult and expensive to construct and as their operation and maintenance require special care, their use is limited to specific research purposes.

Field Experiments

Measurements of water supplies to the field and changes in soil moisture content in the field are sometimes more dependable for computing seasonal water requirement of crops than lysimeter methods. The seasonal water requirements are computed by adding measured quantities of irrigation water, the effective rainfall received during the season and the contribution of moisture from the soil.

The soil moisture depletion method is usually employed to determine the consumptive use of irrigated crops grown on fairly uniform soil when the depth to the groundwater is such that it will not influence the soil moisture fluctuation within the root zone. These studies involve measurement of soil moisture from various depths at a number of times throughout the growth period.

Soil Water Balance

Evapotranspiration can also be determined by measuring the various components of the soil water balance. The method consists of assessing the incoming and outgoing water flux into the crop root zone over some time period. Irrigation (I) and rainfall (P) add water to the root zone. Part of I and P might be lost by surface runoff (RO) and by deep percolation (DP) that will eventually recharge the water table. Water might also be transported upward by capillary rise (CR) from a shallow water table towards the root zone or even transferred horizontally by subsurface flow in (SFin) or out (SFout) of the root zone. In many situations, however, except under conditions with large slopes, SFin and SFout are minor and can be ignored. Soil evaporation and crop transpiration deplete water from the root zone. If all fluxes other than evapotranspiration (ET) can be assessed, the evapotranspiration can be deduced from the changes in soil water content over the time period. Some fluxes such as subsurface flow, deep percolation and capillary rise from a water table are difficult to assess and short time periods cannot be considered. The soil water balance method can give usually only ET estimates over long time periods of the order of week-long or ten-day periods.

Estimation of Evapotranspiration from Pan Evaporation Data

Evaporation from an open water surface provides an index of the integrated effect of radiation, air temperature, air humidity and wind on evapotranspiration. However, differences in the water and cropped surface produce significant differences in the water loss from an open water surface and the crop. The pan has proved its practical value and has been used successfully to estimate reference evapotranspiration by observing the evaporation loss from a water surface and applying empirical coefficients to relate pan evaporation to ET_0. The standard US Weather Bureau Class A open pan evaporimeter or the sunken screen open pan evaporimeter may be used for measurement. The pan evaporation is related to the reference evapotranspiration by an empirically derived pan coefficient.

$$ET_0 = Kp * Epan$$

where,

ET_0: Reference crop evapotranspiration (mm day^{-1})

Kp: Pan coefficient, and

Epan: Pan evaporation (mm day^{-1})

Depending on the type of pan, ground cover in the station, its surroundings as well as the general wind and humidity conditions, pan coefficients will differ. For the Class A evaporation pan, the Kp varies between 0.35 and 0.85 (average Kp = 0.70). For the Sunken Colorado pan, the Kp varies between 0.45 and 1.10 (average Kp = 0.80).

Estimation of Evapotranspiration Using Empirical Formula

Many empirical relations have been developed over the years for estimating ET from meteorological data. The more commonly used methods are Blaney-Criddle method, Thornthwaite method, Penman and Modified Penman methods, Radiation method and Christiansen method.

(a) Blaney-Criddle Method

Blaney-Criddle (1950) developed a simple method correlating evapotranspiration for different crops with the monthly temperature, percentage day time hours and length of growing season.

$$U = KF = \Sigma kf = \Sigma u = \sum \frac{ktp}{100}$$

where,

U: Seasonal consumptive use of water by the crop for a given period (inches)

u: Monthly consumptive use (inches)

K: Empirical seasonal consumptive use coefficient for the growing season

F: Sum of monthly consumptive use factors (f) for the growing season

k: Empirical consumptive use crop coefficient for the month, u/f

f: $\dfrac{t * p}{100}$

t: Monthly temperature (°F)

p: Monthly daylight hours expressed as per cent of daylight hours of the year

Doorenbos and Pruitt (1977) modified the Blaney-Criddle formula, which became known as the FAO Blaney-Criddle formula.

$$ET_0 = C[P(0.46\,T + 8)]$$

where,

ET$_0$: Reference crop ET for the month considered (mm day^{-1})

T: Mean daily temperature over the month considered (°C)

P: Mean daily percentage of total annual day time hours of a given month and latitude (from standard table)

C: Adjustment factor depends on minimum relative humidity, sunshine hours and daytime wind estimates

(b) Thornthwaite Method

Thornthwaite (1948) assumed that an exponential relationship exists between monthly temperature and evapotranspiration.

$$e = 1.6\,(10t/I)^a$$

where,

e: Unadjusted potential evapotranspiration (cm month^{-1})

t: Mean air temperature (°C)

I: Annual or seasonal heat index, the summation of 12 values of monthly heat indices (*i*) when, $i = (t/5)^{1.514}$

a: An empirical exponent.

(c) Penman and Modified Penman Methods

Penman (1948) proposed an equation for evaporation from open water surface, based on a combination of energy balance and sink strength. The method was based on the principle that the incoming solar radiation reaches earth's surface, which warms up the air and soil, provides energy for plant growth and causes evapotranspiration.

Doorenbos and Pruitt (1975) proposed a modified Penman equation, based on intensive studies of the climatic and measured grass evapotranspiration data from various research stations in the world. Though the modified Penman equation is complex, it gives satisfactory estimate of ET$_0$.

$$ET_0 = C[W.R_n + (1-W).f(u).(e_a - e_d)]$$

where,

ET$_0$: Reference crop evapotranspiration (mm day^{-1})

R_n: Bet radiation (R_{ns}-R_{nl}); where R_{ns} is the incoming net short wave radiation and R_{nl} is net long wave radiation

W: Temperature and altitude related weighting factor

f (u): Wind related function

$e_a - e_d$: Vapour pressure deficit (mbar)

e_a: Saturation vapour pressure (mbar) at the mean air temperature

e_d: Mean actual vapour pressure of the air (mbar)

C: The adjustment factor to compensate for the effect of day and night weather conditions

(d) Radiation Method

Radiation method is useful for the estimation of ET_0 in areas where data on air temperature and sunshine, cloudiness or radiation are available. Estimated general levels of humidity and wind velocity (not measured) are also needed. The relationship is:

$$ET_0 = c(W.Rs)$$

where,

ET_0: Reference crop evapotranspiration for the period considered (mm day^{-1})

Rs: Solar radiation in equivalent evaporation (mm day^{-1})

W: Weighting factor which depends on temperature and altitude.

c: Adjustment factor, which depends on mean humidity and daytime wind conditions

(e) FAO Penman-Monteith Method

Though many direct and empirical methods have been developed to estimate ET_0 from different climatic parameters, they were not found to have universal applicability. Food and Agricultural Organization (FAO) had earlier recommended Blaney-Criddle method, radiation method, modified Penman method and pan evaporation method for the estimation of ET_0. But, of late, the FAO Penman-Monteith equation is recommended by FAO as the new standard for reference evapotranspiration (Allen *et al.*, 1998). This is based on a review of the FAO methodologies on crop water requirements by experts and researchers organized by FAO in collaboration with the International Commission for Irrigation and Drainage, and the World Meteorological Organization. The method overcomes the shortcomings of the previous FAO Penman method and provides values which are more consistent with actual crop water use data worldwide. The FAO Penman-Monteith equation uses standard climatic data that can be easily measured or derived from commonly measured data. All calculation procedures have been standardized according to the available weather data and the time scale of computation. The equation requires radiation, air temperature, air humidity and wind speed data. Procedures have been developed for estimating missing climatic parameters. The method is physically based and incorporates both physiological and aerodynamic parameters.

FAO Penman-Monteith equation has been derived from the original Penman-Monteith equation and the equations of aerodynamic and canopy resistance.

The FAO Penman-Monteith equation is:

$$ET_0 = \frac{0.408\Delta(R_n - G) + \gamma \dfrac{900}{T+273} U_2 (e_s - e_a)}{\Delta + \gamma \ (1 + 0.34u_2)}$$

where,

ET_0: Reference crop evapotranspiration (mm day^{-1})

R_n: Net radiation at the crop surface (MJ m^{-2} day^{-1})

G: Soil heat flex density [MJ m^{-2} day^{-1}]

T: Mean daily air temperature at 2m height (°C)

u_2: Wind speed at 2 m height (ms^{-1})

e_s: Saturation vapour pressure (kPa)

e_a: Actual vapour pressure (kPa)

e_s-e_a: Saturation vapour pressure deficit (kPa)

Δ: Slope vapour pressure curve [kPa °C^{-1}]

γ: Psychrometric constant [kPa °C^{-1}]

ET_0 is computed using the FAO software CROPWAT 8 (FAO, 2009) by employing the FAO Penman-Monteith equation. As an example, the ET_0 for different agro-ecological units of Palakkad district in Kerala State, computed by FAO CROPWAT 8 software, employing FAO Penman-Monteith equation is given in Table 10.3.

Table 10.3: Reference Crop Evapotranspiration (ET_0) for Different Agro-ecological Units (AEU) of Palakkad District in Kerala State

Month	AEU Decade (10 days)	Reference Crop Evapotranspiration (ET_0) (mm decade^{-1})							
		AEU 2.3	AEU 3.2	AEU 4.1	AEU 4.2	AEU 4.5	AEU 4.6	AEU 5.1	AEU 5.2
June	I	38	30	33	40	41	44	36	50
	II	38	30	33	40	40	44	36	50
	III	38	30	33	40	40	44	36	50
July	I	31	31	26	37	36	40	36	44
	II	31	31	26	37	36	40	36	44
	III	31	31	26	37	36	40	36	44
August	I	38	26	28	38	37	40	34	43
	II	38	26	28	38	37	40	34	43
	III	38	26	28	38	37	40	34	43
September	I	38	33	32	39	39	42	38	44
	II	38	33	32	39	39	42	38	44
	III	38	33	32	39	38	42	38	44
October	I	38	34	30	39	51	39	38	46
	II	38	34	30	39	51	39	38	46
	III	38	34	30	39	50	39	38	46
November	I	36	36	28	37	35	36	35	44
	II	36	36	28	37	34	39	35	44
	III	36	36	28	37	33	39	35	44

Contd...

Table 10.3–Contd...

Month	AEU Decade (10 days)	Reference Crop Evapotranspiration (ET_0) (mm decade^{-1})							
		AEU 2.3	*AEU 3.2*	*AEU 4.1*	*AEU 4.2*	*AEU 4.5*	*AEU 4.6*	*AEU 5.1*	*AEU 5.2*
December	I	43	32	28	37	32	37	34	42
	II	43	32	28	37	32	37	34	42
	III	43	32	28	37	32	37	34	42
January	I	49	34	29	37	32	34	39	45
	II	49	34	29	37	32	34	39	45
	III	49	34	29	37	33	34	39	45
February	I	50	44	34	41	37	43	48	52
	II	50	44	34	41	38	43	48	52
	III	50	44	34	41	39	43	48	52
March	I	54	47	40	50	41	49	55	61
	II	54	47	40	51	42	49	55	61
	III	54	47	40	52	43	49	55	61
April	I	52	46	42	47	48	52	52	61
	II	52	46	42	47	48	52	52	61
	III	52	46	42	48	48	52	52	61
May	I	49	46	40	45	44	50	46	56
	II	49	46	40	45	44	50	46	56
	III	49	46	40	45	44	50	46	56
Total		**1548**	**1317**	**1170**	**1465**	**1419**	**1524**	**1473**	**1764**

Source: CWRDM (2010).

Crop Evapotranspiration

The crop evapotranspiration (ETc) is the evapotranspiration from disease-free, well-fertilized crops, grown in large fields, under optimum soil and water conditions and achieving full production under the given climatic conditions. The amount of water required to compensate the evapotranspiration loss from the cropped field is defined as crop water requirement. Crop evapotranspiration can be calculated from climatic data and by integrating directly the crop resistance, albedo and air resistance factors in the Penman-Monteith approach. As there is still a considerable lack of information for different crops, the Penman-Monteith method is used for the estimation of the standard reference crop to determine its ET_0. Experimentally determined ratios of ETc/ET_0, called crop coefficients (Kc), are used to relate ETc to ET_0.

$$ETc = Kc * ET_0.$$

Differences in leaf anatomy, stomatal characteristics, aerodynamic properties and even albedo cause the crop evapotranspiration to differ from the reference crop evapotranspiration under the same climatic conditions.

Crop Coefficient

The crop coefficient (Kc) integrates the effect of characteristics which distinguish a typical field crop from the grass reference, which has a constant appearance and a complete ground cover. Consequently, different crops will have different crop coefficients. Due to variations in the crop characteristics throughout its growing season, Kc for a given crop changes from sowing till harvest. Finally, as evaporation is an integrated part of crop evapotranspiration, conditions affecting soil evaporation will also have an effect on Kc. The Kc values of selected horticultural crops/crop combinations are given in Tables 10.4 and 10.5.

Table 10.4: Crop Coefficient (Kc) of Selected Horticultural Crops for
Initial, Middle and Late Crop Growth Stages

Sl.No	Crops	Crop Coefficient (Kc)			Sources
		Initial Stage	Middle Stage	Late Stage	
1.	Broccoli, Brussel sprouts	0.70	1.05	0.95	Allen *et al.* (1998)
2.	Cabbage, Carrot, Cauliflower	0.70	1.05	0.95	Allen *et al.* (1998)
3.	Celery	0.70	1.05	1.00	Allen *et al.* (1998)
4.	Garlic	0.70	1.00	0.7	Allen *et al.* (1998)
5.	Lettuce	0.70	1.00	0.95	Allen *et al.* (1998)
6.	Onion	0.70	1.00	1.00	Allen *et al.* (1998)
7.	Spinach	0.70	1.00	0.95	Allen *et al.* (1998)
8.	Radish	0.70	0.90	0.85	Allen *et al.* (1998)
9.	Brinjal, Bell Pepper, Tomato	0.60	1.05	0.90	Allen *et al.* (1998)
10.	Cucumber	0.60	1.00	0.75	Allen *et al.* (1998)
11.	Pumpkin, Winter squash	0.50	1.00	0.80	Allen *et al.* (1998)
12.	Sweet melons	0.50	1.05	0.75	Allen *et al.* (1998)
13.	Watermelon	0.40	1.00	0.75	Allen *et al.* (1998)
14.	Tapioca	0.30	0.80	0.30	Allen *et al.* (1998)
15.	Sweet potato	0.50	1.15	0.65	Allen *et al.* (1998)
16.	Potato	0.50	1.15	0.65	Allen *et al.* (1998)
17.	Turnip	0.50	1.10	0.95	Allen *et al.* (1998)
18.	Beans	0.50	1.05	0.90	Allen *et al.* (1998)
19.	Cow pea	0.40	1.05	0.60-0.35	Allen *et al.* (1998)
20.	Peas	0.50	1.15	1.10	Allen *et al.* (1998)
21.	Soybeans	0.50	1.15	0.50	Allen *et al.* (1998)
22.	Artichokes	0.50	1.00	0.95	Allen *et al.* (1998)
23.	Asparagus	0.50	0.95	0.30	Allen *et al.* (1998)
24.	Tea	1.10	1.15	1.15	Allen *et al.* (1998)
25.	Coffee	0.90	0.95	0.95	Allen *et al.* (1998)
26.	Palm trees	0.95	1.00	1.00	Allen *et al.* (1998)
27.	Cacao	1.00	1.05	1.05	Allen *et al.* (1998)

Contd...

Table 10.4–Contd...

Sl.No	Crops	Crop Coefficient (Kc)			Sources
		Initial Stage	Middle Stage	Late Stage	
28.	Rubber	0.95	1.00	1.00	Allen *et al.* (1998)
29.	Strawberries	0.40	0.85	0.75	Allen *et al.* (1998)
30.	Banana	0.50	1.10	1.00	Allen *et al.* (1998)
31.	Pineapple	0.50	0.30	0.30	Allen *et al.* (1998)
32.	Grapes	0.30	0.70	0.45	Allen *et al.* (1998)
33.	Almonds	0.40	0.90	0.65	Allen *et al.* (1998)
34.	Apples, Cherries, Pears	0.60	0.95	0.75	Allen *et al.* (1998)
35.	Apricots	0.55	0.90	0.65	Allen *et al.* (1998)
36.	Avocado	0.60	0.85	0.75	Allen *et al.* (1998)
37.	Citrus	0.70	0.65	0.70	Allen *et al.* (1998)
38.	Coconut	0.75	0.75	0.75	Menon and Pandalai (1967)
39.	Nutmeg	0.94	0.94	0.94	CWRDM (1997)
40.	Black Pepper	0.70	0.70	0.70	CWRDM (1997)
41.	Cardamom	1.00	1.00	1.00	Not available
42.	Arecanut	0.94	0.94	0.94	Bavappa *et al.* (1982)

Table 10.5: Crop Coefficient (Kc) Values of Crops in Selected Mixed Cropping Systems

Sl.No.	Crop	Crop Coefficient (Kc)	Source
1.	Coconut (Coconut-Arecanut-Pepper)	0.78	CWRDM (1997)
2.	Arecanut (Coconut-Arecanut-Pepper)	0.36	CWRDM (1997)
3.	Pepper (Coconut-Arecanut-Pepper)	0.09	CWRDM (1997)
4.	Nutmeg (Coconut-Nutmeg)	0.87	CWRDM (1997)
5.	Coconut (Coconut-Nutmeg)	0.83	CWRDM (1997)

Factors Determining the Crop Coefficient

Crop Type

Due to differences in albedo, crop height, aerodynamic properties and leaf and stomata properties, the evapotranspiration from full grown, well-watered crops differs from ET_0.

Crop Growth Stages

As the crop develops, the ground cover, crop height and the leaf area change. Due to differences in evapotranspiration during the various growth stages, the Kc for a given crop will vary over the growing period. The growing period can be divided into three or four distinct growth stages: initial, crop development/mid-season and late season.

Soil Cover

When the soil is covered with mulches, evaporation will be altered and hence Kc values will change. Kc values will also be influenced by cover crops and intercrops/mixed crops.

Soil Evaporation

Differences in soil evaporation and crop transpiration between field crops and the reference surface are integrated within the crop coefficient. The Kc for full-cover crops primarily reflects differences in transpiration as the contribution of soil evaporation is relatively small. After rainfall or irrigation, the effect of evaporation is predominant when the crop is small and scarcely shades the ground. For such low-cover conditions, the Kc is determined largely by the frequency with which the soil surface is wetted. Where the soil is wet for most of the time from irrigation or rain, the evaporation from the soil surface will be considerable and Kc may exceed 1. On the other hand, where the soil surface is dry, evaporation is restricted and Kc will be small and might even drop to as low as 0.1.

Crop Water Requirement

Water requirement of a crop is the quantity of water regardless of source, needed for normal crop growth and yield in a period of time at a place and may be supplied by precipitation, or by irrigation, or by both. Crop water requirement is one of the basic needs for crop planning on a farm and the planning of an irrigation project.

Water is needed by a plant mainly to meet the demands of evaporation (E), transpiration (T) and metabolic needs of the plants, altogether known as consumptive use (CU). Since water used in the metabolic activities of a plant is negligible, being only less than one per cent of quantity of water passing through the plant, evaporation (E) and transpiration (T), *i.e.* evapotranspiration (ET) is directly considered as equal to consumptive use (CU). Under field conditions, it is difficult to determine evaporation and transpiration separately, and hence, they are estimated together as evapotranspiration. In addition to ET, water requirement (WR) includes losses during the application of irrigation water to the field (percolation, seepage, and run off) and water required for special operation such as land preparation, transplanting, leaching of salts, frost protection, crop cooling etc.

WR = ET or CU + application losses + water needed for special operations.

Water requirement is therefore a demand, and the supply would consist of contribution from irrigation (IR), effective rainfall (ER) and soil profile contribution from groundwater (S).

$$WR = IR + ER + S$$

Although the values for crop evapotranspiration and crop water requirement are identical, crop water requirement refers to the amount of water that needs to be supplied, while crop evapotranspiration refers to the amount of water that is lost through evapotranspiration. The irrigation water requirement basically represents the difference between the crop water requirement and the sum of effective precipitation and soil profile contribution from groundwater. Water requirement of any crop depends on crop factors such as variety, growth stage, and duration of plant, plant population and growing season. Soil and climatic factors, and crop management practices such as tillage, fertilization, weeding etc influence crop water requirement. Water requirement of crops varies from area to area and even field to field in a farm depending on the above-mentioned factors.

The water requirement for the cropping season computed for major horticultural crops in different agro-ecological units of Palakkad district in Kerala State using ET_0 and Kc values of Tables 10.3–10.5 is given in Table 10.6 as an example. Similar exercise can be done for any locality.

Table 10.6: Water Requirement of Major Horticultural Crops in Different
Agro-ecological Units (AEU) of Palakkad District in Kerala State

AEU Crops and Seasons	Crop Water Requirement (mm)							
	AEU 2.3	AEU 3.2	AEU 4.1	AEU 4.2	AEU 4.5	AEU 4.6	AEU 5.1	AEU 5.2
Cowpea								
Jan to May	480	480	370	430	410	440	475	535
Sept to Dec	350	270	260	340	370	360	325	400
Tapioca								
Sept to May	810	665	590	725	695	740	755	880
Feb to Oct	735	635	585	735	730	785	735	885
Sweet Potato	545	410	375	480	475	485	475	560
Bitter gourd								
Jan to May	545	485	420	490	470	510	540	615
Sept to Dec	415	315	295	395	410	410	375	460
Pumpkin								
Jan to Jul	690	620	540	660	635	665	690	820
Aug to Feb	645	505	450	565	545	545	575	675
Brinjal								
Sept to Jan	420	320	300	390	390	400	370	450
May to Sept	380	330	310	410	400	445	385	485
Tomato								
Oct to Mar	550	430	375	460	420	460	475	555
Mar to Jul	530	495	440	540	530	585	550	675
Bhindi								
Sept to Dec	355	275	265	345	375	365	335	405
Feb to May	485	430	375	440	420	455	485	550
Cabbage	N.A.	N.A.	455	N.A.	N.A.	590	N.A.	N.A.
Cauliflower	N.A.	N.A.	365	N.A.	N.A.	485	N.A.	N.A.
Chilli								
Sept to Jan	485	370	345	450	450	460	430	520
May to Sep	445	385	360	475	465	515	445	560
Banana								
Sept to Aug	1310	1115	1005	1255	1220	1315	1270	1520
May to Apr	1330	1095	990	1230	1200	1270	1255	1485
Coconut	1155	960	875	1095	1075	1145	1100	1320
Nutmeg	1305	1090	990	1240	1215	1295	1250	1495
Pepper	1075	895	815	1025	1000	1065	1030	1230
Arecanut	1455	1205	1095	1375	1345	1435	1380	1650
Cardamom	N.A.	N.A.	1165	N.A.	N.A.	N.A.	N.A.	N.A.
Coffee	N.A.	N.A.	1225	N.A.	N.A.	N.A.	N.A.	N.A.
Tea	N.A.	N.A.	1280	N.A.	N.A.	N.A.	N.A.	N.A.
Rubber	1385	1155	N.A.	1315	N.A.	N.A.	1320	N.A.
Mango	N.A.	N.A.	N.A.	N.A.	N.A.	N.A.	N.A.	1580

Source: CWRDM (2010).

Effective Rainfall

Effective rainfall means useful or utilizable portion of the total rainfall. Effective rainfall has been interpreted differently by different workers and hence the reported values of effective rainfall in a place may vary. FAO has defined annual or seasonal effective rainfall as part of the total annual or seasonal rainfall which is used for crop production at the site where it falls, but without pumping. For agricultural production, effective rainfall refers to that portion of rainfall which can be effectively used by the plants. This is to say that not all rainfall are available to the crops as some of it is lost through runoff (RO) and deep percolation (DP). How much water actually infiltrates into the soil depends on amount, intensity and duration of rainfall, other climatic parameters, land and soil characteristics, soil water content, depth to groundwater table and the nature as well as management of crops. Therefore, the effective rainfall varies with place, land use/land cover systems, and management practices. Rainfall is highly effective when little or no runoff takes place. Small quantities of rainfall are not very effective as water from these rainfalls is quickly lost through evaporation. The evaluation of effective rainfall involves the measurement of rainfall and/or irrigation, losses by surface runoff, percolation beyond root zone and soil moisture use by crops. Precise measurements are often done by weighing type lysimeters. In general, effective rainfall can be calculated using FAO/AGLW formula,

$$Peff = 0.6 * Pdec - 10/3, \text{ when } Pdec < \text{or} = 70/3 \text{ mm}$$

$$Peff = 0.8 * Pdec - 24/3, \text{ when } Pdec > 70/3 \text{ mm}$$

where,

Peff: Effective rainfall for 10 days

Pdec: Rainfall for 10 days

The effective rainfall computed for different agro-ecological units of Palakkad district in Kerala State is given in Table 10.7 as an example.

Table 10.7: Effective Rainfall of Different Agro-ecological Units (AEU) of Palakkad District in Kerala State

Month	AEU Decade (10 days)	Effective Rainfall							
		AEU 2.3	AEU 3.2	AEU 4.1	AEU 4.2	AEU 4.5	AEU 4.6	AEU 5.1	AEU 5.2
Jun	I	125.8	85.5	98.0	152.0	48.0	2.7	117.5	112.1
	II	185.0	143.9	192.8	168.0	13.8	0	108.8	44.0
	III	180.2	171.8	160.0	152.0	10.5	5.1	66.7	17.5
Jul	I	183.8	159.9	175.6	192.0	47.0	6.9	198.6	28.2
	II	154.8	151.8	269.6	232.0	22.7	8.7	158.0	0
	III	162.8	110.7	273.6	110.4	7.6	5.5	38.8	222.8
Aug	I	95.8	99.4	194.0	152.0	24.2	5.7	56.7	76.6
	II	96.6	59.8	146.2	72.0	14.6	0	137.6	28.0
	III	60.9	63.0	199.2	43.2	5.9	7.8	74.2	50.4
Sep	I	61.3	82.0	47.2	88.0	5.7	8.2	32.2	26.8
	II	68.6	94.1	20.8	56.0	18.2	8.2	44.2	6.4
	III	49.4	70.1	44.8	35.2	1.4	8.2	17.9	0

Contd...

Table 10.7–Contd...

Month	AEU Decade (10 days)	Effective Rainfall							
		AEU 2.3	AEU 3.2	AEU 4.1	AEU 4.2	AEU 4.5	AEU 4.6	AEU 5.1	AEU 5.2
Oct	I	65.3	80.6	72.2	72.0	16.0	54.4	84.8	40.0
	II	70.4	101.0	76.0	152.0	69.6	58.4	63.6	4.6
	III	83.2	80.4	44.0	110.0	147.6	50.6	0	51.2
Nov	I	44.0	60.5	68.0	36.0	76.6	54.6	8.7	26.9
	II	24.5	24.7	7.5	32.8	69.4	55.7	36.0	44.2
	III	3.5	2.8	11.2	40.0	29.5	54.7	0	0
Dec	I	0	3.7	5.4	6.3	40.2	24.0	0.8	2.7
	II	0	0.7	0.6	0.3	0	16.2	0.4	0
	III	0	0	3.4	0	2.7	32.0	0	3.1
Jan	I	7.5	0	0	0	0	0	0	0
	II	0	0	0	0	0	0	0	0
	III	0	0	0	0	2.8	0	0	0
Feb	I	0.1	3.2	0	0	0	0	0	0
	II	0	0	0	0.4	0.7	0	0	0
	III	0	0	4.6	0	0	0	0	0
Mar	I	0	4.7	0	0	1.6	0.3	1.3	0
	II	1.3	17.2	4.7	0.1	2.8	0	0	1.3
	III	1.0	3.1	5.1	0	8.7	0	0	1.0
Apr	I	10.0	20.9	13.4	26.4	11.2	8.7	13.9	17.3
	II	10.4	30.8	11.2	25.6	16.6	8.7	16.2	3.4
	III	28.6	31.1	4.9	27.2	7.0	8.9	4.0	0
May	I	25.0	38.6	25.5	32.0	14.4	0	47.4	0
	II	29.3	26.4	33.0	32.0	16.7	2.7	0	0
	III	56.4	81.7	46.6	36.0	40.3	5.7	27.6	44.7

Source: CWRDM (2010).

Irrigation Requirement

Net Irrigation Requirement

Irrigation requirement indicatively represents the fraction of the crop water requirement that needs to be satisfied through irrigation in order to guarantee the optimal growing conditions of the crop. The net irrigation requirement (NIR) is the irrigation water which is to be delivered to the field for the crop to use. It is expressed as depth of water in mm, and is the quantity of water just required to bring the soil moisture level in the root zone depth of the crops to field capacity. Thus, NIR is the difference between field capacity and soil moisture content in the root zone prior to the application of irrigation water. It does not include water applied for leaching of salts, frost protection, crop cooling, or other purposes, even though water for these purposes is required for crop production and is applied through an irrigation system. NIR is computed over a certain period of time and expresses the difference between the crop evapotranspiration and the effective rainfall contributions over the same time step.

$$NIR = WR - ER - S$$

Thus, NIR of a crop is the water requirement of crops exclusive of effective rainfall (ER) and contribution from groundwater (S). Groundwater contribution from water table is considered as negligible for the dry season, especially in the case of garden land crops. The net irrigation requirement for the cropping season computed for major horticultural crops in different agro-ecological units in Palakkad district of Kerala State is given in the Table 10.8.

Table 10.8: Net Irrigation Requirement of Major Horticultural Crops in Different Agro-ecological Units (AEU) of Palakkad District in Kerala State

AEU Crops and Seasons	Net Irrigation Requirement (mm)							
	AEU 2.3	AEU 3.2	AEU 4.1	AEU 4.2	AEU 4.5	AEU 4.6	AEU 5.1	AEU 5.2
Cowpea								
Jan to May	415	355	260	350	345	395	430	515
Sept to Dec	95	70	45	65	90	45	140	205
Tapioca								
Sept to May	545	390	320	465	400	455	580	700
Feb to Oct	230	125	85	245	375	635	280	490
Sweet Potato	370	280	230	270	205	140	365	375
Bitter gourd								
Jan to May	440	310	380	385	385	480	490	590
Sept to Dec	162	119	81	120	138	86	202	266
Pumpkin								
Jan to Jul	420	285	230	360	420	590	450	605
Aug to Feb	450	345	275	355	260	285	430	505
Brinjal								
Sept to Jan	205	155	120	130	130	360	220	275
May to Sept	0	0	0	20	180	100	25	150
Tomato								
Oct to Mar	475	365	305	380	290	265	445	470
Mar to Jul	310	150	110	215	310	510	295	450
Bhindi								
Sept to Dec	90	70	40	80	90	65	140	210
Feb to May	405	280	230	325	240	425	405	525
Cabbage	N.A.	N.A.	0	N.A.	N.A.	275	N.A.	N.A.
Cauliflower	N.A.	N.A.	0	N.A.	N.A.	160	N.A.	N.A.
Chilli								
Sept to Jan	240	185	145	190	140	100	260	325
May to Sep	5	1	0	60	230	425	30	185

Contd...

Table 10.8–Contd...

AEU Crops and Seasons	Net Irrigation Requirement (mm)							
	AEU 2.3	*AEU 3.2*	*AEU 4.1*	*AEU 4.2*	*AEU 4.5*	*AEU 4.6*	*AEU 5.1*	*AEU 5.2*
Banana								
Sept to Aug	670	495	400	610	755	1035	715	1025
May to Apr	790	580	485	695	695	815	830	1055
Coconut	550	380	300	505	585	785	605	840
Nutmeg	640	450	355	580	690	925	695	975
Pepper	210	340	125	155	255	555	355	320
Arecanut	720	520	415	620	790	1040	780	1105
Cardamom	N.A.	N.A.	460	N.A.	N.A.	N.A.	N.A.	N.A.
Coffee	N.A.	N.A.	490	N.A.	N.A.	N.A.	N.A.	N.A.
Tea	N.A.	N.A.	520	N.A.	N.A.	N.A.	N.A.	N.A.

Source: CWRDM (2010).

Gross Irrigation Requirement

Gross irrigation requirement (GIR) is the total quantity of water to be extracted (by diversion, pumping) and applied in reality to the field/irrigation scheme, taking into account water losses. It is obtained by adding the losses in water application to the field and other losses to the net irrigation requirement. Some water is lost while transporting it from its source to the crop root zone. Losses occur due to such causes as leakage from pipelines, seepage and evaporation from open farm channels, and evaporation from droplets sprayed through the air. Because of these losses, more water must be pumped than that required to be stored in the crop root zone. Information on irrigation efficiency is necessary to be able to transform NIR into GIR.

$$GIR = \frac{NIR}{E}$$

where,

GIR: Gross irrigation requirement (in field) (mm)

NIR: Net irrigation requirement (mm)

E: Field application efficiency of the system (decimal fraction)

The field application efficiency depends upon the irrigation system/method adopted. The application efficiency ranges from 65 to 85 per cent for basin irrigation, 60 to 85 per cent for border irrigation, 50 to 85 per cent for furrow irrigation, 60 to 85 per cent for surge irrigation, 60 to 90 per cent for sprinkler irrigation and 70 to 95 per cent for micro/drip irrigation. The gross irrigation requirement can be determined for a field, for a farm, for an outlet command area or an irrigation project, depending on the need, by considering the appropriate losses at various stages of the crop.

Irrigation Scheduling

Irrigation scheduling is the process by which the timing and quantity of irrigation water application are determined. Excess application results in loss of plant nutrients, water logging and salinity. Lower application results in less than optimum production even when other inputs are given. Application of the right quantity of water at the right time requires a basic knowledge of soils, crops and the climatic conditions. There are several techniques of determining the timing (frequency) and quantity of irrigation application required at farm level. They can be broadly grouped under: 1) Depth/Interval irrigation approach, 2) Soil moisture depletion approach, 3) Climatological approach, and 4) Plant water status approach. The traditional, arbitorily chosen depth/interval for irrigation often results in the supply of excess or less of water than required at different stages of crop growth. Determination of soil moisture status for irrigation scheduling by gravimetric method or by means of instruments like tensiometer, gypsum block, neutron probe, soil moisture meter etc. is scientific, but expensive and require the use of equipment. Data from evaporimeter and lysimeters are location-specific and require trained manpower for measurement and application. Plant parameters like leaf temperature, leaf resistance to vapour diffusion etc. are good indices for irrigation scheduling. But, expensive equipment like infrared thermometer and porometer are needed for the measurement of leaf temperature and leaf diffusion resistance.

Management Allowable Depletion

Management allowable depletion (MAD) is the quantity of soil moisture depletion allowed without affecting the growth and yield of the crop. The maximum management allowable soil moisture depletion factor (P^1) depends mainly on the type of crop and evaporative demand. This fraction of available water content which can be used by the crop without affecting its evapotranspiration and/or growth is actually the depth of water readily available to the crop. It is dependent on the crop's sensitivity to soil moisture deficit, soil type and the irrigation method adopted. MAD is given by the function:

$$MAD = P^1 * AWC * D$$

where,

P^1: Crop sensitivity factor to soil moisture deficit (decimal fraction),

AWC: Available water content (mm m^{-1}), and

D: Depth of root zone (m)

The depth of root zone and management allowable soil moisture depletion (MAD) in selected horticultural crops is given in Table 10.9.

Table 10.9: Depth of Root Zone and Management Allowable Soil Moisture Depletion Factor (P^1) in Selected Horticultural Crops

Crops	Depth of Root Zone (m)	Management Allowable Soil Moisture Depletion Factor (P^1)
Cow pea	0.5-0.7	0.45
Tapioca	0.5-0.8	0.35
Sweet potato	1.0-1.5	0.65
Potato	0.4-0.6	0.35
Bitter gourd	0.6-1.0	0.50
Pumpkin	1.0-1.5	0.35
Watermelon	0.8-1.5	0.40
Chillies	0.5-1.0	0.30

Contd...

Table 10.9–Contd...

Crops	Depth of Root Zone (m)	Management Allowable Soil Moisture Depletion Factor (P')
Brinjal	0.7-1.2	0.45
Tomato	0.7-1.5	0.40
Onions	0.3-0.6	0.30
Bhindi	0.5-1.0	0.40
Cabbage	0.5-0.8	0.45
Carrots	0.5-1.0	0.35
Cauliflower	0.4-0.7	0.45
Banana	0.5-0.9	0.35
Pineapple	0.3-0.6	0.50
Apples, Cherries, Pears	1.0-2.0	0.50
Citrus	1.2-1.5	0.50
Mango	1.0-2.0	0.60
Coconut	0.7-1.1	0.65
Pepper	0.3-0.5	0.45
Cardamom	0.3-0.5	0.50
Nutmeg	0.7-1.0	0.50
Arecanut	0.5-0.8	0.50
Coffee	0.9-1.5	0.40
Tea	0.9-1.5	0.40
Cocoa	0.7-1.0	0.30
Rubber	1.0-1.5	0.40

Source: Allen *et al.* (1998).

Depth of Irrigation

Depth of irrigation application is the depth of water that can be stored within the root zone between field capacity (FC) and the management allowable soil moisture depletion (MAD) level for a given crop, soil and climate. Thus, the depth of irrigation water applied in a single irrigation since the previous irrigation varies with soil type, crop, depth of root zone, crop growth stages, irrigation method and irrigation efficiency. It is given by the function:

$$d = \frac{P^1 * AWC * D}{E}$$

where,

d: Depth of irrigation application (mm)

P^1: The crop sensitivity factor to soil moisture deficit (decimal fraction)

AWC: Available water content (mm m^{-1})

    D: Depth of root zone (m), and

    E: The irrigation application efficiency factor

Since water storage capacity of clayey soils is high, more quantity of irrigation water can be applied per irrigation at wider intervals. But, in the case of sandy soils, less quantity of water has to be applied per irrigation at closer intervals. The deeper the root zone of the crop, more water can be stored in the extractable region of the soil. Since the root system is shallow and transpiration is low in early growth stages of a crop, less quantity of water may be applied at closer intervals. Depending upon the irrigation method and the irrigation efficiency, the total depth of water to be applied varies.

Irrigation Interval

Irrigation interval refers to the frequency of applying water to a particular crop at a certain stage of growth and is expressed in days. Timing of irrigation should conform to soil water depletion requirements of the crop which are shown to vary considerably with evaporative demand, rooting depth and soil type as well as with stages of crop growth. Irrigation interval is given by the function,

$$I = \frac{MAD}{Tc - ER - S}$$

where,

       I: Irrigation interval (days)

  MAD: Management allowable depletion (mm)

    ETc: Mean crop evapotranspiration per day during a time period considered (decade or month) (mm)

    ER: Effective rainfall (mm)

     S: Contribution from groundwater (mm)

Conclusion

The availability of water in the right quantity at the right time is essential for good crop growth and yield. The required amount of water as well as the time of requirement may differ in different crop plants. One of the prerequisites relating to water use efficiency in crop production is the knowledge of the basic soil-plant-water relationships and methods of predicting the water requirement of the crop. This chapter gives basic information in this field. The development planning for agricultural production based on the appreciation of the characteristics of the natural resources like climate, soil and water will go a long way in the rational management of the resources.

References

Allen, R.G., Pereira, L.S., Raes, D. and Smith, M., 1998. Crop evapotranspiration: Guidelines for computing crop water requirements. *FAO Irrigation and Drainage Paper*. Food and Agricultural Organization, Rome, Italy, 56: 300.

Bavappa, K.V., Nair, M.K. and Premkumar, 1982. *Arecanut Palm*. Central Plantation Crops Research Institute, Kasargod, p. 340.

Blaney, H.F. and Criddle, W.D., 1950. Determining water requirement in irrigated areas from climatological and irrigation data. *USDA*, SCS–TP 96.

CWRDM, 1997. Water requirement of multiple cropping system with spices. Final report of the project submitted to ICAR, New Delhi. Centre for Water Resources Development and Management, Kozhikode, p. 102.

CWRDM, 2010. Crop water requirement for different agroecological zones of Palakkad district. Centre for Water Resources Development and Management, Kozhikode, p. 170.

Doorenbos, J. and Pruitt, W.C., 1975. Guidelines for Predicting Crop Water Requirements. *FAO Irrigation and Drainage Paper*. Food and Agricultural Organization, Rome, 24: 197.

Doorenbos, J. and Pruitt, W.C., 1977. Guidelines for Predicting Crop Water Requirements. *FAO Irrigation and Drainage Paper*, Food and Agricultural Organization, Rome, 24(Revised): 144.

FAO, 2009. *User Guide on CROPWAT 8.0 for Windows*. Food and Agricultural Organization, Rome, Italy.

ICAR, 1977. *Water Requirement and Irrigation Management of Crops in India*. Water Technology Centre, Indian Agricultural Research Institute, New Delhi, p. 402.

Menon, K.P.V. and Pandalai, K.M., 1967. *The Cococnut Palm: A Monograph*. Indian Central Coconut Committee, Ernakulam, p. 384.

Michael, A.M., 1978. *Irrigation: Theory and Practice*. Vikas Publishing House Pvt Ltd., New Delhi, p. 801.

Penman, H.L., 1948. Natural evapotranspiration from open water, bare soil and grass. *Proc. Royal Soc. London*, pp. 120–145.

Plaut, Z. and Moreshet, S., 1973. Transport of water in plant-atmosphere system. In: *Arid Zone Irrigation*, (Eds.) B. Yarcon, E. Danfors and Y. Vaadia. Chapman and Hall Ltd., London, Springer-Verlag, Berlin, Hiedelberg, New York, pp.123–141.

Varadan, K.M., 1996. *Water Module for Upland Crops of Kerala*. Centre for Water Resources Development and Management, Kozhikode, p. 59.

Varadan, K.M., Madhava Chandran, K. and Lakshmanan, K., 1990. *User Guide on Farm Irrigation Scheduling for Upland Crops of Kerala*. Centre for Water Resources Development and Management, Kozhikode, Kerala, p. 42.

Thornthwaite, C.W., 1948. An approach toward a rational classification of climate. *The Geographical Review*, 38(1): 55–94.

2012, Nutri-Horticulture   *Pages 195–220*
Editor: Professor K.V. Peter
Published by: DAYA PUBLISHING HOUSE, NEW DELHI

Chapter 11

Rainwater Harvesting Methods

K.P. Visalakshi

Professor (Agricultural Engineering),
Kerala Agricultural University, P O KAU – 680 656, INDIA
E-mail: visalam2009@yahoo.co.in

Introduction

Water is undoubtedly the most precious and unique natural resource that exists on our planet. This provides life support system for human beings, vegetation and animals. The phenomenal growth in population during last few decades has resulted in excessive use of water resources. Improper management or over exploitation of ground water and its inadequate replenishment has led to gradual decline of water table in many parts of the country. Proper planning, development, management and optimal utilization of water resources are of paramount importance at this juncture. Rainwater harvesting appears to be one of the most promising alternatives for supplying freshwater in the face of increasing water scarcity and escalating demand. Identification of simple, location specific, reliable and environmental friendly technologies of rainwater harvesting and the promotion of these technologies are urgently needed.

Methods of Rainwater Harvesting

Rain is considered as the first form of water in the hydrological cycle, and is the primary source of water. Rivers, lakes and ground water are all secondary sources of water. We depend mainly on such secondary sources of water, forgetting that rain is the ultimate source that feeds all these seconday sources. Water crisis situation occurs only because, effective collection and storage of rainwater has been ignored. If only we save each and every drop of water and recharge the under ground aquifer, we can rescue ourselves from this perpetual problems of water scarcity. Though water harvesting is an age old practice, it is emerging as a new paradigm in water resources development and management due to the recent efforts to promote water harvesting and groundwater recharge in urban and rural areas.

The basic principle of rainwater harvesting is to 'catch the water where it falls'. It involves collection, storage and recycling of rainwater for domestic, agricultural, or industrial purposes.

Rainwater harvesting does not imply harvesting of water received directly from rains only, but also from all other natural resources like rivers, streams, lakes, ponds, wells, water springs, groundwater aquifers etc. since all such resources draw water from the rain (and snowfall) itself.

Components of Rainwater Harvesting

☆ Collection and storage of rainwater

☆ Effective utilization of stored water

☆ Augmentation of natural water resources.

Rainwater harvesting is mainly done for direct use and for augmenting groundwater storage.

Rainwater Harvesting for Direct Use

The technique of rainwater harvesting for direct use involves catching the rain from localized catchments such as rooftops of buildings, plain and sloping ground surfaces etc. The rainwater that falls on these catchments is diverted into dugout ponds or tanks to use during dry periods.

Rainwater Harvesting from Roof Top Catchments

In most basic form, it is the collection and storage of rainwater from rooftop of buildings and utilization during summer season. Rooftop rainwater harvesting can be used either for storage of rainwater runoff in tanks for domestic needs or for recharging the ground water or both. The components of the system are the roof catchments, collection device, the conveyance system, first flush valve, filter unit, storage tank and overflow pipe to recharge pit (Figure 11.1).

As rooftop is the main catchment, the amount and quality of rainwater collected depends on the area and type of roofing material. Reasonably pure rainwater can be collected from roofs constructed with RCC slab, galvanized / corrugated iron, aluminium or asbestos cement sheets, tiles and thatched roofs. In case of thatched roofs a covering using silpaulin sheet will make the water collection easier. Roof catchments should be cleaned regularly to remove dust, leaves and bird droppings so as to maintain the quality of water. The amount of water that is received in the form of rainfall over an area is called the rain water endowment of that area. Out of this, the amount that can be effectively harvested is called the water harvesting potential. The collection efficiency or the coefficient of runoff accounts for the fact that all the rain water falling over an area cannot be effectively harvested. The runoff coefficients vary from 0.7 to 0.9 with the type of roofing materials.

The storage capacity needed should be calculated taking into consideration the size and type of catchments, the length of dry spells, the amount of rainfall, the per capita water consumption rate, the cost of the system and its reliability for assured water supply.

The storage capacity is calculated as:

$$Q = (n \times q \times t) + e$$

where,

n: Number of persons in the family

q: Per capita consumption, l/day

t: Number of non rainy days or dry period for which water is needed and

e: Evaporation losses from the tank, l (negligible if tank is covered at top)

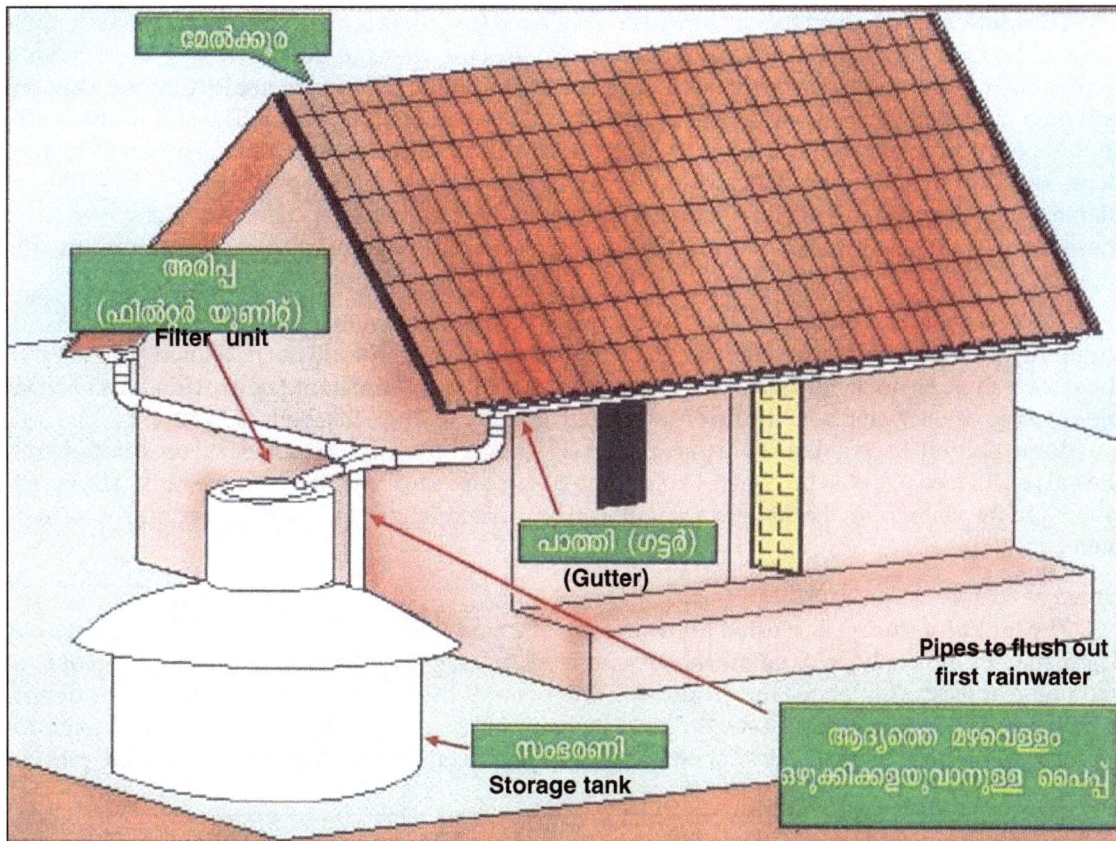

Figure 11.1: Components of Roof Top Rainwater Harvesting System

For Example

The average annual rainfall of Kerala being 3m, the quantity of rain water falling on the roof top of area 50m² is equal to

Area (m²) x annual rainfall (m) x runoff coefficient

= 50 m² x 3m x 0.90

= 135 m³

= 135,000litres

If a five-member family requires 8 litres per head per day for drinking and other domestic purposes, then water requirement for 200 non rainy days is equal to

200 x 5 x 8 = 8000 litres.

Therefore a tank of 10000 litre capacity will be suitable for that family to meet the essential domestic requirements for 200 dry days.

The excess water from the catchments can be diverted to a recharge pit or diverted to a nearby well which will help to raise the level of groundwater table.

The rainwater falling on the catchments is conveyed to storage tank by gutters and pipes. Gutter collects the rainwater run off from the roof and conveys the water to the down pipe. Gutters with semicircular cross section can be made by cutting longitudinally large diameter PVC pipes. They are laid on a mild slop (0.5%) to avoid the formation of stagnant pools of water. The size of the gutter should be according to the flow during the highest intensity rain .A semi circular section of 150mm diameter is enough to carry away most of the intense rainfall. A vertical down pipe of 75 – 100mm diameter may be required to convey the harvested rainwater to the storage tank. An inlet screen (wire mesh) may be fitted at the inlet of down pipe to prevent entry of dry leaves and other debris into the pipe.

The first collection of water during initial one or two rains from the roof is likely to contain dirt, droppings and debris present on the roof. This contaminated water should not be allowed to enter into the storage tank. Hence a first flush diversion system or a by- pass line should be provided in the down pipe to dispose the water from the first few rainfall. In order to drain this polluted water, a pipe and a valve assembly are fixed to the down pipe (Figure 11.2). After the first rain is washed out through this, the valve is closed to allow the water to enter the down pipe and reach the storage tank. The excess water and the water from the first few rainfalls can be utilized for groundwater recharging by various methods.

Storage Tanks

The tanks for storing harvested rainwater may be made of RCC, ferro cement, fiber, bricks etc. These may be either above or below the ground level. In case of underground tanks the top of tank should be at least 30cm above the ground level to prevent the entry of sand, dirt and other debris. Precautions required in the use of storage tanks include provision of an adequate enclosure to minimize contamination from human, animal or other environmental contaminants, and a tight cover to prevent

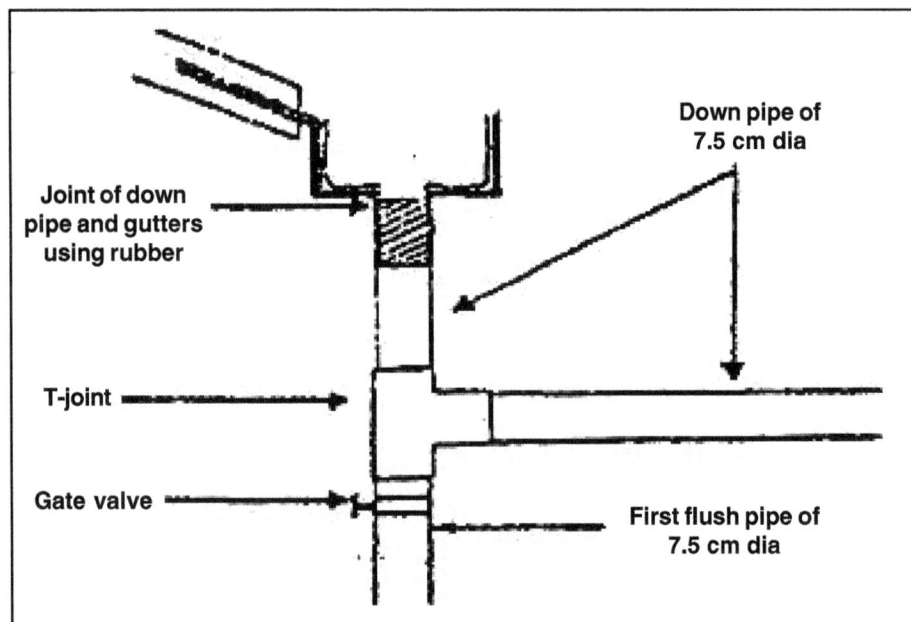

Figure 11.2: Down Pipe and First Flush Arrangement.

algal growth and breeding of mosquitoes. The storage tank is provided with pipe fixtures at appropriate places to draw the water, to clean the tank and to dispose off the excess water.

Among the different types of storage tanks the ferro cement tank has gained wide popularity due to its better performance and comparatively low cost. The tanks are constructed over a suitable saucer shaped concrete or random rubble foundation (Figure 11.3). In the skeletal cage method of its construction, a cylindrical shaped skeletal cage is prepared at the site using mild steel rods(6mm - 8mm) based on the required size and dimensions of the storage tank. This is used as a frame work for casting the storage tank at the site. The cage is then wrapped with two layers of chicken mesh (22 – 20 gauge, 25mm opening), one inside the cage and the other outside (Figure 11.4). Plastering with cement mortar 1:2 is done carefully to fill the gap between the two mesh layers completely (Figure 11.5) . Depending on the capacity of the tank the wall thickness may vary from 2.5 cm to 10 cm. Ferro cement tanks are much better than concrete tanks in its strength and the chances of root penetration is comparatively lesser in this case. The lid of the tank usually dome shaped is also made of ferro cement.

The dirt and debris if any in the water coming through the down pipe are to be removed before entering the storage tank for which a filter unit is also provided over the storage tank. This unit is a chamber filled with filter media such as coarse sand, charcoal, coconut fiber, pebbles and gravels. A plastic/aluminium bucket or a chamber made of ferro cement can be used as filter unit. The container is provided with perforated bottom to allow the passage of filtered water to the tank.

The filter media is arranged from top to bottom in the chamber as detailed in Table 11.1.

Figure 11.3: Foundation for FC Tank

Figure 11.4: Skeletal Cage Wound Over by Chicken Mesh

Figure 11.5: Plastering the FC Tank

Studies have shown that the water, which has purified through this filter and kept closed, will remain safe for drinking for a period of up to six months. To maintain the purity of water the filter bed must be washed or changed twice in a year. A manhole should be provided at the top of the tank for manual cleaning and that should be covered in order to prevent the entry of insects, dust and other foreign materials (Figure 11.6).

Plastic tanks are also being increasingly used as they provide good water quality. Plastic tanks

Table 11.1: Filter Media

Layer	Materials	Thickness (mm)
1	Gravel of 20mm size	50
2	Charcoal	50
3	Coarse sand	50
4	Coconut fiber	50
5		
	Pebbles of 10mm size	10
6	Gravel of 20mm size	50

Figure 11.6: Manhole and Filter Unit

should be constructed from food-grade plastic material in order to prevent leaching of any potentially harmful compounds into the water. Storage tank materials should prevent or minimize light penetration to reduce algal growth and other biological activity, which helps to maintain water quality. For this reason, clear plastic or fiberglass tanks are not recommended.

Comparison of different types of storage tanks in a study under SWAJAL project in Tehri Garhwal, Uttaranchal during 1998-99 as given in Table 11.2 indicates the choice of ferro cement tanks for storage of rainwater.

Table 11.2: Comparison of Different Types of Storage Tanks

Parameters-Storage Tank Type	Capital Cost/Litre (Rs)	Construction	Maintenance
RCC	2.25-2.75	*In situ* construction needs accurate and expensive frame work and skilled technicians	Repairing of leakages expensive and success rate of repair low
Brick/stone masonry	3.5-4.5	*In situ* construction needs skilled labour	
Synthetic polymers	4	Produced in factory, have to be transported to site, can be damaged during transportation	Repairs of leakage not possible locally
Galvanized iron sheet	2.5	Requires skilled brazing/gas welding jobs for *in situ* construction	Need replacement after 5-7 years due to corrosion
Ferro cement	1.75	*In situ* construction possible, construction is simple and easy to learn	Highly resistant to cracking and easy in repairs

At Chellanam, a coastal Panchayat in Ernakulam district of Kerala, the local people developed a very low cost method of roof water harvesting. This method does not require any tank. If sweet water is slowly released on saline water, the former stands as a separate layer on the top of the later. This is because of the lower specific gravity of sweet water as compared to the saline water. This method was found suitable only in coastal belts where water table is shallow and water is saline.

After storing the harvested water in storage tanks, the excess flow can be utilized for recharging the groundwater. It can be recharged to the existing open wells/tube well, ponds/pits and abandoned tube wells/open wells (Figure 11.7). If there is no drinking water scarcity, there is no need to construct the storage structure and the whole harvested water can be directly fed to the above mentioned water sources. Care must be taken to see that water should pass through filter media before putting into the water sources.

Quality of Harvested Water

In India, about 21 per cent of all communicable diseases (11.5 per cent of all diseases) are water borne in nature. According to an estimate, 73 million working person days are lost every year owing to the people falling ill due to water borne diseases. Diarrhoea, which is the most prevalent water borne disease, is responsible for 25-30 per cent of deaths among children below five years of age. Also, epidemics of infectious hepatitis, food poisoning and typhoid are quite common. India incurs an expenditure of about Rs. 36,600/- crores per year on treatment of water related diseases. (Development Alternatives, April 2001). Only 7 per cent of the Indian population has access to municipal sewage system. This has implications on the sanitary conditions and contamination of water. To prevent the incidence of water borne diseases, there is need to improve the quality of drinking water.

Rainwater collection systems are commonly believed to provide safe drinking water without treatment because the collection surfaces (roofs) are isolated from many of the usual sources of contamination (*e.g.* sanitation systems).Although roofs are higher than the ground, dust and other debris can be blown onto them, leaves can fall from trees, and birds and climbing animals can defecate upon them. The quality of drinking water can be much improved if this debris is not allowed to enter the storage tank. The more do we keep a roof clean, the better the water quality will be.

Chlorination to kill bacteria is widely recommended as sterilization for rainwater collection systems but generally chlorinated water is not well liked by users and the chemicals used can be

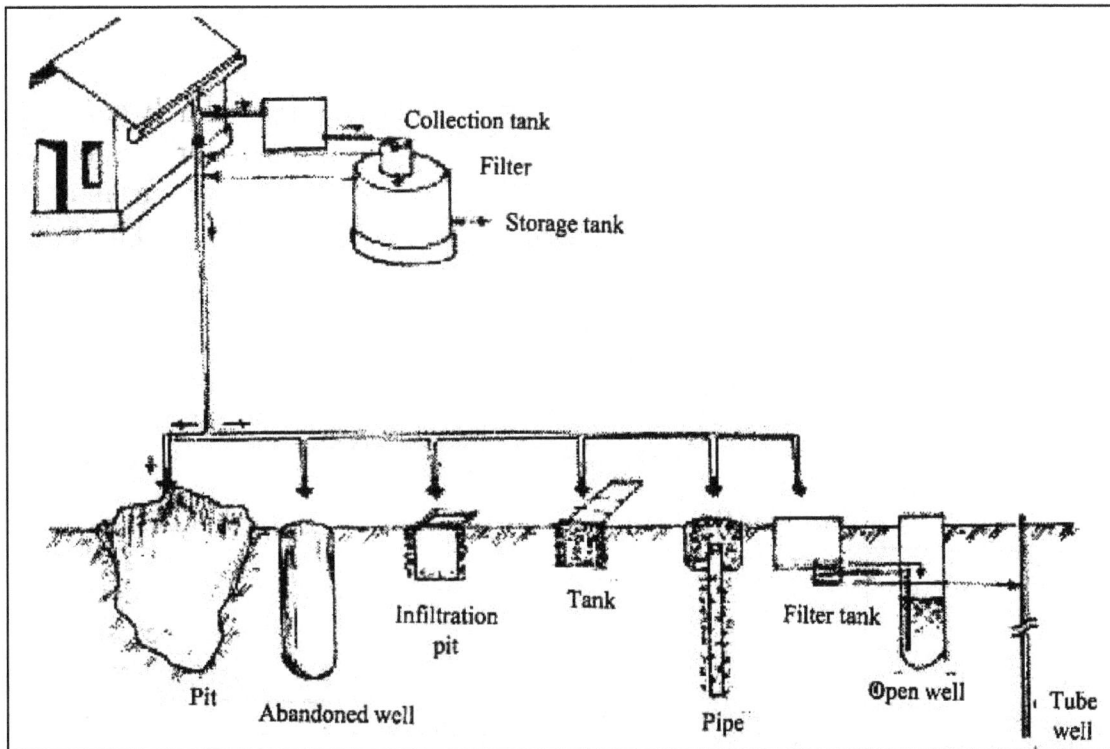

Figure 11.7: Different Methods of Groundwater Recharge Using Excess Flow from Roof Top

dangerous if misused. For this reason chlorination of the water is suggested only where one or more of the following situations are present:

1. A known bacterial risk has been identified through water testing
2. Individuals are getting sick as a result of drinking water
3. It is not feasible to completely empty a tank for cleaning
4. An animal or faecal material has entered the tank.

Chlorination is done with stabilized bleaching powder (calcium hypochlorite- $CaOCl_2$) which is a mixture of chlorine and lime. Chlorination can kill all types of bacteria and make water safe for drinking purposes. About 1 g. of bleaching powder is sufficient to treat 200 liters of water. Chlorine tablets are also available commercially. One tablet of 0.5 g. is enough to disinfect 20 liters of water. Remember to allow 24 hours after the time of chlorination for the chlorine to disinfect the tank before drinking. Chlorine is heavier than water, so it will tend to sink towards the bottom of the tank. Any chlorine smell and taste in the water should dissipate after a short time. If the taste of chlorine is unacceptable, boiling of water for 10-20 minutes before drinking is a suitable alternative to provide safe drinking water.

Water quality testing should be regularly carried out by a relevant agency. If water quality testing is possible, the main focus should be on microbiological testing using tests such as faecal coli forms, Enterococci, and the simple H_2S test. World Health Organization guidelines (WHO 1996) state that faecal bacteria should not be detectable per100ml. of sample. The physical parameters, pH and turbidity should also be measured and compared to WHO guidelines. Rain is considered acidic when the pH is <5.6 and levels below this may cause corrosion of metal roofs and fittings. Heavy metals (*e.g.* lead, copper, cadmium, zinc) should also be monitored periodically, particularly where volcanic or industrial discharges to the air are present.

Water Harvesting from Land Surface

Rain falling on the land surface can be harvested simply by improving the runoff capacity of the land surface through various techniques, including collection of runoff with drain pipes and storage of collected water. Compared to rooftop catchment techniques, ground catchment techniques provide more opportunity for collecting water from a larger surface area and is mainly suitable for storing water for agricultural purposes. Percolation and seepage losses can be controlled by reducing the wet area, self sealing by silting, compaction, natural clay soil lining, bentonite lining, alkali soil lining, soil-cement lining or lining with polythene sheets.

Rainwater and runoff from fields can be harvested and stored in small farm ponds on individual farmer basis or larger reservoirs on community basis. The pond should be located in the farm where catchment is adequate and sufficient runoff is available for collection. Also the command area should be maximum (in case of irrigation) and water should be applied by gravitational flow without involving any external energy source such as water pump. The site should be such that higher storage-excavation ratio (ratio of volume of storage capacity to the volume of excavated earth) can be achieved. Based on the method of construction and the suitability for different topographic features, ponds could be classified into three types, *viz.*, dugout farm ponds suited for flat topography, embankment ponds for hilly and rugged terrains, and dugout cum embankment type ponds.

Trapezoidal farm ponds of suitable size and 1.0 to 5.0 m depth can be constructed with 1:1 side slope at a suitable location in the field. About 200 m³ capacity ponds are sufficient for vegetable cultivation in about 400 m² area. The seepage losses of stored rainwater can be prevented by lining the pond with suitable sealants, the choice of which depends upon the texture of the soil, availability of lining material, durability and cost. Soil cement, cement concrete, brick/stone lining, chemical additives and different types of geo membranes can be used as lining material. HMHDPE film, hot mixed asphaltic concrete, glass fiber, asbestos felt, asbestos fiber etc., are also used for this purpose. However, LDPE black polyethylene sheet and UV- resistant blue silpolin sheet were found the best lining materials. The sheets of required thickness depending on the storage capacity of ponds should be used. Poly sheets are generally available in widths of 1.8 to 14 m. The sheets of required width should be procured to avoid jointing. If needed, the poly sheet pieces can be joined together by heat sealing.

Before laying the sheet, the walls of the pond must be smoothened by removing any protruding objects like pebbles, roots etc and pasting a thin layer of mud on the walls to avoid puncturing of the sheet. A bund of 1m height should be provided around the pond to prevent direct entry of runoff and the sheet should be properly anchored in a shallow trench 30cm x 30cm dug around the outer periphery of the bund (Figures 11.8 and 11.9). Abandoned and unused quarries can also be used for rainwater harvesting by some modifications for easy runoff collection from the surrounding catchments or direct collection from roof tops through pipes and gutters.

If the location permit gravity irrigation, fitting a suitable outlet pipe at the bed level across the wall of the pond may make provision for drawing water from the pond. If the lining material used is not UV resistant, it must be protected by covering with any inert material like bricks or soil, or both, which is

Figure 11.9: Lining of pond

not damaged by sunlight. Although the black LDPE sheet and silpolin sheet are UV resistant, the life span can be increased by proper covering by the above materials.

The runoff from the catchments can be accelerated by use of dense vegetation cover such as grass as it helps to maintain high rate of runoff with minimum soil erosion. Use of geo textiles, plastic sheets, asphalt or tiles along with slop can further increase efficiency by reducing both evaporative losses and soil erosion. Smoothing and compacting of soil surface using graders and rollers will also help to increase surface runoff with minimum erosion. A silt trapping structure should be provided in case of collection of surface runoff.

Rainwater Harvesting for Augmenting the Groundwater Storage

Groundwater is derived primarily from rain and snow melt which infiltrates the land surface and slowly percolates to the water table. Groundwater recharge can take place naturally and by man made constructions/modifications. The recharge that takes place in a natural condition through infiltration of rainfall is the most important source of natural groundwater recharge. Rainfall infiltration primarily depends upon duration and intensity of rainfall, soil moisture characteristics, topographic slopes, land use pattern, agronomic practices, weather conditions preceding, during and succeeding rainfall periods and depth of water table. These conditions further impose a limit on threshold value of rainfall required to affect groundwater recharge. The component of rainfall contribution as infiltration to groundwater varies from 3 to 25 per cent in different hydro geological situations (Table 11.3) and this needs to be supplemented by artificial recharge.

Figure 11.10: Anchoring the Edges

Artificial groundwater recharge is the process by which man fosters the transfer of surface water into the groundwater system. The choice of method depends on local topography, hydro-geological characteristics of aquifers, soil characteristics, availability of recharge water, sediment in recharge water, fluctuation in water levels etc. Artificial recharge projects serve water conservation, overcome problems due to over draft, water management, control sea water intrusion etc.

Table 11.3: Rainfall Contribution as Infiltration in Different Hydro-geological Situations

Hydro-geological Situations	Rainfall Infiltration Factor (Per cent of normal rainfall)
Alluvial areas	
(a) Sandy areas	20 to 25
(b) Areas with higher clay content	10 to 20
Semi consolidated sandstones, friable and highly porous	10 to 15
Hard rock areas	
(a) Granite terrain	
(i) Weathered and fractured	10 to 15
(ii) Un-weathered	5 to 10
(b) Basaltic terrain	
(i) Vesicular and jointed basalt	10 to 15
(ii) Weathered basalt	4 to 10
(c) Phyllites, limestones, sandstones quartzites, shales etc.	3 to 10

Source: Sinha *et al.* (1995).

The methods used for artificial groundwater recharge are as follows:

☆ Water spreading methods like, flooding, percolation ponds/tanks, ditch and furrow system etc.

☆ Subsurface or injection methods like injection wells or recharge wells, recharge pits/trenches, recharge shafts, abandoned/dug wells, percolation well cum bore pit etc.

☆ Stream flow harvesting through check dams, vented cross bars, subsurface dykes etc.

☆ *In-situ* rainwater harvesting through mechanical and agronomic measures.

Water Spreading Methods

Water is allowed to spread over the ground surface which will infiltrate and then percolate to the water table. The basic concept of this method is to increase the area of contact and infiltration opportunity time so as to allow maximum quantity of water to enter into the aquifers. In this method of flooding, water is spread evenly over a large area so that water moves in a thin layer over the land. It is recommended in flat topography and at places where sufficient quantity of water is available.

Percolation Pond/Tank

Percolation ponds/tanks are small water storage structures constructed to harvest runoff from the catchments and to impound for longer time to facilitate percolation of impounded water into the soil substrata both vertically and laterally, thereby recharging groundwater storage in the zone of influence of the ponds (Figure 11.10). This helps to increase the yield of surrounding wells, especially on the downstream side of the pond. The water stored in the percolation ponds can also be used for irrigation for livestock. Besides, it checks soil erosion, reduces flash floods and keeps the drought at bay.

The success of this system mainly depends on the aquifer properties. It is very effective in highly permeable and unconfined aquifers. The natural depressions, existing dry or abandoned wells/tanks

Figure 11.10: Percolation Pond

etc., can be converted into percolation ponds/tanks. Surplus water from reservoirs, storm water, tank, canal etc. can also be diverted into these structures to directly recharge the aquifer. Since these ponds are constructed to recharge the nearby ground, its benefit will be obtained in all the water sources around it.

Ditch and Furrow

Shallow, flat bottomed and closely spaced furrows or ditches are made originating from a supply ditch and descending down the topographic slope to a drainage ditch. Water is distributed through these ditches from the supply channel to the drainage outlet during which infiltration will occur contributing to the groundwater. The width of ditch may vary from 0.3 to 1.8m and should have a mild slope to maintain flow velocity with minimum sediment deposit.

Subsurface or Injection Methods

Injection Wells/Recharge Wells

In this technique, 1 to 2 m. wide and 2 to 3 m. deep trench is excavated, the length of which depends on the site availability and volume of water to be handled. An injection well of 100 to 150 mm dia meter is constructed in the trench, piercing through the layers of impermeable horizons to the potential aquifer reaching about 3 to 5 meters below water levels from the bottom of the trenches. Depending upon the volume of water to be injected, the number of injection wells can be increased to enhance the recharging rate.

This method can be applied for all aquifers confined or unconfined, situated at any depth below ground surface and hence it is more practicable where deep depleted aquifers are to be recharged. These are comparatively costlier and require specialized techniques (Figure 11.11).

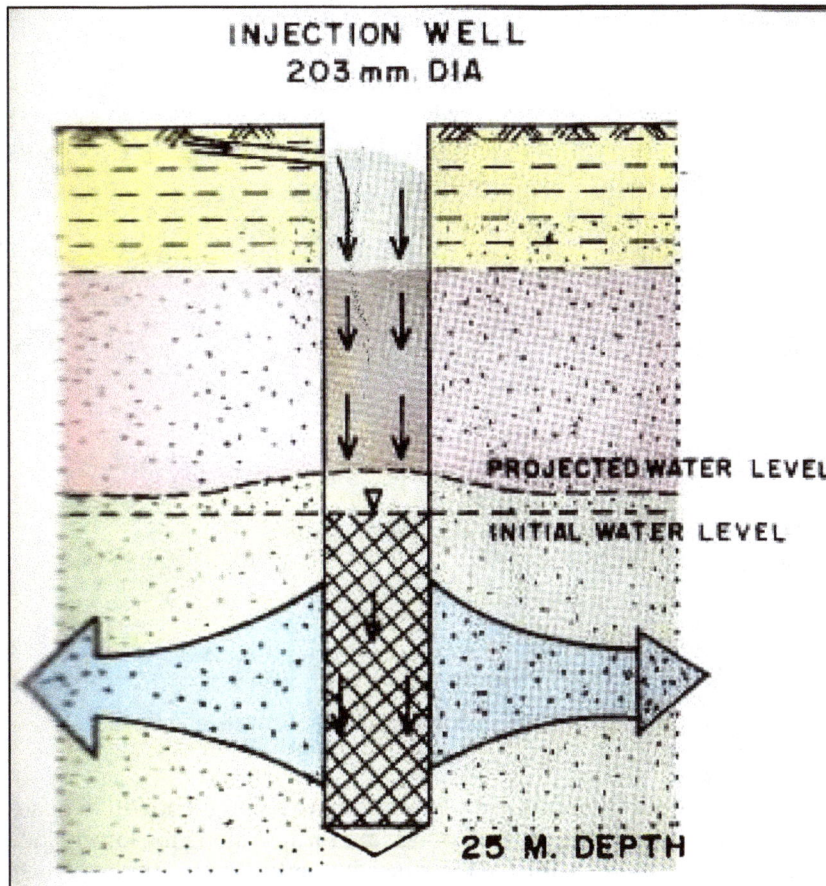

Figure 11.11: Injection Well

Recharge Pits and Trenches

Recharging through pits is practiced in areas where sub-strata restrict the downward passage of water and where aquifer is situated at a moderate depth. Pits penetrating such layers can supply water directly to underlying materials with higher infiltration rates. These pits may be of any shape *i.e.* circular, square or rectangular, usually, 1 to 2m dia/wide and 2-3m deep. These are filled with boulders, gravels and coarse sand (Figure 11.12). Pits and trenches are also used for groundwater recharging through rain harvest from roof tops.

In non-water logged areas, small rain pits can be made at various locations and around wells, which will enhance the percolation of rainwater and increase the level of water table. The dimensions of the pit depend upon soil texture, topography and amount of rainfall received. 1.00m x 1.00m x 0.50m size is suited in all types of terrains. During each rain, water gets collected in these pits and gets absorbed into the earth (Figure 11.13). Care should be taken to dig the pits where slope is more than 16°, since it may cause occurrence of land sides.

Figure 11.12: Pits Filled with Boulders, Gravels and Sand

In sloppy areas, staggered trenches across the slope are recommended. Trenches of 60cm wide and deep and about 2.5m long will be suited. Bunds are formed along the lower side of the trenches using the excavated earth (Figure 11.14). On these bunds plants like pineapple, **vetivar** /fodder grass etc., can be cultivated profitably.

Soak Away or Recharge Shafts

Soak away or recharge shafts are provided where upper layer of soil is alluvial or less pervious. These are bored hole of 30 cm dia. up to 10 to 15 m deep, depending on depth of pervious layer. Bore should be lined with slotted/perforated PVC/MS pipe to prevent collapse of the vertical sides. At the top of soak away required size sump is constructed to retain runoff before it filters through soak away. Sump should be filled with filter media.

Abandoned/Dug Wells

Newly dug wells or existing abandoned well in alluvial as well as hard rock areas can be used as recharge wells. Besides excess water from roof top, the storm water, surplus water from reservoirs, tanks, canals etc., can also be diverted to these dug wells through filter media to directly recharge the aquifer. The water is guided through a pipe to the bottom of the well, below the water level to avoid scouring and entrapment of air bubbles in the aquifers.

Figure 11.13: Water gets Collected in the Pits

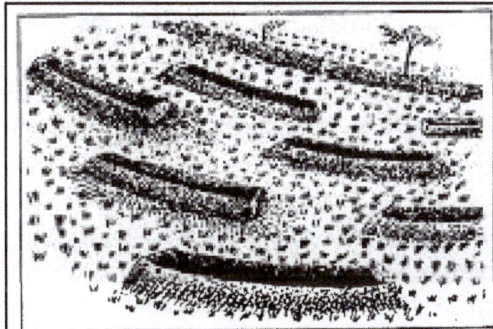

Figure 11.14: Bunds are Formed along the Lower Sides of Trenches

Percolation Well-cum-Bore Pit

Percolation well-cum-bore pit can be constructed as a combination of surface-cum-subsurface recharge structure to recharge deeper aquifers where shallow or superficial formations are highly impermeable. In areas where the soil is likely to be clayey upto say 5m. and more, it is advisable to go in for a percolation well upto 3 or 5m, and a hand bore pit within this well upto a depth of 3 to 5m. from its bottom. The recharge water is guided through a PVC pipe of 6in. diameter from the bottom of the

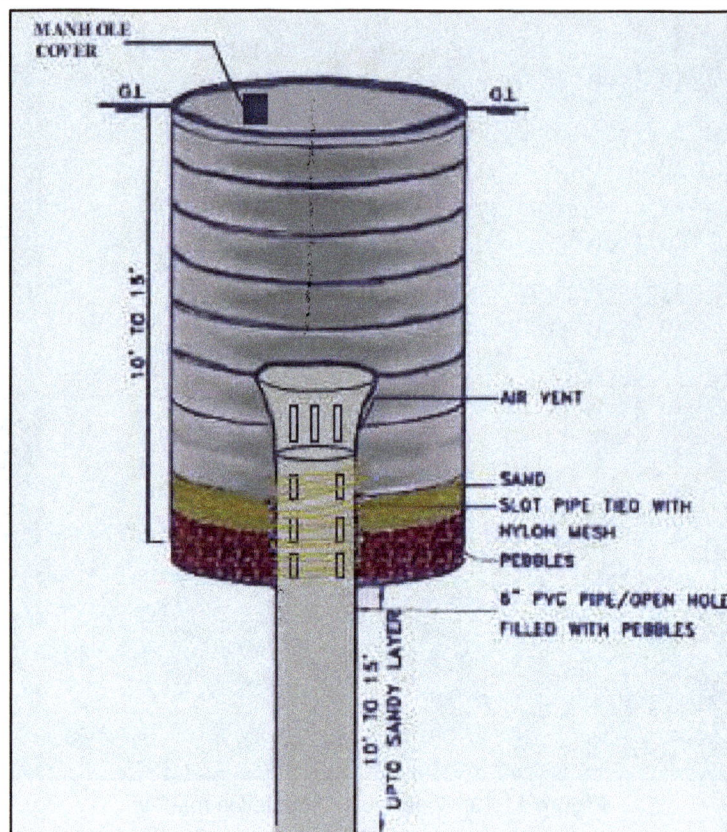

Figure 11.15: Percolation Well-cum-Bore Pit

pond/well, to the water table to avoid scouring and entrapment of air bubbles in the aquifers. Water should pass through filter media before putting into dug wells (Figure 11.15).

Using Natural Termite Hills

Termite hills have several underground tunnels dug out in different directions, some of which usually touches the water table. It is by this way they bring droplets of water on their mouthparts for mound construction. Based on this fact the presence of termite mounds are taken as indicators for water divining. Roof water can directly feed to it. If run off water is used to feed to termite mount, water should be filtered with the help of filtering unit. This is to avoid the silts blocking the termite tunnels in the long run. Termite mount can intake a very large amount of water and can infiltrate to a very large area of subsoil in a short time

Stream Flow Harvesting

Check Dams

Check dams are small barriers built across the direction of flow in shallow rivers and streams. This will reduce the velocity of flow, thus reducing channel scouring and promote sediment deposition. These small dams retain excess water flow during monsoon rains in small catchments behind the structure. Pressure created in the catchments cause the impounded water to infiltrate into the ground. The major environmental benefit is the replenishment of nearby groundwater reserves and wells.

Check dams are built in a range of sizes using a variety of materials. Temporary check dams can be made with locally available wood materials like rubbles, country wood, tree twigs, coconut leaves and fronds, arecanut leaves and bamboo. However, permanent check dams can be constructed using masonry, concrete, steel etc. (Figure 11.16). Small earthen check dams or embankments can easily be constructed by local people. Masonry and RCC structures on the other hand require some degree of advanced construction skill and monetary inputs.

In shallow channels or gullies having depth upto 1m, brushwood dams which are made up of easily available materials are economical and suitable (Figure 11.17). Depending on the size of channels, brushwood dams with single row posts or double row posts can be constructed. In streams of about

Figure 11.16: Permanent Check Dams

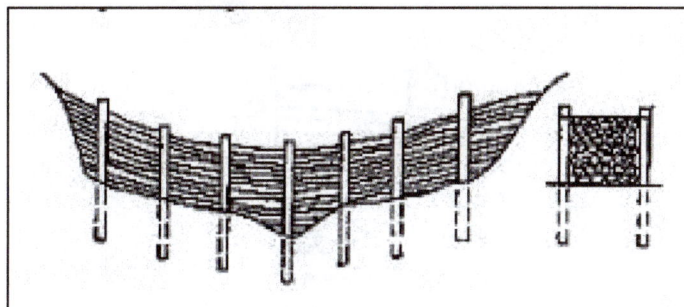

Figure 11.17: Shallow Channels or Gullies

10m width, check dams made up of rocks/large pebbles reinforced by iron wire netting (Gabions) help in better runoff control. During rainy season the excess water flows over the top of the dam. In shallow wide rivers, plants can be grown across the flow, which in due course of time develops as a live bund and by deposition of sand and silt at the root, this will get strengthened.

There are many advantages of storing water under ground as compared to surface storage. The evaporation losses are eliminated and the pollution risks are very low. The water will get self-purified during infiltration through soil and no land surface is wasted for storing water. There is no risk of silting, construction costs are low and maintenance is negligible. Not much skill is required for construction and indigenous materials can be used for the construction. The benefits will be available in the wells in the zone of influence of check dams.

Vented Cross Bar (VCB)

Vented cross bars (locally called *cheerpu*) are similar to check dams except in the pattern of excess flow. In vented cross bars, vents are provided and shutters are fitted on theses vents in the form of cross bars, which can be closed or opened. During heavy rainfall, the shutters are opened to allow water to flow downstream. When the rainfall began to recede, the dam will be closed using the shutters to store the remaining flow of water. Usually, VCBs are closed after the southwest monsoon and water diverted for irrigation. The vent will be opened again by the onset of heavy monsoons.

Sub Surface Dams

The subsurface runoff of water can be arrested by constructing under ground barriers from the bedrock, known as subsurface dam /subsurface dyke, which can augment the recharge of groundwater and thus raising the water table (Figures 11.18 and 11.19). These will benefit the wells in the upstream side of the dam. Subsurface dams can be constructed in rivers, slightly slopping rice fields and in sloping lands having impermeable layer (either hard rock or thick clay) underneath within a depth of 3 to 10m. The top level of the subsurface dam should be 50cm below the ground level, to allow the surface runoff. Sub surface dams can be constructed by masonry, concrete, polyethylene sheet, sand-cement-bentonite mix etc. If a series of subsurface dyke /check dams are constructed across rivers, it can retard the flow of water from rivers to the sea and increase the duration of water flow in the rivers.

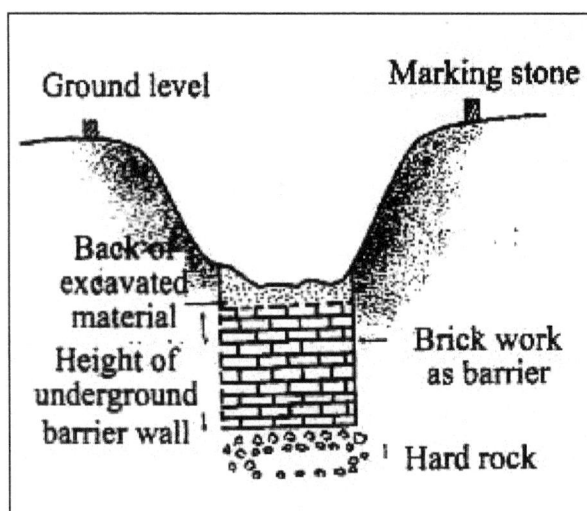

Figure 11.18: Subsurface Dam

Figure 11.19: subsurface Dyke

The selection of suitable artificial recharge method depends upon the hydro-geological characteristics of aquifers, soil characteristics, availability of recharge water, sediment in recharge waters, rate of discharge and availability of funds. A general guideline for selection is given in Table 11.4.

Table 11.4: General Suitability of Recharge Methods

Lithology	Topography	Type of Structure(s) Feasible
Alluvial or hard rock upto 40m depth	Plain area or gently undulating area	Spreading pond, groundwater dams, irrigation tanks, check dams, percolation tanks, unlined canal systems
Hard rock down to 40m depth	Valley slopes	Contour bunds, trenches
Hard rocks	Plateau region	Recharge ponds
Alluvial or hard rock with confined aquifer	Plain area or gently undulating area, flood plain deposits	Injection wells, connector wells
Hard rock	Foothill zones	Farm ponds, recharge trenches
Hard rocks or alluvium	Forested area	Groundwater dams

Source: CGWB, 2000.

In situ **Rainwater Harvesting**

In situ rainwater harvesting is the harvesting of rain, where it falls and conserving it in the soil to retain the soil moisture regimes for a larger duration. Location specific soil and water conservation measures can contribute immensely to groundwater recharge. Depending on the approach, it is divided into mechanical (engineering/artificial) and agronomic or vegetative methods.

Mechanical Methods

Mechanical measures consist of construction of mechanical barriers across the direction of flow of water to retard or retain the runoff and thereby reduce soil and water loss.

(a) Contour Bunds

Contour bunds are small bunds across the slope of the land on a contour so that the long slope is cut into a series of small ones and each contour bund acts as a barrier to the flow of water, thus reducing the velocity of flow (Figure 11.20). Narrow based trapezoidal bunds are constructed along contour to impound runoff water behind them so that all the stored water is absorbed gradually by the soil profile for crop use. Contour bunds are suitable for permeable soil up to a slope of 8 per cent in areas with less rainfall (<600mm). It can be followed up to a slope of 20 per cent, in which case the height of bund is to be increased or small bunds at lesser vertical interval are to be provided.

Figure 11.20: Countour Bunds

(b) Contour Trenches

In areas of heavy rainfall, the bunds may not be sufficient to stop the runoff. Continuous or staggered trenches of 2m length, 0.5m width and 0.5m depth are ideal for these regions. The rainwater stored in these trenches will improve the detention storage and increase the opportunity time for water to infiltrate into the soil profile. This reduces the quantity and velocity of runoff and hence reducing the erosive potential of rainwater. The soil excavated from these trenches can be heaped on the down stream side of the trench to form a bund and deep rooted grasses/pine apple can be planted on these bunds for more effectiveness (Figure 11.21).

(c) Graded Bunding or Channel Terraces

Graded bunding or channel terraces are constructed in high rainfall areas (>600mm). The excess water is removed safely out of the fields avoiding stagnation of water. Graded bunds are recommended even in lesser rainfall areas where the soils are highly impermeable or clayey in nature. In graded bunds, the water flows in a graded channel constructed on the upstream side of the bunds at non-erosive velocities and is led to safe outlets or grassed waterways. The channel grade depends upon

Figure 11.21: Pineapple Planted on Contour Bunds

soil type and length of terrace. Normally the channel grade varies from 0.1 to 0.6 per cent which is much less than the original slope of the land (2-6 per cent). The channels should be wide and shallow so that mean velocity is at minimum. Side slopes should not be steeper than 5:1. The velocity in the channel should not exceed 0.5m/s for sandy soils and 0.75 m/s in clayey soil.

(d) Micro Catchments

For *in situ* moisture conservation for tree crops, micro catchments, saucer basins, crescent bundings, semi circular bunds and catch pits can be introduced (Figure 11.22). These are suitable for flat slopes to slightly slopping lands(less than 5 per cent slope). The size of micro catchments normally ranges from 10 m² to 100 m² and is designed to collect runoff to meet the consumptive use of a single tree. Micro catchment water harvesting is suitable for areas with less rainfall. A bund of 25cm height is raised around each catchment. At the lowest point of each catchment an infiltration pit of 30-40 cm deep is made and a tree is planted in it. The surface runoff from the catchments is collected and stored

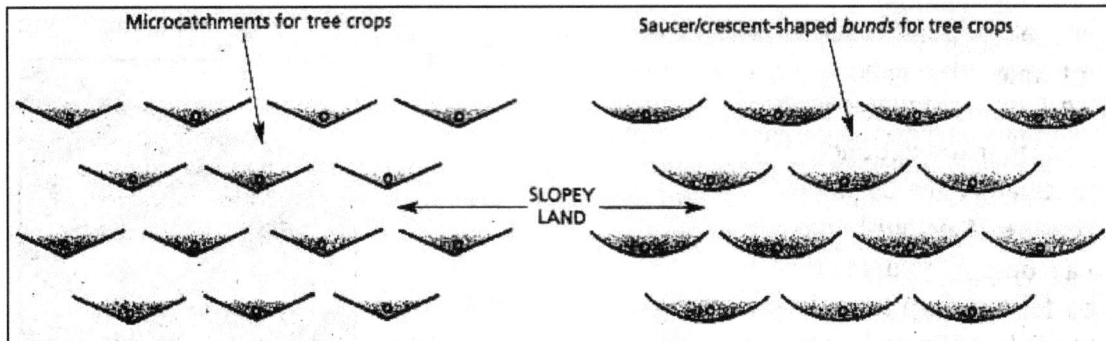

Figure 11.22: Micro catchments

in the soil profile of the plant basin located downstream. Even a single light shower will cause ponding in the basin. Areas where rainfall is as low as 150mm/year and soils that develop crusting on the surface as a result of impact of rain drops are suitable for micro catchment water harvesting.

(e) Terracing

A terrace is an embankment constructed across the slope to intercept the surface runoff.

Terracing is recommended in areas having slopes above 30%. Terracing checks the soil erosion and percolates the runoff. It reduces the degree and length of the slope and velocity of run off so as not allowed to attain the critical value, which initiates scouring of top soil. Terracing helps to conserve rainwater and facilitates tillage operations on sloppy lands. Depending on type of soil, climate and crop requirements, terraces may be constructed level or slopping outward.

Vegetative or Agronomic Methods

Soil conservation through agronomic practices help to reduce the impact of raindrops through interception and thus reduce splash erosion. These practices also help to increase infiltration rates and thereby reduce runoff and overland flow.

(a) Mixed Cropping

Mixed cropping offers a more effective and continuous cover of land, thereby preventing soil erosion. The roots of the various mixed crop components feed at different soil depths leading to more efficient exploitation of soil water and soil fertility. Developing intercrops that do not compete with the main crop is a good soil and water conservation method. Planting coconut, arecanut, mango, jack, papaya etc. as top layer trees; banana, pepper, cocoa, coffee, guava, sapota etc. as second layer plants and turmeric, ginger, vegetables, elephant foot yam, fodder etc., as third layer plants can be practiced. When inter crops spread their canopy, they prevent soil erosion and make more rain to percolate. It not only ensures effective utilization of sunlight and water but also greater productivity of land and water.

(b) Growing Cover Crops

If the depth of top soil is lesser, perennial or tree crops cannot be cultivated and the only crop that can be raised is legumes or grasses. When the plants grow up, their roots check the soil erosion and facilitate infiltration of water to deeper layers. The legumes also help to increase the fertility of soil by fixing atmospheric nitrogen. *Pueraria phaseolides*, *Calopogonium mucunoides*, *Mucuna bracteata* etc., are some of the cover crops used in rubber plantations.

(c) Contour Farming

In this practice, all the normal cultural operations are done across the slope. Thus the contour furrows and plant stems thus created would form multitude of mini barriers across the flow path of the runoff, which improves the in situ detention storage. This in turn increases the infiltration opportunity time, and hence, increased infiltration of rainwater into the soil profile. This will reduce the amount and velocity of runoff and its erosive potential. Contour cultivation remains most effective on moderate slopes of 2 to 7% and deep permeable soil and its effect decreases as the land slope become very flat or very steep.

(d) Mulching

Mulching the land with agricultural wastes reduces soil erosion, evaporation, and excessive heating; controls weeds and increases the water holding capacity and permeability of soil. Since it retains soil moisture, provides organic material and darkness; the microbial activity increases considerably which help to improve the soil health. Mulching with coconut husk/coir pith in 0.50m x 0.50m trenches taken around coconut palms help to conserve water in coconut plantations. The coconut husk can absorb water about 6 times its own weight. This method helps in conservation of rainwater for 7 to 10 years.

(e) Vegetative Barriers

Live fences of suitable crop (vetiver, gliricidia, and hibiscus) will help to check the soil erosion in low slopes. . It also facilitates infiltration of water through air gaps created by its roots. Hence vegetative barriers encourage in situ moisture conservation and ensure silt free overhead flows. Compared to construction of mechanical bunds and trenches, planting vegetative barriers is easy, less laborious and does not require any maintenance.

(f) Micro Irrigation

Judicious use of irrigation water is also a water conservation practice through which the losses occurring during conveyance and application of irrigation water is minimized. This can be achieved by adopting micro irrigation technique, which is the application of right quantity of water and nutrients at right time direct to the root zone of the plant. It includes sprinkler/micro sprinkler and drip irrigation systems. Drip and micro sprinkler irrigation systems are suitable to irrigate horticultural crops where as sprinkler irrigation is used to irrigate field crops. The advantages of these systems include economic and efficient use of various inputs like water, fertilizers (through fertigation), labour and energy in addition to increased productivity and quality of produce.

References

Agarwal, Anil, Sunitha and Khurana, Indira (Eds.), 2005. *Making Water Everybody's Business: Practice and Policy of Water Harvesting*. Center for Science and Environment, New Delhi, p. 456.

CGWB, 2000. *Guide on Artificial Recharge to Groundwater*. CGWB, Ministry of Water Resources, Govt. of India, New Delhi, p. 93.

Dhruvanarayana, V.V., 2002. *Soil and Water Conservation Research in India*. Publication and Information Division, Indian Council of Agricultural Research, New Delhi, p. 454.

Michael, A.M. and Ojha, T.P., 2005 *Principles of Agricultural Engineering, 4th edn.* Vol. II. Jain Brothers, New Delhi, p. 888.

Padre, S., 2002. *Rainwater Harvesting*. Altermedia Publications.

Saifudeen, N., Suresh, P.R. and Marykutty, K.C., (Eds.), 2006. ICAR sponsored Winter School on GIS based watershed planning in Agriculture. Compendium of lecture notes. Department of Soil Science and Agricultural Chemistry, College of Horticulture, KAU. 305p.

Samra, J.S., Sharda, V.N. and Sikka, A.K., 2002. *Water Harvesting and Recycling: Indian Experiences*. Central Soil and Water Conservation Research and Training Institute, Dehradun, p. 347.

Schwab, G.O., Frevert, R.K., Edminster, T.W and Barnes, K.K., 1996. *Soil and Water Conservation Engineering, 4th edn.* John Wiley and Sons, New York, p. 507.

Selvi, V., Singh, D.V. and Madhu, (Eds.), 2004. *Training Manual for Soil Conservation and Watershed Planning*. NATP Sub Project on Watershed Technology.

Singh, Gurmel, Venkataraman, C., Sastry, G. and Joshi, B.P., 1990. *Manual of Soil and Water Conservation Practices*. Oxford and IBH. New Delhi, p. 385.

Singh, K.K., Phogat,V., Tomar, Alka and Singh, Vinod, 2007. *Water Resources Management and Development.* Kalyani Publishers, New Delhi, p. 581.

Sinha, B.P.C. and Sharma, S.K., 1995. *National Groundwater Recharge Estimation Methodologies in India.* INCOH Secretariat, National Institute of Hydrology, Roorkee.

Sudhakaran, V.A. and Ramadas, T.N, 2004. *Mazhavellakoithu – Keralathinu jojicha Jalasamrakshana Samvidhanam (in Malayalam).* Centre of Science and Technology for Rural Development, p. 180.

Thomas, George C., 2010. *Land Husbandry and Watershed Management.* Kalyani Publishers, New Delhi, p. 716.

Visalakshi, K.P., Bridjit, T.K., Archana Chandran, Devadas, V.S., and Abraham, Mini, 2007. *Rainwater Harvesting and Conservation.* Directorate of Extension, Kerala Agricultural University, p. 24.

2012, Nutri-Horticulture
Editor: Professor K.V. Peter
Published by: DAYA PUBLISHING HOUSE, NEW DELHI

Pages 221–230

Chapter 12

Biology of Grafting in Dicot Plants

N.K. Parameswaran* and Renish Jayaraj

Department of Pomology and Floriculture,
College of Horticulture, Vellanikkara – 680 656, INDIA
*E-mail: parameswarannk@yahoo.co.in

Grafting in plants is an age old practice which can be traced back to ancient times. Such an art was known to Chinese as early as 1000 BC. Aristotle (388-322 BC) discussed the method of grafting in his writings with a considerable understanding.

Nomenclature

Grafting

It is the art of joining two pieces of living plant tissue together in such a way that they unite and subsequently grow and develop into a new plant.

Scion

Short piece of detached shoot containing several dormant buds when united with stock forms the upper portion of graft and from which will grow the stem or branches or both of the grafted plant. It would be from a desired cultivar and free from diseases.

Stock (Rootstock, Understock)

The lower portion of the graft which develops into the root system of grafted plant is the stock. It may be a seedling, rooted cutting or a layered plant.

Interstock

The piece of stem inserted by means of two graft union between scion and rootstock is an interstock. An inter stem is used for several reasons such as to overcome incompatibility between scion and rootstock, to make use of winter hardy stem and to take advantage of winter controlling property.

Vascular Cambium

A thin tissue of plant located between bark (phloem) and wood (xylem) is meristematic *i.e.* capable of dividing and forming new cells. For a successful graft union, it is essential that the cambium of scion should be placed in close contact with cambium of stem.

Callus

Mass of parenchymatous cells which develops from and around the wounded plant tissues is the callus. It forms at the junction of graft union arising from the living cell of both scion and stock. The production and interlocking of this parenchyma (callus) cells constitute one of the important steps in the healing process of a successful graft.

Formation of Graft Union

A 'de novo' formed meristematic area must develop between scion and root stock for a successful graft union.

Three basic events which are essentially involved in such processes are:

1. Adhesion of root stock and scion
2. Proliferation of callus at the graft interface- callus bridge
3. Vascular differentiation across the graft interface

The usual sequential steps in the graft union formation are:

Lining Up of Vascular Cambium

Freshly cut scion tissues capable of meristematic activity are brought into secure and intimate contact with similar freshly cut stock tissue in such a way that the cambial region of both are in close proximity. Temperature and humidity condition must be optimal for promoting growth activity in the newly exposed and surrounding cells.

Wound Healing Response

The stage involves formation of necrotic material (one cell deep) from the cell contents and cell walls of cut scion and stock cells.

Callus Bridge Formation

The outer exposed layer of undamaged cells in the cambial region of both scion and stock produces parenchyma cells which soon intermingle and interlock filling up the spaces between scion and stock; this is called the callus tissue.

The callus proliferates for 1-7 days. In most cases, the callus originates from the scion part (probably due to the basal movement of auxins and CHO's etc). The scion and stock cells in the course adhere with a mix of pectin, CHO's and proteins secreted by the dictyosomes which are parts of the Golgi bodies in the cells.

Cambium Formation

Certain cells of this newly formed callus fall in line with the cambium layers of the intact scion and stock and differentiate into new cambium cells.

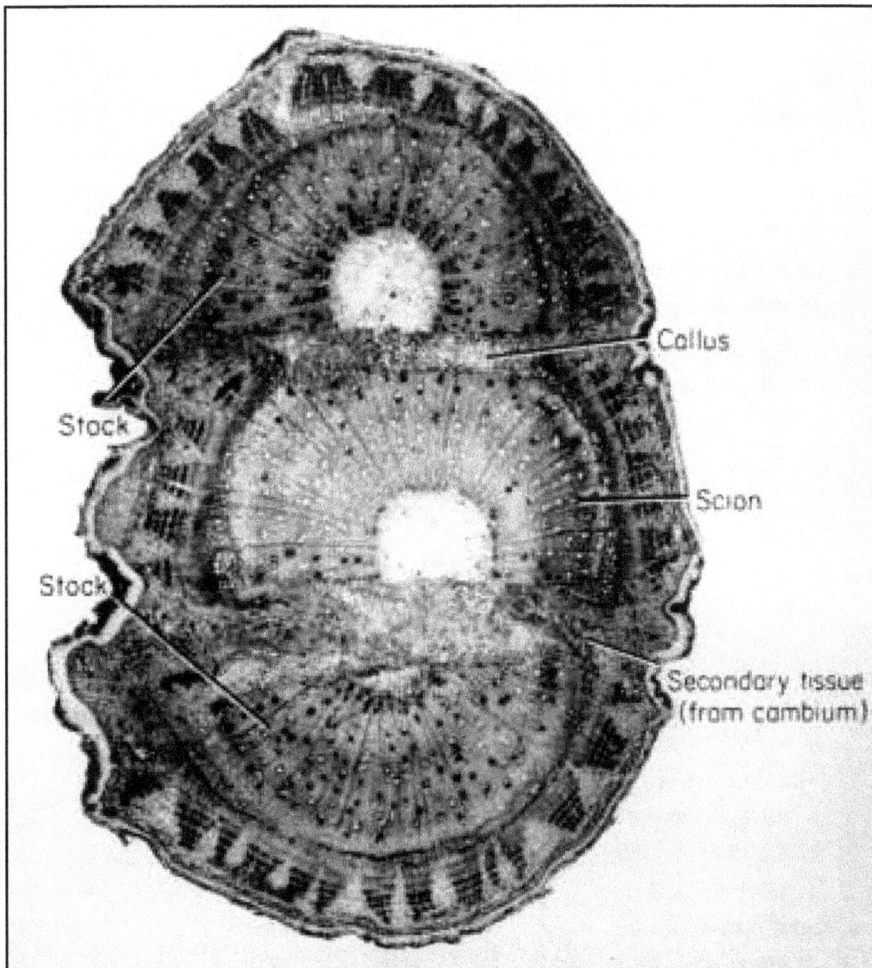

Figure 12.1: Callus Formation as noted during the Graft Union Formation in a Wedge Grafted Plant
(With due to courtesy to Hartmann, Kester and Davies 1990)

Vascular Tissue Formation

These new cambium cells produce new vascular tissues, xylem towards inside and phloem towards outside, thus establishing a secondary vascular connection between the scion and stock, a requisite of a successful graft union.

Healing of graft union is similar to healing of a wound. Such injury to tissue would occur if the cut ends of a branch were split longitudinally and would heal quickly if the split pieces were bound tightly together. New parenchyma cells would be produced by abundant proliferation from cells of the cambium region of both pieces forming callus tissue. Some of the newly produced parenchyma cells differentiate into cambium cells, which subsequently produce xylem and phloem.

Figure 12.2: Salient Anatomical Bases in Plants
(With due to courtesy to Hartmann, Kester and Davies 1990)

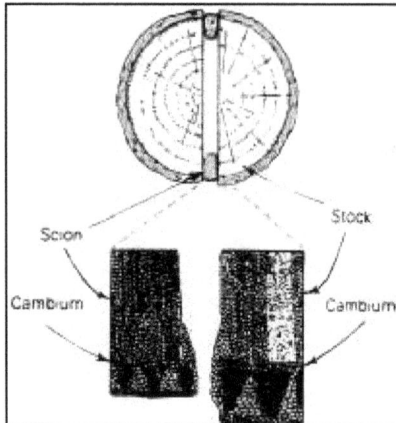

Top view of cleft graft

Phase 1 - Callus formation in region of cambium.

Rapid instant divisions of parenchyma cells in young xylem and phloem around the cambium of both the stock and scion produce undifferentiated callus (parenchyma) cells.

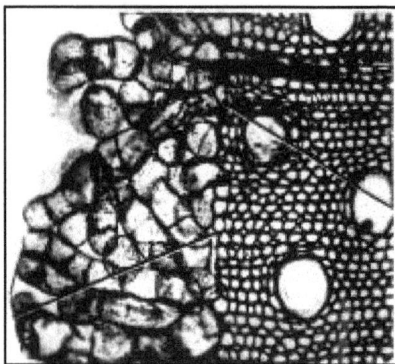

Enlarged view showing callus production from parenchyma cells in the young xylem near the cambium.

Phase 2 - Callus bridge formation.

Proliferation of the callus from the stock and scion filling the gap between the scion and stock, and intermingling to form a callus bridge.

Contd...

Figure 12.2–Contd...

Phase 3 - Differentiation of new cambium.

Parenchyma cells, in the callus between the cambium of the stock and scion, differentiate into cambium cells, thus uniting the cambium of the stock with the cambium of the scion.

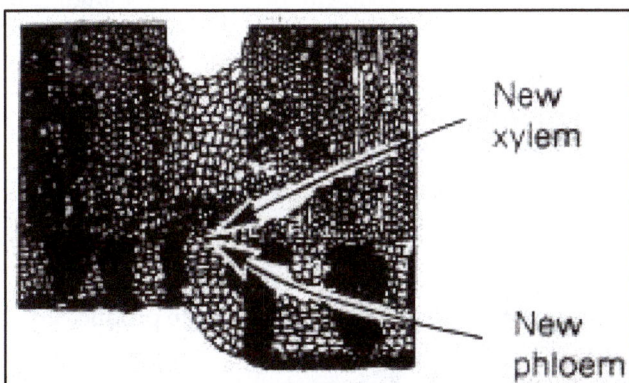

New xylem

New phloem

Phase 4 - Formation of secondary xylem and phloem from new cambium.

The new cambium produces secondary xylem on the inside and secondary phloem on the outside, which form a continuum with the xylem/phloem of the stock and scion. This vascular connection allows translocation between the stock and scion.

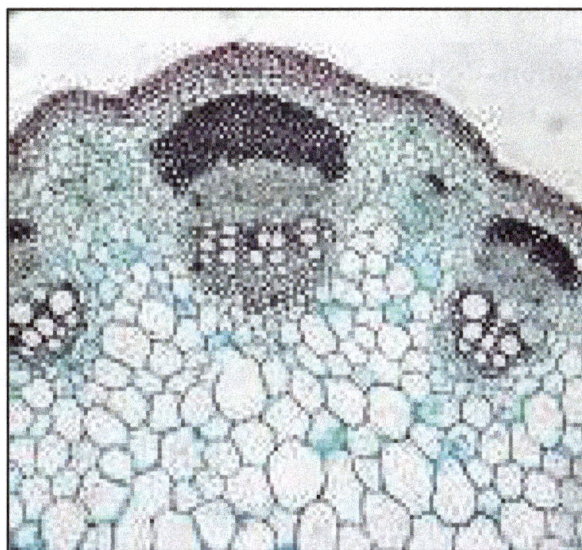

Figure 12.3: Vascular Cambium in a Dicot Plant

Between two split pieces,if one interposed a third detached piece cut, enabling a large number of cells in the cambial region be placed in intimate contact with the cells of the cambial region of the two split pieces, proliferation of parenchyma cells from all cambial areas would soon result in complete healing with the foreign, detached piece joined between the two original split pieces. A graft union is essentially a healed wound, with an additional foreign piece of tissue incorporated into the healed wound.

With the added piece of tissue,the **scion** will not resume its growth normally, until a vascular connection is established so that it obtains water and mineral nutrients. The scion must have invariably a terminal meristematic region, a bud – to resume shoot growth and eventually to supply photosynthates to the root system.

In healing of a graft union, the parts of the grafts, originally prepared and placed in close contact do not spontaneously initiate growth or grow together. The union is accomplished entirely by cells which develop after the actual grafting operation is made.

It is also distinct that the process of graft union formation does not involve any intermingling of the cell contents (*i.e.* no fusion of cell contents or protoplasts). Cells produced by the stock and scion maintain their own distinct identity. Establishment of intimate contact of a considerable amount of cambial region of both stock and scion is vital under favourable environmental conditions.

Factors Influencing the Success of Graft Union

Healing of the graft union is influenced by a number of factors both external and internal. A few of the prominent ones are mentioned below:

Plant Species and Type of Grafts

Fruit plants and other tree crop species differ considerably in the amenability to produce good graft plants

> ☆ *Easy plants*: Apple, grape, pear
> ☆ *Difficult plants*: Mango, hickories, oaks and beaches

Environmental Conditions following Grafting

Environmental conditions during grafting process influence greatly the successful union formation:

Temperature

In grafted apple plants little callus production is formed below 0°C or above 40°C. Such a response varies with plant species and it can be controlled under green house conditions but difficult in the field situations.

Moisture

Optimum moisture level is needed for cell enlargement in the callus bridge. It is maintained usually by using plastic bags or caps over the scion. Rapping with grafting tape, polythene films, grafting wax, etc. are also adopted for maintaining an optimum moisture level during the grafting process. Alternatively, the graft joints are covered with damp peat moss or wood shavings for enabling callusing.

Growth Activity of the Rootstock

Growth stage of rootstock exerts much influence on formation of a successful graft union. The grafting should be preferably done during the periods of high growth activity of rootstocks.

The success of "T-budding" greatly depends on the stage of the bark of the rootstock. It should be at the "slipping" stage; which means the cambial cells are actually dividing and separate easily from each other. "Slipping" usually occurs in late spring or early summer seasons of the year.

Virus Contamination, Insects and Diseases

Virus contamination normally results in delayed incompatibilities manifested as blackline in walnut and brownline in plum. Similarly the bacterial and fungal contamination during wounding while grafting greatly affect the successful graft union formation. Care should be taken to take-up the operations in hygienic conditions minimizing chances for contamination.

Plant Growth Regulators and Graft Union Formation

Exogenous application of auxins to the graft union has given variable response in promoting and subsequent healing and hence not of much practical application.

Polarity in Grafting

Proper polarity is essential if the graft union is to be permanently successful in all large scale grafting operations. Correct polarity is strictly observed and practiced. Generally while grafting two pieces of stem tissue together, the morphologically proximal end of the scion should be inserted into the morphologically distal end of the rootstock.

Genetic Limits of Grafting

A successful graft union essentially requires a close matching of the callus producing tissues near the cambium layers and as such confined to the dicotyledons in the angiosperms having vascular cambium layer existing as a continuous tissue between xylem and phloem. The monocotyledonous plants of the angiosperms will not have a vascular cambium and as such the grafting is more difficult resulting a low percentage of "graft take".

General Rules for Successful Graft Union

1. The more closely related plants are (botanically), better the chances for the graft to be successful
2. Grafting within a clone – the most successful
3. Grafting between clones within a species becomes successful
4. Grafting between species within a genus (50/50 chance of success). Reciprocal interspecies grafts are not always successful
5. Grafting between genera within the same family

   When the plants to be grafted together are in the same family but in different genera, the chances of a successful graft union become rather remote
6. Grafting between families

Successful grafting between plants of different botanical families are usually found impossible except in certain isolated cases.

Graft Incompatibility

Compatibility – It is the ability of two different plants grafted together to produce a successful union and to develop satisfactorily. Graft failure is normally caused by anatomical mismatching/ poor craftsmanship, adverse environment, diseases and graft incompatibility.

Reasons of Graft Incompatibility

1. Adverse physiological responses between grafting partners
2. Virus transmission
3. Anatomical abnormalities of the vascular tissue in the callus bridge

External Symptoms of Incompatibility

Incompatibility in the grafted plants is usually manifested as:

1. Failure of successful graft or bud union in high percentage
2. Early yellowing or defoliation in fall
3. Shoot die-back and ill-health
4. Premature death
5. Marked differences in growth rate of scion and stock
6. Overgrowth at, above or below the graft union
7. Suckering of rootstock
8. Breakage at the graft union

Anatomical Flaws Leading to Incompatibility

Anatomical flaws resulting in unsuccessful or partially successful graft union formation can be attributed mainly to:

1. Poor vascular differentiation during the graft union process
2. Phloem compression and vascular discontinuity resulting in disruption of sap flow translocation of nutrients.

Physiological and Pathogen-Induced Incompatibility

The graft incompatibility can be broadly grouped under:

Non-translocatable or Localized Incompatibility

Problem can be overcome by using mutually compatible interstock in which direct contact between scion and stock is eliminated.

Translocatable Incompatibility

In this situation, the compatibility is not localized and spreads to other parts. In such a situation, the uses of interstock do not solve the problem. It is mainly attributed to involvement of some mobile chemicals causing phloem degradation. Eg: cyanogenic glucosides like prunasin is converted to hydrocyanic acid (Quince grafted on pear)

Pathogen Induced Virus or Phytoplasma Incompatibility

Transmission of Tristeza virus of budded sweet orange when grafted on sour orange rootstock

Predicting Incompatible Combinations

There are different techniques available to predict the incompatible combinations during grafting:

1. Electrophoresis analysis to look for cambial peroxidase banding is widely applied in temperate fruit crops (chestnut, oak and maple). Peroxidases produce specific lignins and

the lignins must be similar for both scion and stock for the graft to be successful on a long-term.

2. Staining of tissues at the graft union and subsequent microscopic examination
3. Magnetic resonance imaging (MRI) checks for vascular discontinuity

Correcting Incompatible Combinations

Correction of incompatible combinations at the field level is not generally cost effective. Such grafted plants can be top-worked. Bridge grafting with a mutually compatible rootstock is also a viable alternative.

Effects of Rootstock on Scion

The components in a grafted plant - the scion and rootstock- mutually exert specific effects on the overall growth and performance of the grafted plant. Some of the salient effects are mentioned here:

Size and Growth Habit

It is the most significant effect manifested in the grafted plant by the influence of rootstock.

The first apple rootstock breeding programme was initiated at the East Malling Research Station, England during 1917. The rootstocks released from the station were designated by "M" for Malling followed by a Roman numeral.

Subsequently, further research was collaborated with the John Innes Horticultural Institute at Merton, UK. These new series of rootstocks have a prefix MM.

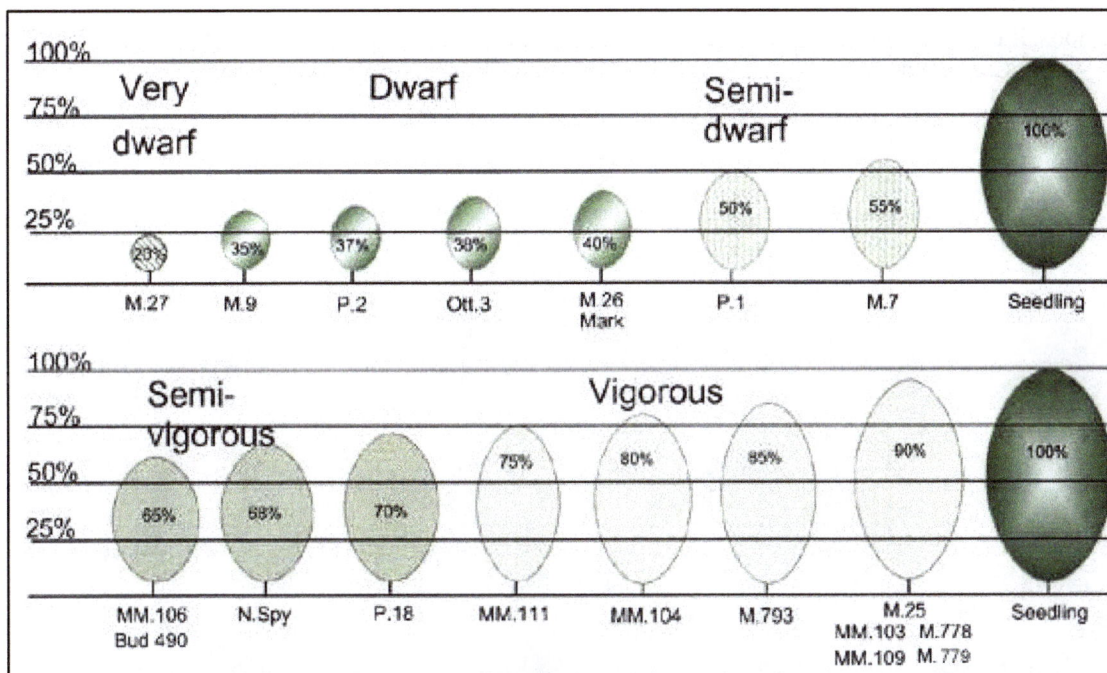

Figure 12.3: Comparison of Apple Rootstocks for Size and Vigour.
The tree size is represented as per cent of seedling.

All the European countries have completely switched over to clonal rootstocks at present. The major advantages of clonal rootstocks are

1. Results in standard size whereas seedling trees exhibit wide variation
2. Tree behaviour and other characters remain true to type
3. Exhibit resistance to diseases, more hardier than the traditional seedling stocks.

Some of the promising dwarfing rootstocks in apple are **M-27, M-9, M-26, M-7, MM-106 and MM-111**

Other Effects of Rootstock on Scion

Rootstocks significantly influence the various fruiting characters of the grafted plants. Some of the prominent effects are:

1. Induction of precocity/early maturity
2. Enhancement of fruit bud formation
3. Enhancement of fruit set
4. Enhancement of yield

References

Esau, K., 1977. *Anatomy of Seed Plants,* 2nd edn. John Wiley, New York., pp. 150–200.

Hartman, H.T., Kester, D.E. and Davis, T.F., 1993. *Plant Propagation: Principles and Practices,* 5th edn. Prentice Hall, India, pp. 305–332.

Simons, R.K., 1987. Compatibility and stock-scion interactions as related to dwarfing. In: *Rootstocks for Fruit Crops,* (Eds.) R.C. Rom and R.F. Carlson. John Wiley, New York, pp. 220–235.

2012, Nutri-Horticulture
Editor: Professor K.V. Peter
Published by: DAYA PUBLISHING HOUSE, NEW DELHI

Pages 231–240

Chapter 13

Basics of Screening Techniques for Disease Resistance Against Fungal and Bacterial Diseases of Vegetables

M.K. Naik, S.V. Manjunatha*, Suresh Patil and Y.S. Amaresh

Department of Plant Pathology, College of Agriculture,
UAS, Raichur, Karnataka, INDIA
*E-mail: manjunaik2000@yahoo.co.in

It would be pertinent to combine resistance to various biotic stresses in vegetable varieties which would minimize hazards of pesticide residues carried on in vegetables which are consumed raw or cooked. Development of resistant varieties entirely depends upon the efficient and basic techniques employed in screening a large number of exotic germplasm, local cultivars, land races and wild species.

A modest attempt is made to review various basics of techniques of screening available at various stages of crop growth like seed, seedling, adult plant and fruit/harvest stage depending upon the nature of host-pathogen system involved.It also highlights the important sources of resistance available in tomato, chilli, bell pepper, cucumber, muskmelon, watermelon, cabbage and cauliflower against major fungal and bacterial diseases. Screening techniques in dieback and anthracnose, *Phytopthora* fruit rot and leaf blight and Fusarium wilt of chilli and capsicum, buck eye rot, early and late blight in tomato, Alternaria leaf spots in crucifers are described methodically. The corresponding sources of resistance to the diseases of these vegetables are mentioned.

The need for management of plant diseases arose out of the domestication of crop plants for providing food to mankind on sustainable basis. Human intervention led to development and deployment of newer varieties of crop plants having higher productivity and amenability to adoption on a large scale. New sets of crop husbandry methods were developed to boost up crop production. These well directed and well intentioned efforts on the part of mankind disturbed the balance between the microbes and their hosts *i.e.* crop plants, in nature, which in turn, triggered the process of disease development in plant populations. Some of these diseases assumed epiphytotic proportion over a period of time threatening the very cultivation of some of the economically important crop plants. Dr J.E Vanderplank very aptly and succinctly reflected that our daily bread comes from crops protected against diseases by genetic resistance. The genetic resistance is seen as pillar of agriculture.

Use of resistant varieties in vegetables has many advantages, since there is obvious economic saving of costs of pesticide application. Seeds of resistant varieties usually do not cost more than that of susceptible cultivars. Further, if the resistance does not protect the crop completely, even partial resistance can markedly reduce the quantity of pesticide to be applied for satisfactory control. There is also a greater advantage of safety for the grower and consumer because of the reduced contact risk and residues. In addition, the resistant varieties are very well compatible with biocontrol measures unlike chemical protection which usually precludes environmentally desirable control methods.

One of the essential features of developing resistant varieties is to identify sources of disease resistance in crops which in turn depend upon efficient and standard screening techniques. Such sources can be identified from the germplasm by following the established screening techniques under natural and artificial conditions. An effort is made to review the various screening techniques developed against major fungal and a few bacterial diseases of solanaceous, cucurbit and cruciferous vegetables. The sources of resistance are also mentioned.

Chilli and Bell Pepper

Dieback and Anthracnose

This disease is caused by *Colletotrichum capsici* and *C. gleosporoides*. The diseases occur both on twigs and ripened fruits and in advanced stages, the seeds are covered by mat of fungal hyphae. Several techniques are standardized under field and laboratory conditions.

Seed Inoculation

Surface sterilized seeds of test genotypes are soaked in conidial suspension of *Colletotrichum capsici* (concentration 1×10^7 spores/ml) and then subjected to blotter test by incubating at $27 \pm 1°C$. Infected seeds of genotypes are easily knocked out in the screening (Naik and Sinha, 1995). Per cent germination and infection are calculated.

Seedling Inoculation

The seedlings are immersed in conidial suspension (1×10^7 spores/ml) of the fungus for about 5 minutes and then transplanted. Dieback of twigs is scored by counting the number of dieback infected twigs out of the healthy twigs in selected group of plants from each genotype.

Pinprick Inoculation of Fruits

Red ripened fruits are surface sterilized and a few pickings are given on the fruits with pin bundles to give slight wounding. Such fruits are then dipped in spore suspension of *C. capsici* and then incubated in humid chamber to create artificial epiphytotic condition. The infected fruits are scored on 0-4 scale for assessing the disease (Naik and Sinha, 1995).

Field Screening

Test genotypes are planted under natural conditions regularly interspersed with susceptible check for every four rows and the infected fruits are scored using 0-4 scale after every harvest of fruits.

Sources of Resistance

CA-960, Pant C-1, K-2, CA-966, PBC-651, PBC-495, PBC-417, PBC-552 and Co-4352 are resistant (Ann., 1994). Out of 127 genotypes Puri Red, Puri Red x S-32, Pant C-1, K-1, K-2-11, D-2-3, D-2-5 and D-

3-1 were resistant (Angadi, *et al.*, 2003). G-4, PMR-5, DS-1, Co-1, IHR-3023 and LCA-301 were resistant (Naik, 2000).

Phytophthora Fruit Rot and Leaf Blight

Screening Technique

Most destructive disease occurs in high rainfall area and is caused by *Phytophthora nicotianae* var *nicotianae*. The culture is prepared by following Kreutzer and Bryant (1994) method. The fungus is grown on steamed barely for 20 to 60 days at 25°C, passed through food grinder and thoroughly mixed into a few cm of unsterilized field soil in a wooden flat. The soil is watered thoroughly with necessary aeration. Three cubic cm of soil mixture is dispensed in 500 ml beaker containing 30 ml water and with 2-3 leaves and incubated at 25°C. The incidence is recorded as percentage of diseased fruits. A slightly modified scale developed by James (1974) is used for leaf scoring.

Sources of Resistance

Bhardwaj (1983) observed more disease severity on sweet type than in hot peppers. Among hot types, Javiott, Russian Yellow, EC-114364, Solan Yellow, G-5 and Perennial have resistance. Seven breeding lines Co-113, Co-2227, Co-2230, Co-2284, Co-2289 and TC-3191 have also resistance to the disease (Ann., 1988).

Fusarium Wilt

A typical soil borne disease and an ideal situation is to look forward to have eventually disease resistant varieties. Various techniques included mixing of three weeks old wheat/maize/sorghum kernel culture to sterilized soil (White, 1927) pouring heavy spore suspension and mycelial suspension to potted plants (Haymaker,1928); dipping injured roots of transplanting in mycelial suspension (Wellman, 1939), inoculating soil with fungus culture and transplanting the seedlings (Schroeder *et al.*, 1953), pouring suspension around the root zone (Henderson and Winstead, 1961) and rapid root-dip transplanting technique (Naik *and* Rawal 2002).

Seed Inoculation Technique

Surface sterilized seeds are dipped in *Fusarium* spore suspension for 12 hours and then subjected to blotter test by inoculating for one week. The seed germination(per cent) and seedling death are considered. Since the fungus is also seed borne, this method works very well for screening at seed stage (Naik *and* Rawal 2002).

Rapid Root Dip–Transplant Technique

Three weeks old seedlings raised in sterile sand are uprooted, roots thoroughly washed in running tap water and 3 mm tip of roots are cut and immersed in *Fusarium* spore suspension and transplanted in sterile soil. Uninoculated plants would serve as check. The plants are observed for wilt symptoms upto 20 days after inoculation. The method is rapid and reproducible and a large number of genotypes can be tested at seedling stage in short time (Naik and Rawal, 2002).

Sick Plot/Pot Technique

Surface sterilized chilli seeds are planted in soil which is made sick by repeatedly incorporating the wilted plant debris and *Fusarium* colonized giant culture. The population of *Fusarium* is monitored by frequent enumeration on *Fusarium* specific medium to ensure adequate colonies in the soil. The plants are observed for wilt symptoms. It is a very good method for adult plant screening.

Sources of Resistance

Masalwadi, a hot pepper cultivar, was resistant (Nayeema *et al.*, 1995) under Srinagar condition. Akilesh Singh *et al.* (1998) identified nine genotypes as moderately resistant. IHR-3018 was resistant and DS-1 and LCA-206 were moderately resistant when thirty genotypes were screened using rapid root dip- transplanting technique (Madhukar *et al.*, 2002).

Tomato

Tomato is an important vegetable grown through out the world and is vulnerable to various diseases which cause huge yield losses. Screening techniques for important diseases are presented below;

Buck Eye Rot

The disease is severe in heavy rainfall areas and is caused by *Phytopthora nicotiane* var *parasitica*.

Artificial Inoculation of Fruits

Healthy green fruits of test genotypes are inoculated with sporangial suspension (9×10^4/ml) and a drop (5μl) is placed on the fruits surface. Fruits are inoculated and incubated at $25 \pm 1°C$ with 95 per cent relative humidity for five days in temperature cum humidity control cabinet and resistance to disease was recorded on 0-5 scale (Dodan *et al.*, 1995).

Field Screening

Twelve plants of each cultivar/line are transplanted in a plot (2.5x2m) alternated with one row of susceptible check during a period so that fruiting should coincide with the period of maximum disease development. The disease incidence in each cultivar or line is recorded periodically and the reaction is categorized on 0-5 scale.

Sources of Resistance

Resistance in this case is available only in small fruited cherry type lines which tend to dilute in advance generations in the process of breeding for desirable fruit size. Resistance is identified in EC-12917/PI-2, EC-128-969, EC-129-166, EC-72901, EC-122062-1, Hybrid -10 (Thareja *et al.*, 1989)and *Lycopersicon esculentum* var *ceresiformae* (Dodan *et al.*, 1995).

Late Blight

Screening Technique

The disease spreads rapidly in wet weather within 3-4 days and plants get blighted. Method adopted by Gallegly and Marvel (1955) is modified by giving disease index of 0, 0.5, 1.0, 2.0, 3.0 and 4.0. The plants with mean disease index upto 1.5 are recorded as resistant, upto 2.0 as moderately resistant and above 2 as susceptible (Kaur *et al.*, 1988).

Sources of Resistance

Tomato cultivars/lines *viz.*, Rossol/VFN, Roma, GP-12 and GP-18 (Erinle and Quinn, 1980) Ottawa 31, PI-270407, Nova and Hybrid TH-2312 (Sokhi *et al.*, 1993) are identified as resistant which can be utilized for further breeding programme.

Early Blight

The disease causes target symptoms on foliage, elliptical shaped lesions on twigs and is also a post harvest disease on fruit.

Field Screening

Plants of each cultivar/line are planted in three rows at 90x30 cm with a border row and single line of susceptible check "Pusa Ruby" between the lines to be screened. After one month of transplanting, plants may be inoculated by atomizing at sunset with conidial suspension of *Alternaria* sp. ($1x10^6$) giving 6-10 conidia/cm^2 leaf surface. The severity of the leaf spots/twigs blight is recorded on three months old plants of each cultivar/line using 0-5 scale (Naik 2000).

Seed Inoculation

Since the disease is also seed borne, screening at seed level is also practiced. Surface sterilized seeds are soaked in *Alternaria* spore suspension ($1x10^6$ spores/ml) and then subjected to blotter test to find out the per cent of germinated and infected seedlings. The method is rapid and a large number genotypes can be screened in a short period.

Fruit Inoculation by Pinprick Method

Red-ripened fruits are surface sterilized and pricked with pin for inducing a small injury. The injured fruits are dipped in spore suspension ($1x10^6$ spores/ml) and incubated for developing symptoms at 27°C. The number of fruits infected and the lesions on each fruit are scored for severity (Prasad 2002)

Sources of Resistance

Cultivar L-25 was resistant while TI-244 had good field tolerance (Erinle and Eroctor, 1992). Lines *viz.*, EC-174076, EC-174057 and Marglobe were resistant (Shahi 1990). Wild species viz, *Lycopersicon pennellii* and *L.peruvianum* are resistant (Biezen *et al.*, 1995). Under field conditions, Arka Alok, Arka Abha, Arka Meghali, Arka Sourbh, IHR-305, IHR-2266, IHR-2288 and IHR-308 were resistant (Naik, 2002). At post harvest stage under artificial inoculation Arka alok, Arka ashish, LE-155 and IHR-2285 were resistant (Naik, 2002).

Bacterial Wilt

It is a serious disease in tropical and subtropical regions of the world. It is the most devastating during hot (30-35°C) and humid weather. It is primarily a soil borne disease and affected plants rapidly wilt and vascular system becomes brown. Seedling inoculation with bacterial suspension (*Ralstonia solanacearum*) and sick plots are used for screening genotypes.

Tomato hybrids Arka Abhijit, Arka Shreshta and cultivars, Arka Abha, Arka Alok are resistant. In Brinjal cultivars Arka Nidhi, Arka Kesav, Arka Neelkant and Swarna Shree are resistant to bacterial wilt.

Cucurbits

The cucumber is grown commercially in mixed cropping and river beds whereas under tropical conditions *Cucurbita* spp and melons are the most commercial ones and are affected by biotic stresses due to downey and powdery mildew, anthracnose and Fusarium wilt. Chemical control is not only expensive and has high residual effects causing health hazards, especially in desert fruits and cucumber, consumed raw. It is essential to concentrate on the development of resistant cultivars.

Foliar and Fruit Diseases

Field Screening

Against powdery mildew, downy mildew and anthracnose, it is usually screened under field condition by planting test genotypes regularly interspersed with a susceptible check and the disease is scored when it is at peak.

Glass House Screening

Against powdery and downy mildew, glasshouse screening can be done by collecting the conidia/sporangia from the infected plants in sterile water and spraying on test plants using atomizer.

Seed Inoculation

Seed inoculation can be followed against anthracnose by soaking the seeds in spore suspension of *Colletotrichum* sp. and followed by blotter test to select uninfected genotypes.

Fusarium Wilt of Cucurbits

Both in muskmelon and watermelon, Fusarium wilt has become a serious problem particularly in riverbed cultivation. Only need of the hour is to develop resistant cultivar(s). Following screening techniques are made use.

Vial Test (Wensley and Mckeen, 1962)

Three weeks old seedlings are uprooted from nursery in plastic trays by flooding water to loosen the sand. The roots are washed thoroughly under running tap-water to remove as much sand as possible. Final washing is done with sterile water. The seedlings are then placed with roots immersed in sterile water. Spore suspension of *Fusarium* is prepared by flooding ten days old culture plates with sterile water. For 100 ml spore suspension, culture growth of three plates resulted in spore concentration of $4\text{-}8\times10^{6}$/ml. The resultant spore suspension is strained through sterile muslin cloth and 40 ml of such spore suspension was transferred into test tubes. Seedlings are then planted in each test tube in such a way that the roots are immersed in spore suspension and cotton is placed around the plant at the mouth of test tube and then incubated at 28 °C for 48 hours before taking observation.

Seed Inoculation Technique

Healthy seeds of test genotypes are surface sterilized and dipped in *Fusarium* spore suspension for overnight and sown on blotter paper lined on Petri plates. Ten seeds can be placed in each plate and incubated for a week. Observations on seed germination and seedling death are recorded.

Sick Pot/Plot Technique

Surface sterilized seeds of test genotypes are planted in sick pot or plot containing sick soil. Sick pots are prepared by placing the giant inoculum of *Fusarium* prepared on sorghum grain. Sick plot is prepared by repeatedly growing the susceptible cultivar and incorporating the wilted plants and debris into the plot. In both the cases the sufficient inoculum of *Fusarium* is monitored by plating the soil on Fusarium specific medium.

Resistant Sources

Cucumber

Powdery Mildew

Poinsette, Yomaki, Spartan salad, PI-197088, *Cucumis. anguria*, *C.denteri* and *C.sagittatus* are resistant and resistance is governed by one, two and three genes.

Downy Mildew

Bangalore, Chinese Long and Poinsette are resistant and governed by single recessive gene.

Anthracnose

Poinsette is highly resistant and PI-175111 and PI-197087 are also highly resistant.

Cucurbita spp.

Powdery Mildew
C.lundelliana and *C.martinezzi* are resistant

Muskmelon

Powdery Mildew
PMR-45, PI-124111, PI-134198 and Edisto Georgia-47 and C-68 are moderately resistant.

PMR-450 is resistant to race 1

Seminole is resistant to race-2

Arka Rajhans and Pusa Sharabati are resistant varieties.

Downy Mildew
PI-124112, Georgia-47, AC-70-154, Perlita, *C.callusus* amd *C.melo var momordica* are resistant.

Fusarium Wilt
Delicious-51, *C.melo var raticulatus, C.melo var indorus, C.melo var chito,* and *C.melo var flexuous* are resistant.

Watermelon

Powdery Mildew
Arka Manik- highly resistant, Asahi Yamato, Sugar Baby- resistant

Anthracnose
Black store, Congo, Charleston Gray, Arka Manik.

Fusarium Wilt
Conquorer, Sumit, Shipper, Calhoun Gray, Smokylee, Charleston gray (Naik, 1990)

Cruciferous Vegetables

The frequent outbreak of disease in cruciferous crops is one of the major constraints in successful vegetable production. A brief account of screening techniques of major disease along with source of resistance is presented

Black Rot

The cells of *Xanthomonas campestris pv campestris* multiply in the vessels, the xylem becomes plugged with extra cellular polysaccharide. Vascular obstruction leads to water stress development of V shaped lesions and blackening of the veins. The pathogen is both seed and soil borne, small, rod shaped, aerobic, gram negative and non spore forming with single polar flagellum.

Screening Techniques
Stab Inoculation
Drops of inoculum are placed on the stems, leaf axils or leaves and then stabbed once or repeatedly through each drop of inoculum with a sterile needle.

Slit Inoculation
An incision is made on the stem or any part of the plant using a sterile razor blade or a pair of scissors. The inoculum is applied to the fresh cut and the plant is kept in moist chamber for a day or so to encourage guttation. Then it is removed and kept outside.

Carborundum Rubs

The leaves are sprinkled lightly with 600 mesh carborundum powder and then gently rubbed with a cotton swab soaked in inoculum suspension.

Syringe Inoculation

A syringe is used to inject inoculum into stems, leaf axils and with great skills into the leaves.

Spray Inoculation

A hand sprayer or pump operated pressure sprayer is used to spray the underside of the leaves until the leaf tissue gets water soaked.

Hydatodes Inoculation

This technique involves introduction of bacteria into guttation droplets, which are subsequently taken up by the host via the hydathodes (Robeson *et al.*, 1989).

Leaf Cut and Dip Method

Five weeks old seedlings at 4-5 leaf stage are inoculated with 7 days old bacterium culture suspenson (1×10^8/ml) by cutting the leaf at the margins and dipped in suspension. The plants are kept at high RH for 12 days and disease progress is recorded after 14 days of inoculation (Sharma *et al.*, 1995)

Reports from India, Australia and various regions of USA indicate that Early Fuji resists a variety of indigenous strains of *Xanthomonas campestris* pv *campestris* (Williams, 1980). Screening and introduction of local germplasm of cabbage and cauliflower revealed resistance in some cultivars like 294 bc reported completely resistant (Chakravarty *et al.*, 173). Pusar Kae, MG-8-2-3 and Sn-445 (Sharma *et al.*, 1977), Ivans and Iglory (Dua *et al.*, 1978),PI-436606 (Hunter *et al.*, 1987), Pusa Mukta (Kadia *et al.*, 1991) and RBS, EC-162587, Sn-445 and Lawanya (Sharma *et al.*, 1995) are also resistant.

Stalk Rot

Stem Inoculation

Genotypes are screened by applying 50 ml of a homogenized fungal suspension adjacent to the stems of two months old plants of each genotype. With this, the plant infection was also the least and leaf lesion was also small (Sharma *et al.*, 1982).

Detached Crucifer Petals

Cauliflower cultivars are screened by using partially colonized crucifer petals (Kapoor *et al.*, 1985)

Inocualtion of Leaf Axil

Leaf axil near collar region is inoculated at curd initiation with 100 mg culture of fungus grown on sand maize meal (1:2w/w). The observations on mortality(per cent) are recorded after 25 days of inoculation. The technique is better than seed inoculation.

Resistant Sources

Early winters, Adams Head, EC-162587, Kn-81, Janavon and EC-131592 (Sharma *et al.*, 1995) are resistant.

Downy Mildew

Plants are sprayed with 10^3 to 10^5 sporangia/ml of distilled water after 3 days of emergence of seedlings. The diseased leaves are also kept between pots to provide enough inoculum of the pathogen.

Inoculated plants are kept under high relative humidity. The disease incidence is recorded after 15 days of inoculation (Sharma *et al.*, 1995).

Resistant Sources

Cauliflower lines like PI-208474, CC, 3-5-1-1, PI, S, PI-1231210, PI-246007 and Early Winter Adam's are resistant (Hoser *et al.*, 1995; Mahajan *et al.*, 1995 and Sharma *et al.*, 1995).

Cabbage

PI accessions 263056, 263057, 357374, 418984, 418984, and 418988 are resistant (Hoser *et al.*, 1995).

Alternaria Blight

The disease caused by *Alternaria brassicae* and *A.brassicola* is seed borne. Once the plantlets start rooting, the rooted plants are spray inoculated with spore suspension of fungus (1×10^4/ml) with atomizer. The disease symptoms appear within a week and scoring of genotypes is done on 0-5 scale. Sharma *et al.* (1991) screened five cauliflower varieties and only KT-9 performed better and it has more deposition of epi cuticular wax on its foliage. Four lines *viz*, IIHR-138, IHR-142, OOHR-217 and IIHR-310 were resistant and IIHR-256, IIHR-264, IIHR-271 and IIHR-318 were moderately resistant among 70 lines (Varalakshmi *et al.*, 1997)

References

Anonymous, 1988. Progress report AVRDC, Shanua, Taiwan Republic of China

Angadi, H.D., Naik, M.K., Patil, M.G. and Patil, R.G., 2003. Evaluation of chilli genotypes against anthracnose disease. *Vegetable Science*, 30(2): 164–165.

Bhardwaj, S.S., 1983. *Ph.D. Thesis*, HPKVV, Solan, 114 pp.

Biezen, E.A., Van der, Glazostskaya, T, Overduin, B., Nijkamp, H.J.J. Hille, J. and Vander, B.E.A., 1995. *Molecular Genetics*, 274: 453–561.

Chakravarthy, B.P., Hegde, S.V. and Gupta, D.K., 1973. *Current Science*, 42: 803–804

Dodan, D.S., Shyam, K.R. and Bhardwaj, S.S., 1995. *Plant Disease Research*, 10: 135– 136.

Dua, I.S., Suman, B.C. and Rao, A.V., 1978. *Indian Journal of Experimental Biology*, 16: 488–491.

Erinle, I.D.and Eroctor, P.G.E., 1992. *Samaru Journal of Agricultural Research*, 9: 63– 67.

Erinle, I.D. and Quirul, J.G., 1980. *Plant Disease*, 64: 701–702.

Gallegy, M.E. and Marvel, M.E., 1995. *Phytopathology*, 45: 109.

Haymaker, II.H., 1928. *Journal of Agricultural Research*, 36.: 675–719.

Henderson, W.R. and Winstead, N.N., 1961. *Plant Disease Reporter*, 45: 272–273.

Hoser, K.J., Lakowska, L.E. and Antosik, 1991. *Cruciferae News Letter*, 16: 145–146.

Hoser, K.J., Lakowska, L.E. and Antosik, 1995. *Journal of Applied Genetics*, 36: 27–33.

Hunter,J.E., Pickson, M.H., Ludwig, J.W., 1987. *Plant Disease*. 71: 263–266.

James, W.C., 1974. *Annual Review Phytopathology*, 12: 27.48.

Kadia, T.S., Tewari, R.N., Singh, R., Singh, N. and Verma, T.S., 1991, *Indian Journal of Horticulture*, 48: 247–250.

Kapour, K.S., Gill, H.S. and Sharma, S.R., 1995. *Phytopathologische Zeitschrifi*, 112: 191–192.

Kaur, S., Singh, Surjan, Kanwar, J.S. and Cheema, D.S., 1988. *Indian Phytopathol.*, 41: 486–487.

Kreutzer, W.A. and Bryant, L.R., 1944. *Phytopathology*, 34: 845–847.

Madhukar, H.M., Naik, M.K. and Sidramaiah, A.L., 2001. *National Symposium of the Society of Soil Biology and Ecology*, UAS, Bangalore.

Madhukar, H.M., Naik, M.K. and Patil, M.G., 2002. *National Symposium in Crop Protection and WTO: An Indian Scenario*, CPCRI, Kasargod.

Mahajan, Vijay, Gill, H.S., More, T.A. and Mahajan, V., 1995. *Euphytica*, 86: 1–3.

Naik, M.K., 1990, *Ph.D. Thesis*, IARI, New Delhi–12

Naik, M.K. and Poonam Sinha, 1995. *Annual Report of IIHR*, Hessarghatta, Bangalore p.43.

Naik, M.K and Rawal R.D., 2002. *Resource Management in Plant Protection*, Vol. 1. PPAI, Hyderabad, 1: 64–84.

Singh, Akhilesh, Singh, A.K. and Singh, A., 1998. *Crop Res.*, 15: 132–133.

2012, Nutri-Horticulture
Editor: Professor K.V. Peter
Published by: DAYA PUBLISHING HOUSE, NEW DELHI

Pages 241–255

Chapter 14

Factors Affecting Shelf Life of Fruits

Ram Asrey* and Kalyan Barman

Division of Postharvest Technology,
Indian Agricultural Research Institute, New Delhi – 110 012, INDIA
*E-mail: ramu_211@yahoo.com

Postharvest product quality develops during growing of the product and is maintained, not improved by postharvest technologies (Hewett, 2006). Fruit quality and shelf life are twin interdependent twin traits; these can not be seen in isolation. Any change in shelf life affects the fruit quality and vice-versa. By and large, postharvest management of fruits is less attended aspect through in most Asian countries and preharvest influence on postharvest shelf life is further a step behind. Scientific findings have shown that combinational use of pre and postharvest management practices give better results in prolonging shelf life of the fruits. So, it is imperative to integrate pre and post production factors to harness their synergy for shelf life extension of fruits. Depending on input, preharvest factors have weak or strong influence on postharvest quality and shelf life of fruits. Preharvest factors alter the production of a number of metabolic products including chlorophyll, carotenoids, xanthophylls, ascorbic acid, tannins, phenolic compounds and other biochemicals. Among the preharvest factors, management of soil fertility (Ca and N), use of plant growth regulators (gibberellins and auxin), selection of suitable variety and irrigation offer much promises. State of ethylene generation, physiological loss in weight, respiration and transpiration are crucial shelf life retention during storage and marketing.

Shelf Life

This term (Shelf life) found its roots in the European culture as the British use to keep the horticultural produce in wooden cabinet in their houses. The period range in which the cosmetic appeal of the produce keeps satisfying to the consumer was called as shelf life. Several criteria like physiological loss in weight, texture, microbial load, taste and overall appearance are used to determine the shelf life of fruits. Generally 8-10 per cent water loss from the fruit indicates the end of shelf life.

The term 'quality' is a complex perception felt by the consumer/viewer in different ways. Generally quality can be viewed as absence of defects or a degree of excellence of produce from consumer view point (Shewfelt, 1999). Quality means different things in the distribution chain. Food quality embraces both sensory attributes that are readily perceived by the human senses and hidden attributes such as safety and nutrition that require sophisticated instrumentation to measure. The quality of the produce/product may be categorized into two groups. Product oriented quality has more scientific support, it is measurable and reproducible. Quality changes can be plotted as a function of time and directly related to physiological changes *viz.* respiratory gas evolution during handling and storage. Product oriented quality can be adjusted to be more responsive to the particular market or a group of consumers by pre and postharvest treatments. This is purely governed by the human mood, region, religion and gender. For example big size kinnow are preferred in Delhi market while Kolkata market prefers medium sized fruits. Red colour onion is preferred in India while gulf countries prefer white onion. Women mostly prefer narrow TSS: acid ration in fruit as compared to their male counter part.

Preharvest Factors Affecting Shelf Life

Climatic Conditions

Climatic factors have an important influence on the quality and shelf life of fruit crops. Apples grown in warm climate, clear days and cool nights were found more conic-elongate than those grown in hot days and warm nights. Likewise differing climatic conditions between two geographical sites had a significant effect on banana fruit size, shape and shelf life.

Light Intensity

Light intensity has profound impact on quality and shelf life of fruits. In guava fruits facing higher light intensity were found inferior in vitamin C content compared to fruits exposed to lower light intensity (Lee and Kader, 2000). In general fruit crops exposed to moderate full sunlight contain more sugars, dry matter and shelf life as compared to shade grown fruits (Nilsson and Gustavsson, 2007). Excess light results in sun scald which is a major problem in most of the fruit crops like persimmon, mandarin, pomegranate, apple, banana, pineapple and berries. Insufficient light results in reduction in colour intensity in red apples, grape and peaches (Solovchenko *et al.*, 2006). It also causes reduction of glossiness in strawberry (Osman and Dodd, 1994).

Temperature

A wide range of appearance disorders are caused by preharvest high temperature. The injury sustained is temperature and time dependant and varies with species, cultivar, stage of development and plant part. High temperature disorders occur due to direct effect of excess water loss which includes damage to cellular membranes, protein and nucleic acids. Indirect effects to quality include inhibited pigment synthesis and thermal degradation of existing pigments and a wide range of sun scald or sun burn symptoms and all these result into poor shelf life (Woolf and Ferguson, 2000).

Freezing/Frost Damage

Radiative freeze damage in strawberry often results in smaller fruits and depending upon the development stage when damage occurs, misshapen fruits obtained from half developed stage. Other examples of degradation and shelf life reduction include discolouration in Kiwi fruits, frost rings on apple and frost circle on peach. Radiative freezing occur under clear sky and calm wind condition. In this, thermal energy moves from air to the vegetation.

Nutrition

Soil Nitrogen

Excessive soil nitrogen has been shown to result in reduced shelf life. High nitrogen fertilizer application can cause imbalances in the level of essential amino acids (poor quality protein) and also increase bitter pits in apples because it affect the Ca absorption. Excess nitrogen dose may delay the pigment synthesis and colour development in apple, cherries and strawberry. High nitrogen application tended to increase ethylene evolution and respiration rate in apple and mango. Excess nitrogen presence decreases fruit firmness in general (Daane *et al.*, 1995). Esters are major contributors to perception of fruit flavour, judicious and split nitrogen application favours flavour synthesis in fruits.

Calcium

Calcium affects fruit senescence and quality by changing intracellular and extracellular processes and the rate of fruit softening depends on fruit Ca status. Ca also plays a regulatory role in various processes that influence cell function and signal transduction. Cell wall degradation results in softening in a fruits, Ca delays softening by delaying in degradation of cell wall polymers. Ca also plays a major role in cell to cell adhesion and this phenomenon is important in textural quality of the fruits. Ca deficiency causes bitter pit, internal breakdown and water core in apple (Rabus and Streif, 2000). In strawberry, if the plants are Ca deficient the fruits are small hard and with seedy patches (Sharma *et al.*, 2006). However optimum and excess level of Ca reduces fruit sweetness in strawberry and cherries.

Other Nutrients

Phosphorus (P) level in soil and plant does not have much direct effect on shelf life and internal fruit quality but it certainly affects the fruit appearance. Research findings show that there is a relationship between pH regulation, organic acids levels and fruit potassium (K) content. Optimum level of K exerts a favorable response on vitamin C content in fruit. Potassium deficiency causes in poor peach fruit colouration and small fruit size on oranges. P deficiency results in large Valencia oranges and excess mask the colour development in cranberry. Fe and Zn deficiency results into reduced fruit size in citrus and both colour and size reduction in peaches. Boron deficiency reduces strawberry fruit size causes external corking of apple fruit. Imbalance in certain nutrient can also have a pronounced impact on fruit shape *e.g.* Zn deficiency alters the shape of peach and cherry fruits. Copper deficiency affects citrus fruit shape (misshape) and kernel filling in walnut. Molybdenum deficiency produce fruit misshape disorder in strawberry. All nutrient based deficiency based disorders tend to decrease fruit shelf life.

Water Management

Water availability is an important factor influencing shelf life and nutritional status of fruits. Drought stress or flooding can result in plant death and photosynthetic reduction. Moisture deficit in soil have been shown to influence ascorbic acid content in fruits. Drought stress and saline water application affect the TSS, sugars and flavour of the fruits. Drought stress and saline water application up to certain improves the quality and shelf life of produce. In peaches and nectarines, trees supplied with optimum amounts of water (100 per cent evapotranspiration, ET) during the fruiting season produces maximum fruit size. However, higher soluble solid level (SSC) and shelf life can be obtained by imposing moderate water stress during fruit growth prior to harvest. This will reduce fruit size, but in some cases only slightly. When peaches were held under moderate water stress (50 per cent ET a month prior harvest), the stored fruit did not develop dry and mealy texture (internal break down symptoms) and shown longer shelf life (Crisosto *et al.*, 1994).

Pruning

Numerous studies show that improving light penetration into the canopy improve the fruit composition of grapes, raspberry, apple, peach and plums. Judicious pruning generally increase soluble solid content (SSC) anthocyanins, total soluble phenols; and reduced titrable acidity, malate, pH and potassium content in fruits. Shading results in significant differences in the aroma of fruits. The greater light interception by an individual fruit and its surrounding leaves the better its shelf life including fruit colour, size and flavour. Peach fruits in the upper position of the canopy have always better shelf life than fruits in the lower, shaded part of canopy. The removal of interior water sprouts can significantly increase light penetration and improve shelf life of lower position fruits (Crisosto *et al.*, 1995).

Plant Growth Regulators

The use of plant growth regulators can positively or negatively affect fruit quality and shelf life. Application of gibberellic acid (GA_3) is commercially practiced in grape, strawberry and citrus fruits to improve their appearance and shelf life (Hewett, 2006). Auxins are also used in various forms to alter maturity and quality improvement in mandarins, berries and some temperate nuts. Pactobutrazole, Dormax, CIPC etc. are also commercially being utilized by the growers as a preharvest treatment in fruit crops to enhance their cosmetic appeal and shelf life during storage. Preharvest application of putrescine (polyamines) in mango exhibited higher firmness, TSS and lower fruit rot during storage (Malik and Singh, 2005). Preharvest foliar application of putrescine also reduces ethylene production of harvested peach fruit and delayed firmness loss (Bregoli *et al.*, 2002). Daminozide (Alar or B_9) preharvest foliar application in apples (2500 mg/l) enhances fruit colour development and firmness. Preharvest applications of paclobutrazol (2000 ppm) have been found effective in minimizing spongy tissue disorder of "Alphanso" mango. Naphthalene acetic acid (NAA) foliar application (300 ppm) in pineapple 30 days before harvesting enhanced shelf life by increasing the potassium ion concentration in mature fruits and minimized internal pulp browning (65 per cent) during cold storage.

Genetics and Cultivar

In addition to the preharvest conditions discussed above genetics and cultivar selection are major factors involved in postharvest quality and shelf life outcomes for the fruits. Since cultivars vary in genetic makeup, they will also vary in characters like size, colour, flavour, texture, nutrition, storage life, processing ability and eating quality. Genetics engineering has made it possible to tailor nutrient, vitamins and pigments rich designer fruit crops with higher shelf life. Such transgenic varieties are now commercially grown in case of mango, papaya, cherry, strawberries and banana.

Postharvest Factors Affecting Shelf Life

Fresh fruits are living tissues subject to continuous change after harvest. Some changes are desirable from consumer point of view but most are not. Postharvest changes in fresh fruits can not be stopped, but these can be slowed down within certain limits to enhance the shelf life of fruits. The postharvest treatments play an important role in extending the shelf life of fruits during storage and retailing. Some of such treatments are discussed briefly hereunder.

Thermal Treatments

Hot Water Treatment

Fruits may be dipped in hot water before storage to control various postharvest diseases and improving peel colour of the fruit. The two main commercial hot water treatments (HWT) are hot water immersion and hot water rinsing and brushing (HWRB). HWT not only remove soil and dust, but also fungal spores from the fruit surface. It is effective in eradicating quiescent infections of the fungi that have become established on and beneath the cuticle and within the pedicel. Additionally it has been also used on fruits to control insect infestation. This treatment causes recrystallization or melting of the wax layer of the fruit surface which sealed barely visible cracks in the cuticle through which water could escape. This sealing of cracks or natural openings significantly limits sites of fungal penetration, reduced weight loss, thus maintaining fruit firmness after prolonged storage. HWT also allows the infected fruit to build up resistance response by the production of lignin-like material production at the inoculation site, followed by accumulation of the phytoalexins and scoparone. HWRB treatment also reduces the respiration rate, ethylene evolution rate and enzymatic activity. As a result of it, there is lower level of fruit softening due to the inhibition of cell wall degrading enzyme and ethylene-forming enzyme (EFE) activity which accelerate the shelf life of fruits. But to attain improved control of postharvest diseases, sometimes certain fungicides to the hot water is also added. Hot water treatment has been applied to various fruit crops, like in mango it is recommended at 50-52 °C for 5 minutes to reduce the fungal infection anthracnose and stem end rot during ripening or storage. This treatment also helps in attaining uniform ripening within 5-7 days. Besides controlling postharvest diseases, hot water treatment at 46 °C for 5 minutes induces tolerance to low temperature injury of avocado fruits during storage at low temperature.

Table 14.1: Recommended Hot Water Treatments for Controlling Different Pathogens in Fruits

Commodity	Pathogens	Temp. (°C)	Time (min)	Possible Injuries
Apple	*Gloeosporium* sp. *Penicillium expansum*	45	10	Reduced storage life
Grapefruit	*Phytophthora citrophthora*	48	3	
Lemon	*Penicillium digitatum* *Phytophthora* sp.	52	5-10	
Mango	*Colletotrichum gloeosporoides*	52	5	No stem rot control
Orange	*Diplodia* sp. *Phomopsis* sp. *Phytophthora* sp.	53	5	Poor degreening
Papaya	Fungi	48	20	
Peach	*Monolinia fructicola* *Rhizopus stolonifer*	52	2.5	Motile skin

Vapour Heat Treatment (VHT)

This treatment proved very effective in controlling infection of fruit flies in fruits after harvest. For this treatment, the boxes containing fruits are stacked in a room, which are heated and humidified by injection of steam. The temperature and exposure time are adjusted to kill all stages of insects (egg, larva, pupa and adult), but fruit should not be damaged. Most difficult stage of insect to control by VHT is the larval stage. A recommended treatment for citrus, mangoes, papaya and pineapple is 43 °C in saturated air for 8 hours and then holding the temperature for further 6 hours. VHT is mandatory for export of mangoes.

Fumigation

The fumigation with SO_2 is successfully used for controlling postharvest diseases of grapes. This is achieved by placing the boxes of fruits in a gas tight room and introducing the gas from a cylinder to the appropriate concentration. Special sodium metabisulphite pads are available, which can be packed into individual boxes of grapes to release SO_2 slowly. The primary function of treatment is to control the *Botrytis cinerea*. The SO_2 fumigation is also used to prevent discolouration of peel of litchis. The fumigation of litchi fruits with SO_2 (600 g/ton of fruit), followed by dipping of fruits in 1M HCl for two minutes help in retaining the red colour of litchi fruits and maintaining the market quality for 3 weeks at 2 °C and 90-95 per cent RH.

Chemical Treatment

Washing with Ozonized Water

Ozone is the most effective natural bactericide of all the disinfecting agents. It is now being used in food processing or storage of perishables as an anti-microbial agent. Ozone is the safe and natural purification and disinfecting agent. It is strong and ideal, germicide, sanitizer, sterilizer, vermicide, anti-microbial, bactericide, fungicide, deodourizer and detoxifying agent (Graham, 2000). Ozone oxidizes the metabolic products and neutralizes the odours generated during the ripening stage in stored fruits. It also enhances the taste of most perishables by oxidizing pesticides and neutralizing ammonia and ethylene gases produced by ripening or decay. The reduction of ethylene gas helps to preserve, reduces shrinkage and increases the shelf life of fruits. Its use does not leave any toxic by-products or residues, does not affect healthy cells or alter its chemistry, and is non-carcinogenic. Ozone always reverts back to its original form-oxygen. Ozone should be constantly consumed and absorbed during the oxidation process. Its effectiveness influenced (lowered) due to the presence of 100 per cent humidity levels. When the humidity level is below 50 per cent, the efficiency of it slows as a bacterial medium (Castillo *et al.*, 2003). Positive effect will show at low and constant levels between 0.05 ppm and 0.1 ppm and it allows workers to enter the storage area and carry out their work.

Washing with Chlorine Solution

Chlorine treatment (100-150 ppm available chlorine) can be used in washing water to help control inoculums build up during packing operations. pH of wash water between 6.5 and 7.5 give best results.

Calcium Application

The postharvest application of $CaCl_2$ or $Ca(NO_3)_2$ play an important role in enhancing the storage and marketable life of fruits by maintaining their firmness and quality (Asrey *et al.*, 2008).The postharvest application of $CaCl_2$ (2-4 per cent) for 5-10 minutes dip, extends the storage life of pear up to 2 months, apple up to 7 months and plum up to 4 weeks at 0-2 °C with excellent colour and quality. Treatment of mangoes with 2-8 per cent $CaCl_2$ solution resulted in delayed softening of fruit during storage at 20 °C by 8-12 days compared to untreated fruits.

Use of Growth Regulator/Fungicides

The growth regulators (GA_3) or fungicide (bavistin) application can be effectively used to extend/enhance the shelf life of fruits. In mandarins, such as kinnow, the application of bavistin/thiabendazol increased the shelf life of fruits. Likewise, the application of cytokinins or GA_3 (20 ppm) is helpful to extend the shelf life of citrus fruits particularly "Baramasi" lemon. To control the mango anthracnose, the postharvest dips of fungicides in cold water rarely reduced level of infections, but when fungicides was applied in hot water, complete control could be achieved. Treatment of litchi fruit with 0.25 per cent SOPP (Sodium-ortho-phenyl-phenate) in 6 per cent wax emulsion provides protection against a number of fungal diseases.

1-Methylcyclopropene (1-MCP)

1-methylcyclopropene (1-MCP) is a synthetic cyclic olefin that inhibits ethylene by binding irreversibly to ethylene receptors (Sisler *et al.*, 1997). The affinity of 1-MCP for the receptor is approximately 10 times greater than that of ethylene. It has a non-toxic mode of action, a negligible residue and is active at very low concentrations. 1-MCP is available as a commercial preparation named Ethylbloc® and SmartFresh® powder. These are stable, complex formulations which releases 1-MCP when dissolved in water. Commercial development of 1-MCP has largely concentrated on the apple fruit. However, Semi-commercial trials have been carried out with a wide variety of fruit crops. In general, the effect of 1-MCP on delaying ripening and senescence rates are associated with the reduced ethylene production, respiration, colour change and rate of softening. Treatment with 1-MCP at 0.1 and 1 il L^{-1} in apples reduces ethylene production, respiration and slower loss of firmness. Softening and respiration rates were decreased when banana treated with 1 il L^{-1} 1-MCP. In strawberry, 1-MCP at 5-15 nL L^{-1} doubled post harvest life when stored at 5 °C.

Polyamines

Polyamines (PAs) are low molecular weight small aliphatic amines that are ubiquitous in living organisms and have been implicated in a wide range of biological processes, including plant growth, development and response to stress (Smith, 1985). Polyamines have been proposed to be a new category of plant growth regulator and are proposed to be involved in large spectrum of physiological and biological processes of fruits. PAs in their free forms have been reported as anti-senescent agent, thus increasing shelf life from both endogenous and exogenous applications (Valaro *et al.*, 1998). The main effects of PAs in postharvest management of fruits are- inhibit biosynthesis of ethylene, increase fruit firmness, reduce respiration rate, reduce chilling injury, retard colour changes and reduce mechanical damage. The most common polyamines are putrescine, spermidine and spermine found in every plant cell. These PAs inhibit biosynthesis of ethylene because polyamine and ethylene biosynthesis are linked through the common precursor S-adenosylmethionine (SAM). PAs and ethylene use the common precursor, SAM, for their biosynthesis. But these two molecules show opposite effects in relation to senescence. Ethylene production is associated with the biosynthesis of ACC. Most of the observations indicate that various PAs can delay senescence by inhibiting ACC synthesis (Lee *et al.*, 1997). Apricot fruit treated with putrescine (1 mM) by low pressure infiltration showed significant reduction in ethylene levels as compared to untreated fruits (Romero *et al.*, 2002). Similar result was found in case of plums, when treated with 1mM and 2 mM putrescine by immersion method (Khan *et al.*, 2007). During storage of fruits at low temperature, incidence of chilling injury (CI) have been found in a number of tropical and subtropical fruits thus reducing its shelf life. Exogenous application of PAs have been reported to reduce CI by protecting the membrane lipid from being conversion in physical state (liquid-crystalline to solid-gel state) and lipid peroxidation due to its antioxidant property. Pre-storage application of polyamines (1 mM putrescine and 1 mM spermidine) in pomegranate fruits reduces chilling injury when stored at 2 °C and 90 per cent RH up to 3 months (Mirdeghan *et al.*, 2007). PAs also maintain firmness of fruit due to their cross-linkage to the carboxyl group of the pectic substances in the cell wall, resulting in rigidification, thus blocks the access of degrading enzymes

(Polygalacturonase, pectin methyl esterase, pectin esterase and cellulase) and reducing rate of softening and less prone to mechanical damage. Mango treated with putrescine (1 mM), spermidine (0.5 mM) and spermine (0.01 mM) maintained significantly higher firmness up to 28 days after storage at 13 °C as compared to untreated fruits (Malik *et al.*, 2005). Similar results were found in case of apples up to 27 weeks of storage at 0 °C.

Non-chemical Treatment

Edible Coating

Edible coatings are thin layers of edible material applied to the product surface in addition to or as a replacement for natural protective waxy coatings and provide a barrier to moisture, oxygen and solute movement. They are applied directly on the surface of fruit by dipping, spraying or brushing to create a modified atmosphere. Unwaxed fruit have pores in their surface through which most the diffusion of O_2 and CO_2 occurs. Water vapour can move through the pores, cuticle and microcracks in the cuticle. During the process of waxing a tightly adhering thin film of the coating substance is applied to the surface of the fruit. Surface coatings have been used extensively on fruits to modify internal atmosphere composition and thereby delay ripening, reduce the water loss, improve the finish of the skin and thus enhancing shelf life. These waxes may be of animal origin, vegetable origin or may be produced synthetically.

Semperfresh

Semperfresh, a food- grade coating used to retard moisture loss, ripening and spoilage of fruit is a mixture of sucrose esters with high proportion of short chain unsaturated fatty acid esters, sodium salts of CMC and mixed mono and diglycerides. Semperfresh significantly reduced water loss and internal CO_2 in several fruits and reduced colour changes, retained acid, increased shelf life and maintained the keeping quality of apples.

Chitosan

Chitosan, a by-product from crustacean shell wastes, is a high molecular weight cationic polysaccharide. Chitosan-based coatings are effective in prolonging the shelf life and improving quality of fruits, by delaying ripening, reducing respiration rate, reducing desiccation, regulating gas exchange, decreasing transpiration losses, modifying the internal atmosphere, maintaining the quality of harvested fruits, retaining fruit firmness, freshness, weight loss, titratable acidity, vitamin C and reducing mould growth. Chitosan could be an ideal preservative coating because of its film forming properties, biochemical properties, inherent antifungal properties, enzyme activity (chitinase) and elicitation of phytoalexins.

Carnauba Wax

It is an edible coating material under the group of lipid, which is mainly impart to reduce water loss and gloss. Carnauba wax is recovered from the underside of the leaves of a Brazilian palm tree (*Copernica cerifera*).

Pro-long

Pro-long is a mixture of sucrose fatty acid esters, sodium carboxymethyl cellulose and mono-and diglycerides. The mode of action of Pro-long involves the creation of a selectively permeable barrier creating internal atmospheres which preserve the fruit by reducing water loss and chilling injury characteristics, which might be utilized both in the storage of fruit and for the maintenance of quality during the marketing period. Coating bananas with Pro- long reduced weight loss, oxygen uptake, ethylene release and chlorophyll loss and modified their internal atmosphere by reducing the permeability of the fruit peel to gases (Banks *et al.*, 1984). Treatment with 0.75 per cent Pro-long also significantly increased the storage life of mangoes, retarded ripening and reduced weight loss and chlorophyll loss.

Table 14.2: Specific Coating Applications for Different Fruits

Coating Materials	Fruits
Prolong	Banana
Semperfresh	Banana
Semperfresh with organic acid	Banana
Ban-seel	Banana and plantains
Tal prolong, Semperfresh and applewax	Apple
Nutri-save	Golden delicious apple
Semperfresh	Granny smith apple
Brilloshine	Apple, avocado, melons and citrus fruits
Nu-coatflo, Brilloshine and Citrashine	Citrus fruits
Semperfresh	Guava
Palm oil	Guava
Vapor gard	Mango
Chitosan	Strawberry and raspberry
N,O-Carboxymethyl chitosan	Fruits

Essential Oils

Essential oils (EOs) are volatile, natural, complex compounds characterized by a strong odour and are formed by aromatic plants in different parts (flowers, buds, seeds, leaves, twigs, bark, herbs, wood, fruits and roots) as secondary metabolites. These are registered food grade materials and have the potential to be applied as an alternative treatment to control postharvest decay of fruits. An important characteristic of EOs and their components is their hydrophobicity, which enables them to partition in the lipids of the bacterial cell membrane and mitochondria, disturbing the structures and rendering them more permeable. As a result, damage of membrane proteins and depletion of proton motive force takes place. Leakage of ions and other cell contents can then occur which after attaining

Table 14.3: Essential Oils Used for the Control of Postharvest Diseases of Fruit

Essential Oil	Major Component	Bacterial spp.	Fruit Crops
Clove oil	Eugenol	*P. expansum, M. fructigena, B. cinerea, P. vagabunda*	Apple, grape
Mint oil	Menthol	*P. italicum, B. cinerea, R. stolonifer*	Orange, strawberry
Thyme oil	Thymol, carvacrol	*B. cinerea, R. stolonifer, M. fructicola*	Strawberry, grape, sweet cherry
Cinnamon oil	Cinnamaldehyde	Natural flora	Kiwifruit
Lemongrass oil	Citral	*C. gloeosporioides, P. expansum, B. cinerea, R. stolonifer*	Mango, peach

extensive loss of cell contents or of critical molecules and ions lead to death of bacterial cells (Denyer and Hugo, 1991). A number of essential oil components has been identified as effective antibacterial, *e.g.* eugenol, carvacrol, thymol, menthol in several fruits like apple, peach, sweet cherry, strawberry, grape, citrus, mango etc.

Biocontrol Agents

Biological control means controlling the number and activity of a pathogen or insect by another member of the community other than man. Among different biological approaches, use of the microbial antagonists like yeasts, fungi, and bacteria is quite promising and gaining popularity. Microbial biocontrol agents possess a number of important advantages over traditional chemical pesticides which make their commercial outlook particularly promising, as, in general, they are considered nonhazardous to humans and animals; biodegradable and environmentally friendly; attack specific target organisms, leaving other beneficial organisms unaffected; are easy to genetically modify; and can be commercially developed with relative ease. These advantages, however, are counterbalanced by a number of limitations which include the sensitivity of most of the currently marketed microbial control agents to adverse environmental conditions such as extreme dryness, heat, and cold; limited shelf life; limited biocontrol efficacy in situations where several pathogens are involved in decay development; and limited effectiveness under high disease pressure. Several modes of action have been suggested to explain the biocontrol activity of microbial antagonists. Amongst these, competition for nutrient and space between the pathogen and the antagonist is considered as the major mode of action by which microbial agents control pathogens causing postharvest decay. In addition, production of antibiotics or anti-fungal metabolites, direct parasitism and induced resistance with reduction of pathogen enzyme activity is other modes of action of the microbial antagonists by which they suppress the activity of postharvest pathogens on fruits (Sharma *et al.*, 2009).

Table 14.4: Microbial Antagonists Used for the Control of Postharvest Diseases of Fruit

Antagonists	Diseases (Pathogen)	Fruit Crops
Bacillus subtilis	Brown rot (*Lasiodiplodia theobromae*), Gray mold (*Botrytis cinerea*), Green mold (*Penicillium digitatum*), Stem end rot (*Botryodiplodia theobromae*), Alternaria rot (*Alternaria alternata*)	Apricot, strawberry, citrus, avocado, cherry, litchi
Bacillus licheniformis	Anthracnose (*Colletotrichum gloeosporoides*), stem end rot (*Dothiorella gregaria*)	Mango
Trichoderma viride	Green mold (*Penicillium digitatum*), Stem-end rot (*Botryodiplodia theobromae*), Gray mold (*Botrytis cinerea*)	Citrus, mango, strawberry
Pseudomonas syringae	Blue mold (*Penicillium expansum*), Green and blue mold (*Penicillium digitatum* and *P. italicum*), Grey mold (*Botrytis cinerea*), Brown rot (*Monilinia laxa*)	Apple, citrus, peach
Candida oleophila	Penicillium rots (*Penicillium digitatum* and *Penicillium italicum*), Crown rot (*Colletotrichum musae*), Anthracnose (*Colletotrichum gloeosporioides*)	Citrus, banana, papaya
Debaryomyces hansenii	Green and blue mold (*Penicillium digitatum* and *Penicillium italicum*), Blue mold (*Penicillium italicum*), Rhizopus rot (*Rhizopus stolonifer*), Sour rot (*Geotrichum candidum*)	Citrus, peach

Irradiation

Irradiation can be applied by exposing the fruit to ionizing radiations from radioisotopes (normally in the form of gamma-rays but X-rays can also be used) and from machines, which produce a high-energy electron beam. The FDA has approved two types of radiation sources for the treatment of foods: gamma rays produced by the natural decay of radioactive isotopes of cobalt-60 or cesium-137, x-rays with a maximum energy of five million electron volts (MeV), and electrons with a maximum energy of 10 MeV. The rays are directed onto the fruit being irradiated, but the fruit itself never comes into contact with sources of radiation. Irradiation may be referred to as a "cold pasteurization" process, as it does not significantly raise the temperature of the treated fruits. Since fruits contain 80-95 per cent water and their intercellular spaces (about 20 per cent of total volume) contain oxygen, the high energy gamma rays generate copious amounts of free radicals from those of water and oxygen in the fruit immediately after radiation. The free radicals in turn bring about the breakage of the genetic material (DNA) of the insects and spoilage microorganisms, thus destroying them. But after a short period (2-3 days) the free radicals get scavenged off or converted into harmless molecules. Gamma radiation also delays ripening and senescence by decreasing the activity of the cell wall degrading enzyme pectin methyl esterase (PME) and the activity of ACC-oxidase involved in ethylene synthesis. Activities of some other enzymes like polygalactosidase, cellulase and 1-aminocyclopropane-1-carboxylate oxidase are also negatively affected. Ripening of bananas is inhibited at irradiation doses of 0.25-0.35 kGy, and the irradiated fruit later can be ripened by treatment with ethylene. Similar results have been reported for mango, papaya, guava and several other tropical and subtropical fruits. But ionizing radiation at doses above 1 kGy can induce various types of physiological disorders in fruits like internal browning in avocados; skin discolouration and stem darkening in grapes; skin damage in bananas; internal cavities in lemon and lime etc.

Table 14.5: Doses of Irradiation for Different Fruits for Postharvest Disease Control

Fruit Crop	Minimum Dose Required (Gy)	Maximum Dose Tolerated (Gy)
Apple	150	100-150
Apricot, peach, nectarine	200	50-100
Avocado	–	15
Lemon	150-200	25
Orange	200	200
Strawberry	200	200
Grapes	–	25-50

For the control of postharvest diseases generally, a minimum dose of 1.75 kGy is required for effective inhibition of postharvest fungi. So to avoid such a high dose, combination treatment such as heat + irradiation is sometime used. Such combination has been shown to be effective for control of brown rot on stone fruits and anthracnose on mango and papaya. Irradiation at doses below 1 kGy is an effective insect disinfestations treatment against various species of fruit flies, mango seed weevil, navel orange worm, codling moth, scale insects and other insect species of quarantine significance in marketing fresh fruits. Most insects are sterilized at doses of 0.05-0.75 kGy; some adult moths survive 1 kGy, but their progeny are sterile. For mango irradiation at a dose between 0.25-0.75 kGy is used for quarantine treatment and shelf-life extension.

Storage Environment

Temperature

Temperature is one of the most important factors affecting the shelf life of fruits. There is an optimum storage temperature for all the fruits. This temperature depends upon the geographical origin of the fruit. For example tropical fruits have evolved in warmer climates and therefore cannot tolerate low temperatures during storage so; they must be stored above 12 °C. In contrast fruits that have evolved in temperate climate can be stored at 0 °C. Tropical fruits like mango, banana and papaya when stored at higher temperature shelf life is terminated because of fruit ripening, while at lower temperatures ripening is not the concern, the limitation to storage being imposed by chilling injury. Chilling injury causes greater susceptibility to disease, as well as skin pitting, skin scald, failure to ripen and flesh breakdown thus reducing shelf life of fruit. However, if fruits are removed from chilling temperature storage to a higher temperature before chilling symptom development occurs, fruits ripen normally; hence they have an overall longer shelf life.

Temperature has a direct effect on the respiration rate of the fruit and this is an indication of the rate of deterioration of the product. The storage life of fruits varies inversely with the rate of respiration. Fruits with a higher rate of respiration generally have a shorter shelf life than those with a lower rate of respiration. The relationship between temperature and rate of metabolic reactions (*e.g.* respiration) may be described by temperature quotient or Q_{10}. Van't Hoff, a Dutch chemist, determined that for every 10 °C increase in temperature the rate chemical reactions are approximately doubles. For example an apple held at 10 °C ripens and respires about 3 times as fast as one held at 0 °C. This increase in respiration has a direct impact on the shelf life of fruits.

Postharvest storage temperature can also influence the growth rate of the microorganisms (bacteria and fungi) that cause postharvest rots. Fruits are generally more susceptible to fungal diseases, rather than bacterial ones. This is because then fruit are quite acidic and this makes them more resistant to bacteria. However as fruit ripen they become increasingly susceptible to invasion by disease organisms. This is because the fruit become less acidic, the skin softens, the sugars increase and the natural defence barriers weaken. These disease-causing organisms grow faster at warmer temperatures. Therefore if storage temperatures are low the rate of disease development can be considerably reduced and the storage life of the fruits can be assured. Storage temperature can also influence which type of disease develops. For example in oranges the fungi that causes blue mould rot grow rapidly at lower temperatures whereas the bacteria that causes the soft rot grows best in warmer conditions.

Relative Humidity (RH)

Relative humidity is a term used for expressing moisture content in air. It is defined as the ratio between the vapour pressure of water in the air to the saturation vapour pressure possible at the same temperature, and it is expressed as percentage (per cent). There is fairly a good relationship between the degrees of water loss from the fruit with the relative humidity of air. When the RH of the air is low, the rate of water loss from the fruit is increased. However, very high RH (more than 95 per cent) can encourage the growth of bacteria and fungi because spores require enough free moisture on the fruit surface for their germination. This occurrence of rot may outweigh the potential benefits of reduced water loss at high RH. So, it has found that 90 per cent RH is usually the best compromise condition for fruit storage. Further moisture loss from pre-climacteric avocado, mango, banana, plantains, and pear fruit hastens ripening process. There is a linear negative relationship between water loss and green life of avocado, banana and mango fruit. For example, the green life of bananas is about 22 days at 20 °C with 95 per cent RH and 16 days at 13 per cent RH. At low RH, 50 per cent and 65 per cent reduction in post harvest life takes place in custard apple and plantain respectively. There are also numerous studies relating reduced chilling injury symptom development in sensitive tropical fruits at high RH. Development of some physiological disorder in fruit is also associated with relative humidity. For example, 'woolliness' in peach, internal breakdown and core blush in apples increased by high RH due to low level of water loss.

Storage Atmosphere (O_2, CO_2 and C_2H_4)

The composition of gases (*e.g.* oxygen and carbon dioxide) in the storage atmosphere can affect the shelf life of fruit. The concentration of O_2 and CO_2 has profound effect on the respiration and other metabolic reactions of fruit. Increases in carbon dioxide and decreases in oxygen concentrations in the storage atmosphere exert reduction in rate of respiration. Further, the activity of decay causing organisms is also reduced at atmospheres containing 10 per cent or more carbon dioxide. But, care must be taken to ensure that sufficient oxygen is retained or carbon dioxide concentration is not too high in the atmosphere so that anaerobic respiration does not initiate which is associated with the development of off-flavours. The critical level of oxygen depends upon the time and temperature of storage. Apart from these, many volatile compounds may also produce from fruit and other sources (*e.g.* internal combustion engines) that may accumulate in the storage atmosphere. Among these, ethylene is the most important compound which if accumulate above certain critical levels reduces storage life of fruit by the commencement of ripening and senescence. So, to enhance the shelf life of fruit, alteration in the concentration of the respiratory gases is done by the use of controlled atmosphere (CA) and modified atmosphere (MA) storage. CA storage generally refers to decreased oxygen and increased carbon dioxide concentration with precise control of these gases whereas, in MA storage the condition is same but the storage atmosphere is not closely controlled. Here higher carbon dioxide and lower oxygen concentration is achieved by the normal process of respiration of fruit as in plastic film packages.

References

Asrey, Ram., Patel, V.B., Singh. S.K., Sagar, V.R., 2008. Factors affecting fruit maturity and maturity standards: A review. *J. Food Sci. Technol.*, 45(5): 381–390.

Banks, N.H., 1984. Some effects of TAL Pro-long coating on ripening bananas. *J. Expt. Bot.*, 35(150): 127–137.

Bregoli, A.M., Scaramagali, S., Costa, G., Sabatini, E., Ziosi, V., Biondi, S. and Torrigiani, P., 2002. Peach (*Prunus persica*) fruit ripening: aminoethoxyvinylglycine (AVG) and exogenous polyamines affect ethylene emission and flesh firmness. *Physiol. Plant.*, 114: 472–481.

Castillo, A., Mckenzie, K.S., Lucia, L.M. and Acuff, G.R., 2003. Ozone treatment for reduction of *Escherichia coli* and Salmonella serotype Typhimurium on beef carcass surfaces. *J. Food Prot.*, 66(5): 775–779.

Crisosto, C.H., Johnson, R.S., Luza, J.G. and Crisosto, G.M., 1994. Irrigation regimes affect fruit soluble solids concentration and rate of water loss of 'O'Henry' peaches. *HortSci.*, 29(10): 1169–1171.

Crisosto, C.H., Mitchell, F.G. and Johnson, S., 1995. Factors in fresh market stone fruit quality. *Postharvest News Inform.*, 6(2): 17–21.

Daane, K.M., Johnson, R.S., Michailides, T.J., Crisosto, C.H., Dlott, J.W., Ramirez, H.T. and Morgan, D.P., 1995. Nitrogen fertilization affects nectarines fruit yield, storage qualities, and susceptibility to brown rot and insect damage. *California Agric.*, 49(4): 13–18.

Denyer, S.P. and Hugo, W.B., 1991. Biocide-induced damage to the bacterial cytoplasmic membrane. In: *Mechanisms of Action of Chemical Biocides*, (Eds.) S.P. Denyer and W.B. Hugo. The Society for Applied Bacteriology, Technical Series No 27. Oxford Blackwell Scientific Publication, Oxford, pp. 171–188.

Graham, D.M., 2000. Ozone as an anti-microbial agent for the treatment, storage and processing of foods in gas and aqueous phases, Direct food additive petition, Electric Power Research Institute, Palo, Alto, California.

Hewett, E.W., 2006. An overview of preharvest factors influencing postharvest quality of horticultural products. *Int. J. Postharvest Technol. Innovation.*, 1(1): 4–15.

Khan, A.S., Singh, Z. and Abbasi, N.A., 2007. Pre-storage putrescine application suppresses ethylene biosynthesis and retards fruit softening during low temperature storage in 'Angelino' Plum. *Postharvest Biol. Technol.*, 46: 36–46.

Lee, M.M., Lee, S.H. and Park, K.Y., 1997. Effects of spermine on ethylene biosynthesis in cut carnation (*Dianthus caryophyllus*) flowers during senescence. *J. Plant Physiol.*, 151: 68–73.

Lee, S.K. and Kader, A.A., 2000. Preharvest and postharvest factors influencing vitamin C content of horticultural crops. *Postharvest Biol. Technol.*, 20: 207–220.

Malik, A.U. and Singh, Z., 2005). Pre-storage application of polyamines improves shelf-life and fruit quality of mango. *J. Hortic. Sci. Biotechnol.*, 80(3): 363–369.

Mirdehghan, S.H., Rahemi, M., Castillo, S., Martínez-Romero, D., Serrano, M. and Valero, D., 2007. Pre-storage application of polyamines by pressure or immersion improves shelf-life of pomegranate stored at chilling temperature by increasing endogenous polyamine levels. *Postharvest Biol. Technol.*, 44(1): 26–33.

Nilsson, T. and Gustavsson, K., 2007. Postharvest physiology of 'Aroma' apples in relation to position on the tree. *Postharvest Biol. Technol.*, 43: 36–46.

Osman, A.B. and Dodd, P.B., 1994). Effects of different levels of preharvest shading on the storage quality of strawberry (*Fragaria x ananassa* Duchesne) cv. Ostara. I. Physical characteristics. *Pertanika J. Trop. Agric. Sci.*, 17: 55–64.

Rabus, C. and Streif, J., 2000. Effect of various preharvest treatments on the development of internal browning disorders in 'Braeburn' apples. *Acta Hortic.*, 518: 151–157.

Romero, D.M., Serrano, M., Carbonell, A., Burgos, L., Riquelme, F. and Valero, D., 2002. Effects of postharvest putrescine treatment on extending shelf life and reducing mechanical damage in apricot. *Food Chem. Toxicol.*, 67(5): 1706–1712.

Sharma, R.R., Patel, V.B. and Krishna, H., 2006. Relationship between light, fruit and leaf mineral content with albinism incidence in strawberry (*Fragaria x ananassa* Duch.). *Sci. Hortic.*, 109: 66–70.

Sharma, R.R., Singh, D. and Singh, R., 2009. Biological control of postharvest diseases of fruits and vegetables by microbial antagonists: A review. *Biological Contr.*, 50: 205–221.

Shewfelt R.L., 1999. What is quality?: A review. *Postharvest Biol. Techol.*, 15: 197–200.

Sisler, E.C. and Serek, M., 1997. Inhibitors of ethylene responses in plants at the receptor level: Recent developments. *Physiol. Plant.*, 100: 577–582.

Smith, T.A., 1985. Polyamines. *An. Rev. Plant Physiol.*, 36: 117–143.

Solovchenko, A.E., Avertcheva, O.V. and Merzlyak, M.N., 2006). Elevated sunlight promotes ripening-associated pigment changes in apple fruit. *Postharvest Biol. Technol.*, 40: 183–189.

Valero, D., Martinez, D., Riquelme, F. and Serrano, M., 1998. Polyamine response to external mechanical bruising in two mandarin cultivars. *Hort Science.*, 33(7): 1220–1223.

Woolf, A.B. and Ferguson, I.B., 2000. Postharvest responses to high fruit temperatures in the field. *Postharvest Biol. Technol.*, 21: 7–20.

2012, Nutri-Horticulture
Editor: Professor K.V. Peter
Published by: DAYA PUBLISHING HOUSE, NEW DELHI

Pages 257–270

Chapter 15

History of Coconut Breeding

R.V. Nair*, K. Samsudeen and M. Shareefa

Central Plantation Crops Research Institute, Kasaragod – 671 124, INDIA
*E-mail: rvncpcri@gmail.com

Probably developed as a source of thirst quenching healthy drink for the early mankind, coconut (*Coconut nucifera* L.) is a palm of many uses today. It is known as tree of life for its diverse uses for every part of the palm with more than 100 different products including for food and drink, fodder for livestock, fibre, cosmetics and building materials. The root has medicinal uses, trunk is used as building material, leaf has many uses from thatching to fuel to composting, leaflet midribs are used for making brooms, tender nut is the source of healthy natural soft drink, mature nut is the source of copra and oil, coconut husk is valued for its fibre, shell has many uses apart from raw material for charcoal making and fresh endosperm has culinary uses. In addition to these uses, coconut is an integral part of the social, cultural and religious traditions of people in many countries. It is very commonly used in temple offerings, religious ceremonies, marriages and auspicious occasions.

Coconut is cultivated in more than 90 countries in about 12 million ha. In most of the countries, it is a small holder's crop and provides livelihood to more than 10 million families. Coconut also has a role in stabilizing farming system and environments as in small islands and coastal lines.

Coconut breeding is very tedious and time consuming. The main reasons for slow progress in coconut breeding work are long juvenile period, long interval between generations, heterozygous nature of the palm, lack of vegetative reproduction, long period of experimentation and large area required for experiments. Much work was done overcoming such limitations and many improved varieties were released.

Organized coconut breeding programme started in the early part of 20th century. In India, breeding work started in 1916 at the erstwhile Central Coconut Research Station, Kasaragod (presently the Central Plantation Crops Research Institute) and Research Stations now under Kerala Agricultural University. In Sri Lanka, research work on coconut began with the inception of the Coconut Research Institute (CRI) in 1928. The work started in Indonesia during the Dutch colonial period when initial research activity was institutionally conducted in 1911.

Coconut, *Cocos nucifera* L., is a monotypic species under the genus *Cocos* with no known wild or domesticated relatives. It is a diploid species with chromosome number 2n = 32 and belongs to family Palmae under monocotyledons. Genetic resources essential for breeding programs are limited to cultivars found in the different geographical regions around the tropical world.

Origin and Distribution

Arecaceae family to which coconut belongs is widely distributed throughout the world. Palms are supposed to have arrived during Cretaceous period or before the fragmentation of Pangaea into Laurasia and Gondwana (Moore 1973, Uhl and Dransfield 1987).

Wide distribution of coconuts in tropics and absence of a wild progenitor made it difficult to assign a place of origin for this tree of life. An American origin was argued based on the fact that most of the botanical relatives are from American continent. Equally vociferous is the argument that coconut originated in some islands of Pacific or Indian Ocean (Purseglove, 1972; Child 1974). *Cocos nucifera* L. belongs to the tribe Cocoeae along with more than 500 other species. Most of these species are endemic to South America which prompted many to suggest an American origin for coconut. Present day distribution and discovery of fossils of Cocoeae members from New Zealand and India lead to the theory of origin of cocoeae tribe in Western Gondwana before fragmentation of Pangaea. It is difficult to ascertain the centre of origin of coconut palm due to long history of domestication and absence of progenitors.

It is proved that coconut nut can take a sea voyage for a period up to 110 days and can travel 4,800 km without loosing the viability (Edmonson 1941). This quality enabled coconut to spread widely all along the tropical coast. Large populations of coconut are found in Pacific Ocean islands, South East Asia, South Asia, Indian Ocean Islands, African coasts, Central and South America and Atlantic Ocean islands. In all these islands, coconut reached even before human occupation.

Coconut palms are well adapted to conditions of high humidity, groundwater availability around the year, well drained soil with good aeration, temperature ranging 20 to 35°C and well distributed rainfall above 1000mm/annum. But coconut palms are susceptible to frost. They do not flower under cold conditions. Generally, coconuts grow very well between 20°N and 20°S of the equator as evidenced by their wide distribution in the tropical world.

Diversity and Genetic Resources

Coconut can be broadly classified into two groups, the Talls and the Dwarfs. Talls are the commonly planted cultivars. Tall palms grow to a height of 20-30m and are slow maturing, first flowering in 6-10 years after planting. They are sturdy and long-lived and may attain an age of 80-100 years. Talls are normally cross-pollinated and hence highly heterozygous and the population is heterogeneous. Tall palms are sometimes referred to as var. typica. The fruit is medium to large in size and shape varies from round, oval to oblong. Colour of the fruit varies from green, yellowish green to brown. The average copra content is usually above 150g/nut and oil content varies from 66 to 70 per cent. The tall palms are hardy and adapted to different soil types and varying environmental conditions. The palms are known by the name of the localities from where they are collected or cultivated. Some of the popular tall cultivars from India are West Coast Tall, East Coast Tall, Tiptur Tall, Benaulim, Andaman Ordinary and Laccadive Ordinary. Some of the popular exotic tall cultivars are Fiji Tall, Philippines Ordinary, Sri Lankan Tall, Rennel Tall, West African Tall, Panama Tall, Malayan Tall and San Ramon.

Dwarf palms are of shorter stature, 8-10m high when 20 years old and flower in 3 to 4 years after planting. They have a short productive life of 35 to 40 years. Some trees may give economic yield up to

50 years. The dwarf palms are more homozygous than talls, due to high degree of self-pollination. Dwarf palms are sometimes referred to as var. *nana* (Griff.) Nar. The fruits are medium to small and colour of fruits may be green, yellow, orange (red) or brown. The palms have a tendency for irregular or alternate bearing. The average copra content in dwarfs ranges 90-120g/nut and oil content is about 65 per cent. The dwarfs are less hardy than the talls, requiring better soil and climatic conditions. The dwarfs are supposed to have originated from talls either through mutation (Menon and Pandalai, 1958) or by inbreeding (Swaminathan and Nambiar, 1961). A few of the popular dwarf cultivars from India are Chowghat Green Dwarf, Chowghat Orange Dwarf and Gangabondam Green Dwarf. Some of the popular exotic tall cultivars are Malayan Yellow Dwarf, Malayan Orange Dwarf and Malayan Green Dwarf.

Diversity in coconut exists for size and shape of the nut, shape of crown, leaf length, inflorescence stalk length, female flower production, bearing habit and resistance to biotic and abiotic stress. Size of the nut varies from very small in Laccadive micro to very large in San Ramon. Nut shape ranges from round to elongate. Crown shape may be spherical, semispherical or X shaped. Collection and conservation of these diverse types in coconut are the major breeding initiatives in all coconut growing countries.

Germplasm Collection and Conservation

Collection and conservation of genetic resources are important aspects of breeding. Characterization and evaluation are essential for utilization of the collected materials. In coconut, many countries are conducting research on these lines. In India, Central Plantation Crops Research Institute (CPCRI) started collection work in 1924 when exotic germplasm was introduced from Philippines, Malaysia, Fiji, Indonesia, Sri Lanka, Vietnam and other Southeast Asian countries. Survey for collecting indigenous germplasm was started in 1958. Today, CPCRI holds the world's largest coconut germplasm with 132 exotic and 229 indigenous collections. The exotic collections are from Pacific Ocean Islands, Southeast Asia, Indian Ocean Islands, African Regions, Central and South America and Atlantic Region. Indigenous accessions are from Kerala, Tamil Nadu, Karnataka, Andaman and Nicobar islands, Lakshadweep islands, Andhra Pradesh, Goa, Orissa, Gujarat and West Bengal.

In Sri Lanka, systematic germplasm conservation programme started in 1984. A dwarf palm block planted in 1940, a collection of varieties planted in 1962, and a collection of palms at the isolated seedgarden (ISG) were the only national resources before 1984. Germplasm materials consisting of 24 cultivars are conserved at two sites, one in Bandirippuwa estate and another in Porthukulama Research Station (PRS).

In Indonesia, history of survey and collection of coconut germplasm dates back to 1926-27, when Dr Tammes collected nuts from high yielding tall coconut population around Mapanget district (North Sulawesi) and planted at the Mapanget Experimental Garden. From 1973 onwards, Central Research Institute for Industrial Crops and its regional institutes conducted many surveys in selected areas of 11 provinces of Indonesia. Collections were planted at the Mapanget Experimental Garden. Some of the recent collections are planted at the research farm at Pakuwon, West Java and at the Bone-Bone Experimental Garden, South Sulawesi. About 100 accessions are in field gene banks at the above three experimental gardens.

In Philippines, prior to 1970 collection activities were done by the Bureau of Plant Industry (BPI), College of Agriculture and Visayas State College of Agriculture (ViSCA), Baybay, Leyte (Santos *et al.*, 1984). Since 1972, PCA (Philippines Coconut Authority) initiated a vigorous campaign of conserving domestic genetic resources. The germplasm collections are planted in the PCA research centers at Davao and Zamboanga (both in Mindanao) and at Albay (Luzon). PCA's genebank in Zamboanga has become one of the most important germplasm depositories of local and foreign coconut ecotypes in the world with around 100 accessions.

Coconut germplasm collection in Thailand was established in 1965. A few varieties from local and foreign sources were collected and conserved in Chumphon Horticulture Research Centre (CHRC). Twenty one varieties (10 dwarfs and 11 talls) are maintained at CHRC. In 1997, a coconut germplasm genebank (under COGENT/ADB project) with 20 local accessions was set up at Kanthuli, Suratthani province.

Germplasm collection and conservation programme are in progress in other coconut growing countries. In Vietnam, a total of 15 local tall cultivars and five dwarfs were conserved in Dong Go Station (Ben Tre province) and Trang Bang Station (Tay Ninh province). In Papua New Guinea (PNG), 42 tall and 11 dwarf ecotypes are planted in conservation blocks at Stewart Research Station (SRS). In Mexico, the Centro de Investigacion Cientifica de Yucatan (CICY) established a coconut germplasm collection in the Yucatan Peninsula. The coconut germplasm collections held by CICY include 37 accessions of exotic and local cultivars. In Côte d'Ivoire, the Institut de Recherches pour les Huiles et Oléagineux (IRHO) introduced, from different tropical areas, 53 ecotypes to the Marc DELORME Research Station. Thirty-six talls and 17 dwarfs are conserved in this field gene bank.

Germplasm Utilization

Breeding New Varieties and Hybrids

Improvement of plantation crops, particularly coconut, is a long drawn and slow process because of its distinct floral biology, heterozygous nature, low rate of sexual propagation with one seedling from one nut, lack of selection procedure for isolation of superior hybrid seedling, lack of reproducible asexual methods for rapid multiplication, prolonged interval between generation and long juvenile phase before flowering. It also requires vast land area because of wide planting distances and long period of experimentation required to obtain results. Inspite of all these limitations, systematic improvement work is in progress and substantial advancement is achieved in coconut.

The organized coconut breeding started for the first time in the world in 1916 at the erstwhile Coconut Research Stations at Kasargod and Nileshwar (Pilicode) of former Madras Presidency, now in Kerala India. The major objectives of breeding coconut are improving the yield by increasing size of the nut and per palm yield, improving the copra and oil content of nuts, production of short statured varieties and resistance to biotic and abiotic stresses.

Methods of Breeding

Selection

A programme of selection for yield was practiced ever since coconut improvement work started in majority of the coconut growing countries in the world. The selection method generally involves selection of high yielding mother palms, selection of seed nuts and finally selection of seedlings in the nursery.

Mother Palm Selection

Mother palm selection is the most important factor for coconut improvement. In India, palms yielding 80 nuts and more with a copra outturn of 20 kg/palm/year consistently over a period are considered as mother palms. Studies on path coefficient and regression analyses indicate that average number of female flowers, number of functional leaves in the crown, internodal distance at fixed mark, total leaf production upto three years after sowing and time taken for flowering are important components showing the largest direct effect on yield and thereby indicating their value in selection programme (Bavappa and Sukumaran, 1983).

Seedling Selection

Nursery selection is based on characters seen to be correlated with high yield of adult palms, like early sprouting, faster growth rate, early splitting of the unexpanded leaf into leaflets, seedling vigour in terms of girth at collar, height and number of leaves besides freedom from pests and diseases (Sathyabalan and Mathew, 1976).

Identification of Prepotent Palm

The concept of prepotency in coconut was first proposed by Harland (1957) working at the Coconut Research Institute, Sri Lanka. He defined prepotent palms as mother palms which, inspite of having been indiscriminately pollinated by miscellaneous males, are sufficiently possessed of dominant yield factors to ensure that their offsprings are also high yielding. Thus a prepotent palm would be one where the stable gene combinations, particularly those controlling polygenic quantitative traits like yield which may be located on different chromosomes, tend to cohere but do not recombine, thus resulting in the *en bloc* transmission of parental characters to progeny even under random mating. It is this coherence mechanism that enables a prepotent palm to function as reservoirs for potential variability. Since the identification of prepotent talls based on progeny yield performance would take several years, nursery selection methods to correlate seedling characters with adult palm potency were attempted. The coconut breeder can use the seedling index such as collar girth and leaf production for locating genetically superior prepotent palms in natural population.

Based on the preliminary evaluation of cultivars available at Kasargod, a multilocational trial started in 1972 in different states and based on the overall superior performance, Lakshadweep Ordinary was selected and released by CPCRI in 1985 for commercial cultivation under the name Chandrakalpa. Another cultivar, Banwali Green Round, from Goa region was selected and released in 1987 by Konkan Krishi Vidyapeeth, Dapoli for cultivation in Konkan coast under the name Pratap. Similarly, the Agricultural Research Station, Ambajipeta, East Godavari district of Andhra Pradesh released another selection of Philippines Ordinary in the name of Double Century at AICRP-Palms held during 1995. Other high yielding coconut selections released recently are Kalpa Pratibha, Kalpa Dhenu and Kalpa Mitra which are having high oil yield, drought tolerance and good tender nut water quality. Kalparaksha and Kalpasree are recently released varieties which are resistant to root wilt and high yielding.

Exploitation of Hybrid Vigour

Tall x Dwarf Hybrids

Discovery of hybrid vigour in coconut by Patel (1937) in West Coast Tall and Chowghat Green Dwarf crosses made in 1932 at Nileshwar Coconut Research Station was a significant landmark in the history of coconut improvement. He found that these hybrid seedlings showed early germination, increased collar girth and higher leaf numbers as compared to progenies of the female parent. This important finding paved the way for successful breeding programmes in coconut not only in India but in other countries like the Philippines, Indonesia, Sri Lanka and Jamaica. The superiority of WCT x COD was further established in the various multilocational trials. The AICRP on palms recommended this hybrid under the name Kera Sankara for release in Kerala, Maharashtra and Coastal Andhra

Pradesh. In another comparative trial involving WCT and LCT as female parent and COD and GBGD as male parents, the superior performance of LCT x COD and LCT x GBGD were reported and recommended for release during 1985 under the name Chandralaksha and Lakshaganga for general cultivation in Kerala. Breeding efforts were made at Pilicode centre with GBGD as male parent with six tall parents and superior hybrids in terms of copra yield/palm and copra out turn/ha were selected. Among these, three hybrids, LCT x GBGD, ADOT x GBGD and WCT x GBGD were released as Lakshaganga, Anandaganga and Keraganga respectively.

Table 15.1: Varieties Released in India

Name	Agency	Area for which Recommended	Nut Yield/ Palm/Year	Copra (g/nut)	Oil Content (per cent)
Chandrakalpa	CPCRI, Kasaragod	Kerala, Karnataka, TN	97	195	70
Kerachandra	CPCRI, Kasaragod	AP, Maharashtra	110	198	66
Chowghat Orange Dwarf	CPCRI, Kasaragod	All coconut growing regions	Tender nut variety		
Kalpa Pratibha	CPCRI, Kasaragod	West Coast region and peninsular India	91	256	67
Kalpa Dhenu	CPCRI, Kasaragod	West Coast region and Andaman and Nicobar Islands	86	242	65.5
Kalpa Mitra	CPCRI, Kasaragod	West Coast region and West Bengal	80	241	66.5
Kalparaksha	CPCRI, Kasaragod	West Coast region and root (wilt) disease tracts of Kerala	65	215	65.5
Pratap	KKV, Dapoli	Konkan region	150	152	59
VPM-3	TNAU	Tamil Nadu	77	191	66
ALR 1	TNAU	Tamil Nadu	126	131	64
Kamrupa	AAU	Assam	101	162	64
Kera Sagara	KAU	Kerala	99	203	67.8
Kalpasree	CPCRI, Kayamkulam	Root (wilt) prevalent areas in South India	90	96.3	66.5

Studies were carried out at Veppamkulam in Tamil Nadu on different Tall x Dwarf crosses involving East Coast Tall with different pollen parents and the research led to the release of Veppamkulam Hybrid Coconut (ECT x Dwarf Green) for large scale cultivation in Tamil Nadu. A second T x D hybrid (ECT x MYD) was also released as VHC-2 for large scale cultivation in Tamilnadu. In a study aimed to develop a superior, high-yielding cross combination, pollination was made between East Coast Tall (ECT) as ovule parent and Malaysian Orange Dwarf (MOD) as pollen parent. The resultant hybrid had higher nut yield, copra yield with high copra outturn compared to VHC 2. The hybrid ECT x MOD was released for wider cultivation as VHC 3 (Veppankulam Hybrid Coconut 3) by the Tamil Nadu State Variety Release Committee in January 2000.

Although the cross combinations between Tall x Dwarf are high yielding and precocious, there are limitations such as large scale production of T x D hybrids is labour intensive and expensive operation since hand pollination by trained climbers is involved and absence of colour marker in T x D hybrids.

Dwarfs x Tall Hybrids

The distinct advantage of this hybrid over T x D is that it could be produced on a large scale by regularly emasculating dwarf mother palms, permitting free natural crossing with pollen from tall palms standing nearby. Use of Dwarf Orange or Yellow as female parents enables identification of hybrid seedlings because of colour marker. Yellow, orange or red petiole colour is recessive to brown and green pigments and hybrids show a greenish brown or brownish colour depending on the colour of talls used in crossing. Among T x D and D x T hybrids, the latter was definitely superior to the former in terms of precocity and tree to tree variation was minimum in this hybrid. During the VII Workshop of AICRP on palms, COD x WCT hybrid was released as Chandrasankara, for general cultivation in Kerala. Another hybrid Kalpasankara (CGD x WCT) was released for cultivation in the root (wilt) prevalent tracts of South India.

Tall x Tall Hybrids

The T x T hybrids are produced by intervarietal hybridization of tall cultivars under controlled conditions. Individual palms of high breeding value are identified and these genotypes are grown on isolated seed gardens and utilized for production of T x T hybrids. Though late in bearing, the yield potential of T x T hybrids is good.

Table 15.2: Coconut Hybrids Released in India

Hybrids	Parentage	Agency for Release	Area for which Recommended	Annual Nut Yield/palm	Copra Content (g/nut)	Oil Content (Per cent)
Chandra Sankara	COD x WCT	CPCRI, Kasaragod	Kerala, Karnataka, Tamil Nadu	110	208	68
Kera Sankara	WCT x COD	CPCRI, Kasaragod	Kerala, Karnataka, Maharashtra, Andhra Pradesh	106	198	68
Chandra Laksha	LCT x COD	CPCRI, Kasaragod	Kerala, Karnataka	109	195	69
Laksha Ganga	LCT x GBGD	KAU	Kerala, Tamil Nadu	108	195	70
Kera Ganga	WCT x GBGD	KAU	Kerala	100	201	69
Kera Sree	WCT x MYD	KAU	Kerala	112	216	66
Kera Sowbhagya	WCT x SSAT	KAU	Kerala	130	195	65
Ananda Ganga	ADOT x GBGD	KAU	Kerala	95	216	68
Godavari Ganga	ECT x GBGD	APAU	Andhra Pradesh	140	150	68
VHC-1	ECT x MGD	TNAU	Tamil Nadu	98	135	70
VHC-2	ECT x MYD	TNAU	Tamil Nadu	107	152	69
VHC-3	ECT x MOD	TNAU	Tamil Nadu	156	161	64.5
Kalpasamrudhi	MYDxWCT	CPCRI, Kasaragod	West Coast region	117.2	219.6	67.5
Kalpa Sankara	CGDxWCT	CPCRI, Kayamkulam	Root(wilt) prevalent areas in South India	84	170	67.5

Breeding for Special Characteristics

Disease Resistance

Among the various diseases, root (wilt) is the most devastating in coconut. The characteristic symptoms of the disease are flaccidity, yellowing and marginal necrosis of leaflets followed by a progressive decline in yield. The screening of the available germ plasm has not yielded any resistant type. Hence hot spot areas were surveyed and disease free palms were identified. These are now being utilized in the hybridization programme (Jacob *et al.*, 1998). Healthy WCT and CGD palms standing in the midst of root(wilt) diseased palms were identified (Nair *et al.*, 2000). Kalpa Raksha (selection from Malayan Green Dwarf) was released for root (wilt) disease affected tracts due to its capacity to resist the disease to a greater extent. This is the first coconut variety released for cultivation in root (wilt) diseased tract of Kerala. Resistance breeding programme involving WCT and CGD palms identified in farmer's plots located in hot spots of root(wilt) disease resulted in the release of two more coconut varieties for cultivation in root (wilt) prevalent areas of India *viz.*, Kalpasree (selection from CGD) and Kalpasankara (Hybrid between CGD and WCT).

Drought Tolerance

The four southern states of India (Andhra Pradesh, Tamil Nadu, Kerala and Karnataka) which account for 90 per cent of the area under coconut are periodically exposed to low rainfall or delayed onset of monsoon, both resulting in poor yield during drought. The impact of drought on coconut yield is well documented and the effect persists even for subsequent 2 to 3 years. Under these circumstances, evolving drought tolerant varieties/hybrids is the only solution towards increasing coconut production in drought affected areas. The desirable traits of drought tolerant cultivars are accumulation of epicuticular wax on the leaf surface, low stomatal frequency, low stomatal resistance and leaf water potential. The activities of certain enzymes like glutamate oxaloacetic transaminase (GOT) and acid phosphatase (APH) have their influence on drought tolerance (Rajagopal *et al.*, 1988). Based on these characters, the drought tolerant cultivars identified are Federated Malay States, Java Giant, Fiji, Laccadive Ordinary and Andaman Giant. All the identified drought tolerant varieties are currently being utilized in the breeding programme to evolve high yielding hybrids possessing drought tolerance (Rajagopal *et al.*, 1990).

Nut Water Quality

During the study conducted to find out a cultivar for tender nut purpose, different cultivars were organoleptically screened for nut water quality and twelve cultivars were selected. The biochemical evaluation indicated that Chowghat Orange Dwarf had maximum content of total sugars, reducing sugars and low contents of sodium and potassium. The AICRP on palms recommended this cultivar for release as tender nut variety in Kerala(Anonymous, 1991).

Coconut Breeding in some Important Coconut Growing Countries

Coconut Breeding in Indonesia

Hybridization work commenced around 1930 with crosses between selected palms from the first germplasm survey. Coconut germplasm imported from Ivory Coast were crossed with local and Malayan Yellow Dwarf at two centres- Lampung and Sumatra. Production of hybrids commenced in 1978 with pollen imported from Ivory Coast. The first hybrid trial planted at Pakuwan in West Java indicated that local hybrids showed precocity for flowering, more copra content/nut and yield of copra in the first year of bearing and was exceptionally good particularly in Nias Yellow Dwarf x Tenga Tall.

In 1975, the Government introduced coconut hybrid PB 121 from Ivory Coast. PB 121 was distributed to the farmers through the SCDP (Smallholder Coconut Development Project). In 1984, the Government released coconut hybrids KHINA-1 (GKN x Tenga Tall or DTA), KHINA-2 (GKN x Bali Tall or DBI) and KHINA-3 (GKN x Palu Tall or DPU). Tall x Tall hybrids were also released, namely: KB-1 (32 x 32), KB-2 (32 x 2), KB-3 (32 x 83) and KB-4 (32 x 99). These local hybrids, especially KHINA-1, are now planted on a small scale by farmers and private growers.

Coconut Breeding in Sri Lanka

A coconut breeding programme is in operation since the inception of the Coconut Research Institute (CRI) of Sri Lanka in 1928. Being a predominantly cross-pollinating crop with a long generation interval, and in the absence of a proven method of vegetative propagation, mass selection and hybridization are the major tools used in coconut breeding. The pollen of pre-potent palms was initially used in the production of tall x tall (CRIC 60) and dwarf x tall (CRIC 65) hybrids through controlled hand pollination. Hand pollination was a tedious process and since 1955, the Isolated Seed Garden (ISG) technique was adopted for assisted natural pollination for mass production of hybrids. Organized multilocation evaluation trials for selected cultivars commenced in 1984. Along with the SLT variety (progeny from 'plus palms'), four other cultivars, namely, tall x tall (CRIC 60); green dwarf x tall (CRIC 65); yellow dwarf × tall and Moorock Tall (an accession which had undergone at least three generations of selection for high copra weight), were tested in five locations with varying soil and rainfall conditions. The first three trials were set up in 1984. They were managed under the general guidelines of the CRI, with no extra inputs. In all the three sites, the hybrid CRIC 65 was proved superior to the tall cultivars with respect to leaf production, flowering and fruiting.

Coconut Breeding in Philippines

Coconut germplasm conservation was started by Bureau of Plant Industry in the early fifties. Introduction of foreign genetic materials consisted of tall populations namely Rennel from Solomon Islands; Gazelle, Markham valley and Karkar (all from Papua New Guinea), West African Tall(from Ivory Coast) and dwarf population namely Malayan Red Dwarf, Malayan Yellow dwarf, Cameroon Red Dwarf, and Green Dwarf(Brazil) all introduced from Ivory Coast.

Coconut Breeding in Thailand

The first hybrid trial was planted in 1975. The preliminary results suggest that Malayan Yellow Dwarf x West African Tall is the most precocious and has the highest early yield with the Thai Tall x West African Tall hybrid ranking second for yield. The hybrid between Malayan Yellow Dwarf x West African Tall (Sawi 1) is precocious, uniform and consistent in yield as well as adapted to a wide range of environmental conditions. The Chumphon 1 (Thai Tall x West African Tall) is also early in bearing, consistent in yield, tolerant to drought and produces copra of high quality. This was released in 1987.

Coconut Breeding in Jamaica

The coconut is not indigenous to Jamaica. The first cultivar, the Jamaica Tall, was introduced in the 15th century. Lethal Yellowing Disease (LYD) was first reported in the island in 1884 (Been 1992a), with periodic outbreaks since. The Jamaica Tall cultivar was highly susceptible to this disease. The search for LYD-resistant cultivars became important for the survival of the coconut industry in Jamaica. The plant breeding programme of the Coconut Industry Board began in 1961 with this search as a priority. Of the local cultivars screened, the Malayan Dwarf (a cultivar introduced in the late 1930s) had high LYD resistance. Several cultivars were introduced from major coconut-growing regions, but none possessed greater LYD resistance than the Malayan Dwarf. F_1 hybrids were made using combinations of the Malayan Dwarf and selected tall cultivars from the introductions, the Jamaica Tall and Panama Tall. The progenies were incorporated into hybrid trials. From these trials, the Maypan (Malayan Dwarf × Panama Tall) was selected and released for commercial planting in 1974 (Charles 1961). It has since superceded the parent dwarf as the most widely planted coconut cultivar in Jamaica.

Scope of Biotechnology in Coconut Breeding

Anther Culture

Problems encountered with conventional breeding of coconut are its long life span and high heterozygosity which make plant breeding a long, difficult and expensive process. Production of homozygous lines will have a tremendous impact on generating new cultivars through breeding programs. Using homozygous lines, it is possible to produce pure F1 hybrids. At least 60 years of backcrossing and selfing would be required to get homozygous coconut lines by breeding. Generation of dihaploid (DH) plants would result in rapid production of homozygous lines, shortening the process to 1 or 2 years. Recessive alleles in the parents are easily uncovered in DH plants and therefore desirable mutations can be easily detected. They provide a wide spectrum of traits for selection in breeding programs. A few studies were conducted on coconut anther culture (Kovoor (1981), Iyer (1981), Thanh-Tuyen and de Guzman (1983), Monfort (1985)) and the first successful anther culture protocol via pollen embryogenesis was reported by Perera *et al.,* in 2008.

Embryo Culture

The commercial application of embryo culture is rescue and culture of Macapuno variety of coconut in Philippines. Macapuno is a high value coconut due to its relative rarity and importance in the food industries. The nut is filled with a jelly like endosperm and is in great demand in the ice cream and pastry industries. The tree is propagated by planting the nut from the Macapuno bearing trees resulting in about 2 to 25 per cent Macapuno yield. The embryo does not germinate due to its abnormal surroundings, it is in. The embryo is excised and grown *in vitro* with proper nutrients to develop into a normal tree and the yield of *in vitro* grown Macapuno palms goes upto 75 per cent Macapuno nuts.

Germplasm Storage/Exchange

Germplasm conservation in palm is expensive, labour intensive and often impractical due to lack of vegetative propagation system. Coconut germplasm exchange in particular is again hindered by size of nut and absence of dormancy resulting in quick germination *in situ*. The transport of large nuts would be costly because of its bulkiness, weight, spoilage and moreover quarantine regulations. At present, embryo culture has become an important tool for safe germplasm exchange as it reduces the cost of transportation and would meet the phytosanitory regulations. The techniques for aseptic collection of embryo under the palm itself, its storage, transport, *in vitro* conservation and retrieval and *ex vitro* establishment are already standardized.

In vitro Screening for Biotic and Abiotic Stress

Coconut palms experience moisture stress of different magnitudes depending upon the extent of dry spell. There is need to breed specifically drought-tolerant coconut germplasm, practical steps are neglected because of the technical problems particular to the coconut tree. Selection methods adopted for other plant species such as withholding irrigation, germination in mannitol (Ashrof and Abu-Shakra, 1978) or survival through desiccation (Sullivan and Ross, 1979) could not be applied to the coconut because of seed-nut and seedling size which discourage greenhouse experimentation. Field experimentation is, besides, hindered by the low planting density (about 200 palms/ha) and the difficulty of controlling the physical environment and nutritional status of plants grown in the open. Using tissue-culture approach, tolerance to an array of stresses may be selected from among calluses and cells grown *in vitro* in a medium simulating the respective stresses. The germplasm may be screened for stress tolerance by progressively increasing the stress with each passage of culture and apprising the survivors. Water stress conditions were stimulated during *in vitro* stage and subsequent plant development using mannitol, poly ethylene glycol (PEG) and NaCl are applied in the medium.

A working clonal propagation system for coconut could also be used to produce disease free coconut palms and to provide ideal homogeneous test plants to screen for resistance against many dreaded diseases.

Micropropagation

Development of tissue culture techniques is aimed at rapid multiplication of elite planting materials which can greatly save time, space and resources. Success in coconut micropropagation is reported from leaf tissues, inflorescence tissues and regeneration also is reported.

Cryo Perseveration

Traditionally coconut germplasm is conserved in *ex situ* gene banks. Use of *in vitro* culture techniques like slow growth and cryopreservation are alternative viable options for safe medium and long term conservation of coconut germplasm.

Molecular Biology

DNA-based markers possess the potential to significantly increase the efficiency of coconut genetic improvement programmes, especially in the areas of germplasm management, genotype identification and marker-assisted selection of economically important traits.

Study of Genetic Diversity and Characterization of Germplasm

Reliable assessment of the genetic relationships between varieties/populations of coconut and the accurate estimation of the genetic diversity present in coconut are pre-requisites for sustainable future coconut breeding and genetic resources conservation programme. This is particularly important for coconut due to its long generation time and associated costs of maintaining mature palms. Assembly of elite breeding populations and their thorough evaluation are therefore a priority. Morphological traits have the disadvantages of being influenced by both environmental and genetical factors and may therefore not provide an accurate measure of genetic diversity. Methods independent of environmental factors for establishment of these parameters are of great importance to the plant breeders. At present the evaluation and characterization of coconut germplasm are based exclusively on morphological and reproductive traits. Attempts to characterize coconut populations based on isozymic, polyphenol and carotenoid differences are reported in many countries with different coconut populations. In this regard, molecular marker techniques are advantageous as they directly reflect variations in the DNA sequence and therefore of independent of environment.

Various molecular markers are used for studying the genetic relationships in coconut germplasm *viz*, RAPD (Ashburner *et al.*, 1997; Upadhyay *et al.*, 2004), RFLP (Lebrun *et al.*, 1998), AFLP (Perera *et al.*, 1998; Teulat *et al.*, 2000) and SSR (Perera *et al.*, 2003; Meerow *et al.*, 2003).

Marker Assisted Selection

Marker Assisted Selection refers to use of DNA markers which are tightly-linked to target loci as a substitute for or to assist phenotypic screening. It involves selection of plants carrying genomic regions, involved in the expression of traits of interest through molecular markers. With the development and availability of an array of molecular markers and dense molecular genetic maps in crop plants, MAS has become possible for traits both governed by major genes as well as quantitative trait loci (QTLs).

For crop species with long generative cycles such as coconut, linkage mapping and molecular markers will allow early selection of traits in breeding programmes. The time required for bringing new improved varieties can be greatly reduced. Moreover, these technologies will play crucial roles in the identification and isolation of any gene of interest in the future.

References

Ashburner, G.R., Thompson, W.K. and Halloran, G.M., 1997. RAPD analysis of South Pacific coconut palm population. *Crop Sci.*, 37: 992–997.

Ashraf, C.M. and Abu-Shakra, S., 1978. Wheat seed germination under low temperature and moisture stress. *Agron. J.*, 70: 135–139.

Bavappa, K.V.A. and Sukumaran, C.K., 1983. Coconut improvement by selection and breeding: A review in the light of recent findings. In: *Coconut Research and Development*, (Ed.) N.M. Nayar. Wiley Eastern Ltd., New Delhi, pp. 44–55.

Been, B.O., 1992. Lethal Yellowing: The Jamaican experience. In: *Lethal Yellowing: Research and Practical Aspects*. Kluwer Academic Publishers, Dordrecht.

Child, R., 1974. *Coconuts, 2nd edn.* Longman, London.

CPCRI, 1985. *Annual Report 1984–2005*. Central Plantation Crops Research Institute, Kasaragod, India.

CPCRI, 1987. *Annual Report 1986–87*. Central Plantation Crops Research Institute, Kasaragod, India.

CPCRI, 1988. *Annual Report 1987–88*. Central Plantation Crops Research Institute, Kasaragod, India.

CPCRI, 2008. *Annual Report 2007–08*. Central Plantation Crops Research Institute, Kasaragod, India.

Edmondson, C.H., 1941. *Viability of Coconut after Floating in Sea*. Occasional Papers B.P. Bishop Museum, Hawaii 16: 293–304.

Harland, S.C., 1957. *The Improvement of Coconut Palm by Breeding and Selection*. Circulation paper No. 7/57. Coconut Res. Instt. Bull.No. 15, Ceylon.

Iyer, R.D., 1981. In: *Proceedings COSTED Symposium on Rissue Culture of Economically Important Plants*, (Ed.) A.N. Rao. National University Singapore, Singapore, pp. 219–230.

Jacob, P.M., Nair, R.V. and Rawther, T.S.S., 1998. Breeding for root (wilt) resistance. In: *Coconut Root (Wilt) Disease*, (Eds.) K.U.K. Nampoothiri and P.K. Koshy. Codeword Process and Printers, Mangalore, 1998, pp. 97–104.

Kovoor, A., 1981. Palm tissue culture: State of art and its application to the coconut. *FAO Plant Production and Protection Paper* 30, FAO, Rome.

Lebrun, P., N'cho, Y.P., Seguin, M., Grivet and Baudouin, 1998. Genetic diversity in coconut (*Cocos nucifera* L.) revealed by restriction fragment length polymorphism (RFLP) markers. *Euphytica*, 101: 103–108.

Meerow, A.W., Wiser, R.J., Brown, J.S., Kuhn, D.N., Schnell, R.J. and Broschat, T.K., 2003. Analysis of genetic diversity and population structure within Florida coconut (*Cocos nucifera* L.) using microsatellite DNA, with special emphasis on the Fiji dwarf cultivar. *Theor. Applied Genet.*, 106: 715–726.

Menon, K.P.V. and Pandalai, K.M., 1958. *The Coconut Palm: A Monograph*. Indian Central Coconut Committee, Ernakulam, Kerala, India, p. 384.

Monfort, S., 1985. Androgenesis of coconut: Embryos from anther culture. *Z Pflanzenzuchtg*, 94: 251–254.

Moore, H.E. Jr., 1973. The major groups of palms and their distribution. *Genetics Herbarum*, 11: 27–141.

Nair, M.K., Nampoothiri, K.U.K. and Damodaran, S., 1991 Coconut breeding: Past achievements and future strategies. In: *Coconut Breeding and Management*, (Eds.) E.G. Silas, M. Aravinakshan and A.I. Jose. Kerala Agricultural University.

Nair, M.K. and Nampoothiri, K.U.K., 1993. Breeding for high yield in coconut. In: *Advances in Coconut Research and Development*, (Eds.) M.K. Nair, H.H. Khan, Gopalasundaram and E.V.V. Bhaskara Rao, pp. 61–70.

Nair, R.V., Jacob, P.M., Thomas, R.J. and Sasikala, M., 2004. Development of varieties of coconut (*Cocos nucifera*) resistant/tolerant to root (wilt) disease. *J. Plantation Crops*, 32(suppl): 33–38.

Patel, J.S., 1937. Coconut breeding. *Proc. Assoc. Econ. Biol.*, 5: 1–16.

Perera, P.I.P., Hocher, V., Verdeil, J.L., Bandupriya, H.D.D., Yakandawala, D.M.D. and Weerakoon, L.K., 2008. Androgenic potential in coconut (*Cocos nucifera* L.). *Plant Cell Tiss. Organ Cult.*, 92: 293–302.

Perera, L., Russel, J.R., Provan, J. and Powell, W., 2003. Studying genetic relationship among coconut varieties/populations using microsatellite markers. *Euphytica*, 132: 121–128.

Perera, L., Russel, J.R., Provan, J., NcNicol, J.W. and Powell, W., 1998. Evaluating genetic relationship between indigenous coconut (*Cocos nucifera* L) accessions from Sri Lanka by means of AFLP profiling. *Theor. Applied Genet.*, 96: 545–550.

Purseglove, J.W., 1972. *Tropical Crops: Monocotyledons*. Longmans, London.

Rajagopal, V., Kasturibai, K.V. and Voleti, S.R., 1990. Screening of coconut genotypes for drought tolerance. *Oléagineux*, 45(5): 215–223.

Rajagopal, V., Shivishankar, S., Kasturibai, K.V. and Voleti, S.R., 1988. Leaf water potential as an index of drought tolerance in coconut *Cocos nucifera*. *Pl. Physiol. Biochem.*, 15(1): 80–86.

Santos, G.A., Cano, S.B., dela Cruz, B.V., Ilagan, M.C. and Bahala, R.T., 1984. Coconut germplasm collection in the Philippines. *Philippine Journal of Coconut Studies*, 9(1–9).

Satyabalan, K., 1993. *The Coconut Palm: Botany and Breeding*. Asian and Pacific Coconut Community.

Sullivan, C.Y. and Ross, W.M., 1979. Selecting for drought and heat resistance in grain sorghum. In: *Stress Physiology in Crop Plants*, (Eds.) H. Mussel and R.C. Staples. John Wiley and Sons, New York, pp. 263–281.

Swaminathan, M.S. and Nambiar, M.C., 1961. Cytology and origin of the dwarf coconut palm. *Nature* 192: 85–86.

Teulat, B., Aldam, C., Trehin, R., Lebrun, P., Barker, J.H.A., Arnold, G.M., Karp, A., Baudouin, I. and Rognon, F., 2000. An analysis of genetic diversity in coconut (*Cocos nucifera* L.) population from across the geographical range using sequence tagged microsatellites (SSRs) and AFLPs. *Theor. Applied Genet.*, 100: 764–771.

Thanh-Tuyen, N.T. and De Guzman, E.V., 1983. Formation of pollen embryos in cultured anthers of coconut (*Cocos nucifera* L.). *Plant Sci. Lett.*, 29: 81–88.

Uhl, N.W. and Dransfield, J., 1987. *Genera Palmarum*. The L.H. Bailey Hortorium and The International Palm Society, Lawrence.

Upadhay, A., Jayadev, K., Manimekalai, R. and Parthasarathy, V.A., 2004. Genetic relationship and diversity in Indian coconut accessions based on RAPD markers. *Sci. Hortic.*, 99: 353–362.

2012, Nutri-Horticulture   *Pages 271–308*
Editor: Professor K.V. Peter
Published by: DAYA PUBLISHING HOUSE, NEW DELHI

Chapter 16

History of Nematology Research

Jiji Rajmohan

Professor, Department of Entomology,
College of Agriculture and Research Institute,
P.O. Vellayani, Trivandrum, INDIA
E-mail: jijirajmohan2004@yahoo.co.in

Nematodes are triploblastic, bilaterally symmetrical, unsegmented, pseudocoelomate and vermiform animals. They exist almost everywhere in nature. The different groups of nematodes are fungal feeders, bacterial feeders, predators, animal parasites, algal feeders, omnivores and plant parasites. There are approximately 4,000 known plant-parasitic nematode species. Depending on the species and their population level, damage to plants ranges from mild to severe. Plant-parasitic nematodes are recognized as one of the greatest threats to crops throughout the world. Nematodes alone or in combination with other soil microorganisms attack almost every part of the plant including roots, stems, leaves, fruits and seeds. The name "nematode" comes from the Greek words: nema, which means "thread", and toid, which means "form". Typically the nematode body is elongate, spindle shaped or fusiform, tapering towards both ends and circular in cross section. The length may vary from 0.2mm (*Paratylenchus*) to about 11mm (*Paralongidorus maximus*). In a few plant parasitic nematodes, their body may be pear, lemon, reniform or irregularly saccate shaped as in the females of *Meloidogyne*, *Heterodera*, *Nacobbus*, *Rotylenchulus* and *Tylenchulus*.

The damages caused by nematodes are often overlooked since they are hidden enemies and the associated symptoms are also attributed to nutritional and water related disoders. The epidemics of sugarbeet nematode sickness due to cyst nematode, yellow diseases of black pepper, molya disease of wheat and barley and the earcockle disease of wheat have led to the recognition of the science of plant nematology as an imoportant branch of Agricultural science.

The wheat seedgall nematode *Anguina tritici* discovered by Needham in 1743 was the first plant parasitic nematode to come to the attention of early investigators. In 1855, Berkeley detected the root-knot nematode, *Meloidogyrre* spp. to be the cause of root galls on cucumber plants in England. In 1857,

Kuhn reported the stem and bulb nematode, *Ditylenchus dipsaci* infesting the heads of teasel. In 1859, Schacht reported the sugarbeet cyst nematode, *Heterodera schachtii* from Germany.

Land Marks in the History of Nematology Research

☆ 1873 - Butschli - Description of the morphology of free-living nematodes.

☆ 1884 - de Man - Taxonomic monograph of soil and fresh water nematodes of the Netherlands.

☆ 1889 - Atkinson and Neal - Publication about the root-knot nematodes in the United States.

☆ 1892 - Atkinson - First report of root-knot nematode and *Fusarium* complex in vascular wilt of cotton.

☆ 1907 - Cobb,N.A. - Joined the USDA and considered to be the Father of American Nematology.

☆ 1914 - Cobb,N.A. - Contributions to the Science of Nematology.

☆ 1918 - Cobb,N.A. - Development of methods and apparatus used in Nematology.

☆ 1933 - Goodey,T. - Book on "Plant parasitic nematodes and the diseases they cause".

☆ 1934 - Filipjev - Book on "Nematodes that are important for Agriculture" translated from Russian to English in 1941 by Stekhovan,S under the title "A Manual of Agricultural Helminthology".

☆ 1943 - Carter - Description of nematicidal value of D-D used for soil fumigation.

☆ 1945 - Christie - Description of the nematicidal value of EDB.

☆ 1948 - Allen - Taught the World's first formal university course in Nematology at the University of California, Berkeley.

☆ 1950 - Oostenbrink - Wrote a Book on "The Potato Nematode, A dangerous parasite to Potato Monoculture."

☆ 1951 - Christie and Perry - Role of ectoparasites as plant pathogens.

Goodey, T - Wrote a book on "Soil and fresh water nematodes".

'Food and Agriculture Organisation (FAO) the United Nations organised the first International Nematology course and Symposium at Rothamsted Experiment Station, England.

☆ 1955 - European Society of Nematologists founded.

☆ 1956 - *Nematologica* - The first journal published exclusively for Nematology from The Netherlands.

☆ 1961 - Society of Nematologists was founded in the United States.

☆ 1967 - Organization of Tropical American Nematologists was founded.

☆ 1969 - Journal of Nematology was first published by the Society of Nematologists, USA.

☆ 1973 - *Nematologia Mediterranea* - published from Italy.

☆ 1978 - Revue de Nematologie published from France.

☆ 1930s - 1990s - Barron, Duddingeon, Mankau, Linford, Sayre and Zuckerman-provided an insight on the Biological control of plant-parasitic nematodes. Enhanced understanding of antagonists and related bioagents enhancing the potential for practical biocontrol.

☆ 1940s - 1990s - Van Gundy-Advancement in survival mechanisms of nematodes which provided fundamental knowledge and facilitated practical control.

☆ 1950s - 1990s – Triantaphyllou- Provided advancement in Cytogenetics, mode of reproduction/sexuality-and information data base for genetics/molecular research. Enhanced understanding of evolution and taxa interrelationship.

☆ 1960s - 1990s - Caveness, Jones, Oostenbrink, Sasser and Seinhorst -International programme such as International *Meloidogyne* Project - They expanded educational base of nematalogists worldwide and provided ecological - taxonomic data base.

☆ 1960s - 1990s - Nickle, Poinar and Steiner- Biological control of insects with nematodes.

☆ 1960s - 1990s - Brenner, Dougherty and Nicholas- *Caenorhabditis elegans* developmental biology and genetics - model system - provided fundamental information on cell lineage, behaviour, gene function, ageing and overall genome for the model biological system

In addition to the above, the research advancements are in progress in the following areas:

Molecular markers for resistance genes which provide efficiency of breeding for resistance.

Molecular analysis of host-parasitic interactions which provide fundamental knowledge on mechanisms of pathogenesis.

Cloning of resistance genes - Elucidation of the molecular mechanism of resistance-Transgenic host resistance to plant parasitic nematodes.

Land Marks of Nematology Research in India

In India, origin of nematology research can be traced to the report of Barber, the then Economic Botanist working at Coimbatore on root knot infesting tea in South India in 1901. Butler (1906)identified root knot nematode in black pepper in Kerala. Further, Butler reported a disease of rice caused by *Ditylenchus angustus* in 1919. Reports of Ayyar (1926, 1933 and 1944) on *Meloidogyne* spp infesting vegetables and other crops, Dastur (1936) on *Aphelenchoides besseyi* on rice and Jones (1961) on golden nematode of potato *Heterodera* (*Globodera*) *rostochiensis* Nilgiris were other remarkable landmarks in the country (Seshadri, 1986). Organised research in nematology started in early sixties. The pioneering work of nematode survey and taxonomy by workers at Aligarh, Hyderabad and New Delhi by Siddiqi, Jairajpuri Das and Khan during 1959-1965, not only laid a stable foundation for the growth and development of nematolgy in the country but also brought Indian nematology in the international map. Six hundred and forty species of Tylenchids, 78 of Aphelenchids, 72 of Longidorids, and 8 of Trichodorids were recorded (Bajaj, 1999). Twenty three species belonging to major genera like *Meloidogyne, Radopholus, Ditylenchus, Heterodera, Rotylenchus, Hoplolaimus* and *Helicotylenchus* were identified as important pests, causing 7.2 to 100 per cent loss in several cash crops in India (Patel *et al.*, 1999).

Organisation of International Nematology course at IARI in 1964 and the South East Asian Post graduate Nematology courses between 1967 and 1979 and initiation of nematolgy teaching at the State Agricultural Universities during the same period laid foundation for nematolgy teaching in India. Establishment of the Nematolgical Society of India in 1969 and the starting of publication of the Indian Journal of Nematology together with the organization of a series of national symposia were instrumental in creating awareness on the relevance of the subject. The establishment of the AICRP on nematode pests of crops and their control at 14 centres in 1977 strengthened the research activities in the country. Further nematological units were established at several traditional and agricultural universities and ICAR Institutes and the Institutions made significant advances in basic and applied nematology

The history and development of Nematology in India are given below in chronological order.

☆ 1901 - Barber first reported root-knot nematode on tea from Tamil Nadu, South India.

☆ 1906 - Butler reported root-knot nematode on black pepper in Kerala.

☆ 1913, 1919 - Butler reported Ufra disease on rice in Bengal due to the infestation of *Ditylenchus angustus* in 1919.

☆ 1926, 1933,1934- Ayyar reported root-knot nematode infestation on vegetables in 1933 and other crops in India.

☆ Dastur 1936 - *Aphelenchoides besseyi* on rice

☆ Jones 1961- Golden nematode of potato *Heterodera* (*Globodera*) *rostochiensis* Nilgiris

☆ 1959 - Prasad, Mathur and Sehgal reported cereal cyst nematode for the first time from India.

☆ 1961 - Nematology laboratory established at Agricultural College and Research Institute, Coimbatore, with the assistance of Rockefeller Foundation and Indian Council of Agricultural Research.

☆ 1961 - Jones reported the potato cyst nematode for the first time from Uthagamandalam, Tamil Nadu.

☆ 1961 - Nematology unit established at the Central Potato Research Institute, Simla.

☆ 1963 - Laboratory for potato cyst nematode research established at Uthagamandalam with assistance of Indian Council of Agricultural Research.

☆ 1964 - First International Nematology Course held at I.A.R.I., New Delhi.

☆ 1966 - Nair, Das and Menon reported the burrowing nematode on banana for the first time from Kerala.

☆ 1966 - Division of Nematology established at I.A.R.I. New Delhi.

☆ 1968 - First South-East Asian Post-Graduate Nematology course held in India.

☆ 1969 - Nematological Society of India founded and first All India Nematology Symposium held at I.A.R.I, New Delhi.

☆ 1969 - Third South-East Asian Nematology course conducted in 1970 at New Delhi.

☆ 1971 - Indian Journal of Nematology published.

☆ 1971 - Fourth South-East Asian Nematology course at New Delhi.

☆ 1972 - First All India Nematology Workshop held at IARI, New Delhi

☆ 1973 - Fifth South-East Asian Nematology course at New Delhi.

☆ 1975 - Sixth South-East Asian Nematology course at New Delhi.

☆ 1976 - Summer Institute in Phytonematology held at Allahabad.

☆ 1977 - Department of Nematology established at Haryana Agricultural University, Hisar.

☆ 1977 - All India Co-ordinated Research Project (AICRP) on nematode pests of crops and their control started functioning at 14 centres in India with its Project Co-ordinator at I.A.R.I, New Delhi.

☆ 1979 - M.Sc.(Ag.) Plant Nematology course started at Tamil Nadu Agricultural University, Coimbatore.

☆ 1979 - All India Nematology Workshop and Symposium held at Orissa University of Agriculture and Technology, Bhubaneshwar.

☆ 1979 - Seventh South-East Asian Nematology course at New Delhi.

☆ 1981 - Department of Nematology established at Tamil Nadu Agricultural University, Coimbatore.

☆ 1982 - Department of Nematology established at Rajendra Agricultural University, Pusa, Bihar.

☆ 1983 - All India Nematology Workshop and Symposium held at Solan, Himachal Pradesh.

☆ 1985 - All India Nematology Workshop and Symposium held at Udaipur, Rajasthan.

☆ 1986 - National Conference on Nematology held at I.A.R.I, New Delhi.

☆ 1987 - All India Nematology Workshop at Govt. Agricultural College, Pune.

☆ 1987 - Group Discussion on "Nematological problems of Plantation crops" held at Sugarcane Breeding Institute, Coimbatore.

☆ 1992 - Silver Jubilee Celebration of Division of Nematology, I.A.R.I., New Delhi.

☆ 1992 - Summer Institute on "Management of Plant Parasitic nematodes in different crops" organised by ICAR at Haryana Agricultural University, Hisar.

☆ 1995 - All India Nematology Workshop and National symposium on Nematode problems of India held at IA.R.I, New Delhi.

☆ 1997 - Summer School on "Problems and Progress in Nematology during the past one decade" was organised by ICAR at I A.R.I, New Delhi.

☆ 1998 - Afro-Asian Nematology Conference held during April 1998 at Coimbatore.

☆ 1999 - National seminar on "Nematological Research in India: Challenges and preparedness for the new millennium" at C.S. Azad University of Agriculture and Technology, Kanpur.

☆ 2000 - National Nematology Symposium on "Integrated Nematode Management" held at OUAT, Bhubaneshwar, Orissa.

☆ 2001 - National Congress on "Centenary of Nematology in India: Appraisal and Future plans" at LA.R.I, New Delhi.

☆ 2002 - "Centenary of Nematology in Tamil Nadu" Celebrated at Department of Nematology, Tamil Nadu Agricultural University, Coimbatore

☆ 2003 - Winter School on "Biological control of plant parasitic nematodes" from December 2–22, 2003 at Department of Nematology, Tamil Nadu Agricultural University, Coimbatore

☆ 2004 - ICAR sponsored Summer School on "Recent technologies in the Management of Phytoparasitic nematodes for sustainable agriculture" -September 8-28, 2004 at CCS, Haryana Agricultural University.

- National Symposium on "Paradigms in Nematological Research for biodynamic farming" held at Bangalore from November, 17-19, 2004

☆ 2005 - National Symposium on "Recent advances and Research priorities in Indian Nematology" held at IARI, New Delhi from December, 9-10, 2005.

Extent of Crop Loss

Nematodes are becoming an increasing threat to many crops. Worldwide, nematodes are responsible for loss up to 14 per cent of annual crop. Nematodes damage crops by directly feeding on plant roots, transmitting viruses and/or facilitating bacterial and fungal infections. Plant parasitic nematodes pierce plant cell walls with their stylet and feed on plant cells. This may deprive the host of nutrients and water for their growth. The conducting tissues responsible for the translocation of nutrients and water are often blocked. Recent projections estimated that in India, nematodes may be causing loss of about 5 per cent in oil seeds, 8 per cent in cereals and pulses, 10 per cent in fruit crops and 12 per cent in vegetables. The annual crop loss is estimated Rs. 242 billion (Sheshadri and Gaur, 1999). A reduction of root mass, a distortion of root structure or enlargement of the roots may indicate presence of nematodes. Symptoms of nematode damage intensify during stressful conditions like excessively dry or wet weather. Furthermore, nematode damage provides an opportunity for other plant pathogens to invade the root and thus further weaken the plant.

Table 16.1: Genera of Most Common Plant Parasitic Nematodes

Root knot nematode	*Meloidogyne* spp.
Cyst nematode	*Globodera* spp., *Heterodera* spp.
Reniform nematode	*Rotylenchulus* spp.
White tip nematode	*Aphelenchoides* spp.
Spiral nematode	*Helicotylenchus* spp.
Dagger nematode	*Xiphinema* spp.
Awl nematode	*Dolichodorus* spp.
Lance nematode	*Hoplolaimus* spp.
Lesion nematode	*Pratylenchus* spp.
Needle nematode	*Longidorus* spp.
Pin nematode	*Paratylenchus* spp.
Ring nematode	*Criconemella* spp.
Sheath nematode	*Hemicycliphora* spp.
Sting nematode	*Belonolaimus* spp.
Stubby root nematode	*Paratrichodorus* spp., *Trichodorus* spp.
Stunt nematode	*Tylenchorynchus* spp.
Rice root nematode	*Hirscmaniella* spp.
Burrowing nematode	*Radopholus similis*

Symptoms

Nematode damage in crops is often confused with drought, root feeding by other pests, malnutrition or disease. Typical symptoms are wilting despite sufficient soil moisture, yellowing of the foliage, uneven or stunted growth and reduced root mass with or without galling.

Symptoms Produced by Above Ground Feeding Nematodes

Leaf Discolouration

The leaf tips become white in rice due to ice white tip nematode, *Aphhelenchoides besseyi*, yellowing of leaves on chrysanthemum due to chrysanthemum foliar nematode, *Aphelenchoides ritzemabosi*.

Dead or Devitalized Buds

In straw berry plants the nematode *A. fragariae* affects the growing point and kills the plants and results in blind plant.

Seed Galls

In wheat, *Anguina tritici* larvum enters the flower primordium and develops into a gall. The nematodes survive for longer period (even upto 28 years) inside the cockled wheat grain.

Twisting of Leaves and Stem

The basal leaves become twisted in onion when infested with *Ditylenchus dipsaci*. In rice crop, the top leaves become twisted when infested with *D. angustus*.

Crinkles or Distorted Stem and Foliage

The wheat seed gall nematode *A. tritici* infests the growing point and as a result, distortions in stem and leaves take place.

Necrosis and Discolouration

The red ring disease of coconut is caused by *Radinaphelenchus cocophilus*. Due to the infestation, red coloured circular area appears in the trunk of infested palm.

Lesions on Leaves and Stem

Small yellowish spots are produced on onion stem and leaves due to *D. dipsaci*. Leaf lesions are caused by *Aphelenchoides ritzemabosi* in chrysanthemum.

Symptoms Produced by Below Ground Feeding Nematodes

The nematodes which infest and feed on the root portion and exhibit symptoms on below ground plant parts and the above ground plant parts are classified as:

1. Above ground symptoms due to root feeding
2. Below ground symptoms

Above Ground Symptoms

1. *Stunting*: Reduced patches of stunted plants appear in the field. Example: in potato due to *Globodera rostochiensis*, in gingelly due to *Heterodera cajani* and in wheat by *Heterodera avenae*.

2. *Discolouration of foliage*: Patchy yellow appearance in coffee is due to *Pratylenchus coffeae*. *G. rostochiensis* infested potato plants show light green foliage. *Tylenchulus semipenetrans* induces fine mottling on the leaves of orange and lemon trees.

3. *Decline and die back*: In banana, decline, die back and toppling down are caused by *Radopholus similis*. Spreading decline in citrus is due to *R. citrophilus* and slow decline of citrus is due to *Tylenchulus semipenetrans*. In grapevine, slow decline is caused by *Meloidogyne* spp.

4. *Wilting*: Wilting during day time is due to *Meloidogyne spp*. In hot weather, the root - knot infested plants tend to droop or wilt even in the presence of enough moisture in the soil. Severe damage to the root system due to nematode infestation leads to day wilting of plants.

Below Ground Symptoms

1. *Root galls or knots*: The characteristic root galls are produced by *Meloidogyne spp*. False root galls are produced by *Nacobbus batatiformis* on sugar beet and tomato. Small galls are

produced by *Hemicycliophora arenaria* on lemon roots. *Ditylenchus radicicola* causes root galls on wheat and oats. *Xiphinema diversicaudatum* causes galls on rose roots.

2. *Root lesions*: The penetration and movement of nematodes in the roots cause typical root lesions. Example: necrotic lesions are induced by *Pratylenchus* spp. on crossandra; and the burrowing nematode *Radopholus similis* in banana. Similarly, *Pratylenchus coffeae* and *Helicotylenchus multicinctus* cause reddish brown lesions on banana root and corm. The rice root nematode also causes brown lesions on rice root.

3. *Reduced root system*: Due to nematode feeding the root tip, growth is arrested and the root produces branches. This may be of various kinds like oars root, stubby root and curly tip.

   (a) *Coarse root*: Infestation by *Paratrichodorus* spp. arrests growth of lateral roots and leads to an open root system with only main roots.

   (b) *Stubby roots*: the lateral roots produce excessive rootlets (example: *P. chistei*).

   (c) *Curly tip*: In the injury caused by the nematode *Xiphinema* spp., it retards the elongation of roots and causes curling of roots known as 'Fish hook' symptom.

4. *Root proliferation*: Increase in the root growth or excessive branching occurs due to nematode infestation. The infested plant root produces excessive root hairs at the point of nematode infestation. (example: *Trichodorus christei*, *Nacobbus* spp., *Heterodera* spp., *Meloidogyne hapla* and *Pratylenchus* spp.).

5. *Root rot*: The nematodes feed on the fleshy structure and result in rotting of tissues (example: *Scutellonema bradys* in yam and *Ditylenchus destructor* in potato cause root rot.

6. *Root surface necrosis*: The severe injury caused by nematode may lead to necrosis of tissues.

Nematodes of Global Importance

Nematodes form important threat to agriculture world wide (Siddiqi, 1996; Luc *et al.*, 1993, 2005).

Heterodera (Cyst nematodes): Female nematode is swollen or obese, lemon shaped, 300-600 um in diameter with a distinct neck. Females produce several hundred eggs, and after death, the female cuticle forms a protective cyst. Eggs are retained within the cyst. Cysts are either partially enclosed in root tissue or in the soil. It is called a cyst nematode because the greatly swollen, egg-filled adult female is referred to as the "cyst stage". Male vermiform (*ie.* wormlike) is found in soil. Juvenile vermiform is 450-600 um long.

Major species of *Heterodera* are *H. glycines, H. avenae, H. schachtii, H. trifolii, H. goettingiana, H. cajani* and *H. zeae*.

Globodera (Cyst nematodes): Similar to *Heterodera* but the cyst is globose. Species confined to the cooler places. Major species: *G. rostochiensis, G. pallida, G. tabacum*.

Meloidogyne (Root-knot nematodes): Females embedded in root tissue, globose, 0.5-0.7 mm in diameter with slender neck. Male vermiform 1-2 mm long, free living in soil. Juveniles slender, vermiform about 450 um long. Most of the females are within the galls on the roots. World-wide distribution.

Major species: *M. arenaria, M. incognita, M. javanica, M. hapla, M. chitwoodi*.

Ditylenchus (Stem and bulb nematode): Slender vermiform nematodes. Ectoparasites of plant stems, leaves and within the tissues. Potato rot nematode (*D. destructor*) is one of the five nematodes, listed on the EPPO quarantine list A-2 (Zero tolerance required in countries in which the pests are imported by reasons of prevailing ecological conditions).

Major species of *Ditylenchus* are *D. destructor, D. dipsaci* and *D. angustus.*

Anguina (Seed gall nematodes): Typical gall forming endoparasites of seeds, stems and leaves of cereals, grasses and other plants. Adult stages are found only in plant galls, juveniles are found in galls, plant tissues or soil. As the gall matures and dies, the infective juveniles can survive many years in a quiescent state. Major species of *Anguina* are *A. tritici, A. agrostis* and *Anguina wevelli.*

Pratylenchus (Lesion nematodes): An important group of migratory endoparasites and ectoparasites of roots. They cause serious damage to many economic plants world-wide. They are small nematodes (less than 1 mm long). Major species are *P. penetrans, P. brachyurus, P. coffeae, P. zeae, P. goodeyi, P. thornei* and *P. vulnus.*

Radopholus (Burrowing nematodes): These small nematodes (less than 1 mm long) constitute an important group of endoparasitic nematodes of plant roots and tubers. The major species is *R. similis* with two host races which differ in parasitism of citrus.

Hirschmanniella (Root nematodes): Medium size to long, slender migratory endoparasites, many on roots (1-4 mm). *H. oryzae* is a major pest of rice in several countries. Major species are *H. oryzae, H. mucronata* and *H. spinicauda.*

Hoplolaimus (Lance nematodes): An important group of basically migratory ectoparasites which feed on roots of many kinds of fruits and other economic plants world-wide. Medium length (1-2 mm). Major species are *H. columbus, H. seinhorsti* and *H. indicus.*

Rotylenchulus (Reniform nematodes): Immature females establish permanent feeding sites in roots, become semi-swollen and protrude from roots. They are 0.23-0.64 mm long and have a kidney shaped body. Males are vermiform. Eggs are laid in gelatinous matrix.

The major species is *R. reniformis,* found in both tropical and warm temperate soils.

Tylenchulus (Citrus nematode): Immature females are in soil and are vermiform. Mature female anterior part is embedded in root tissues, the slender posterior part protrudes from roots and is swollen. Males and juveniles are vermiform and slender. The major species is *T. semipenetrans* found every where in citrus growing areas.

Helicotylenchus (Spiral nematodes): Small to medium sized nematodes (0.4-1.2 mm), usually in spiral shape. Ectoparasitic, semi-endoparasitc or endoparasitic nematodes of roots. The most damaging species is *H. multicinctus.* Major species are *H. multicinctus, H. mucronatus, H. dihystera* and *H. pseudorobustus.*

Criconemella (Ring nematodes): Migratory ectoparasites. Females are 0.2-1mm long, stout with prominent retrorse annules. Males are slender and short; juveniles are like females with annules. Major species are *C. xenoplax, C. axesti* and *C. spharocephalum.*

Xiphinema, Longidorus, Trichodorus and *Paratrichodorus* (Dagger, needle and stubby root nematodes): Slender, virus transmitting nematodes 0.8-5 mm long. Ectoparasites on roots of perennial and woody plants. World-wide distribution.

Major species are *X. americanum, X. elongatum, Longidorus africanu* are *Paratrichodorus minor.*

Aphelenchs (Bud and Leaf and Pine wood nematodes): They have a world-wide distribution. *A. fragariae* and *A. besseyi* feed on and damage strawberry plants; the later species also damages rice. *A. ritzemabosi* causes necrosis on leaves of chrysanthemums and other ornamentals. Pine wood nematode (*Bursaphelenchus xylophilus*) is implicated in a serious disease of pine trees (pine wilt), which devastated pine forests in Japan and occurs in North America on various pines. In 1997, white pine trees in Maryland were devastated due to heavy infestation of this nematode. This is a serious quarantine pest and all pine wood chips or wood products for import and export purposes need to be checked for this nematode.

Major Crops Affected by Nematodes

Rice

Plant-parasitic nematodes associated with rice can be divided into two groups depending on their parasitic habits: the foliar parasites, feeding on stems, leaves and panicles and the root parasites. Some of the nematodes of economical importance are the white tip nematode (*Aphelenchoides besseyi*); the rice stem nematode (*Ditylenchus angustus*); the root nematode (*Hirschmanniella* spp). and the rice cyst nematode (*Heterodera oryzae*). The white tip nematode is seed borne and occurs in many rice growing areas. Earlier symptoms are emergence of chlorotic tips of new leaves with a white splash pattern. The grain is small and distorted and kernel may be discolored and cracked. The rice stem nematode is the cause of "ufra" disease in several countries, mainly in deep water rice areas. A root-knot nematode (*M. graminicola*) damages rice in several countries. Four cyst-nematode species infect rice roots (*H. oryzae*, *H. oryzicola*, *H. elachista* and *H. sacchari*), and the infected roots turn brown to black. Lemon shaped white females and brown cysts can be seen on infected roots. Also, lesion nematode species cause severe damage to rice. Recently rice root nematodes were reported from Kerala (Sheela *et al.*, 2006).

Wheat and Other Cereals

Several plant-parasitic nematodes are associated with wheat and barley and the most economically important ones are cereal cyst nematode (*Heterodera avenae*), the seed gall or ear-cockle nematode(*Anguina tritici*), root-knot and the lesion nematodes. The cereal cyst nematode is present in many countries. The seed gall nematode, cause of the "ear-cockle" disease of wheat and barley, was the first described plant-parasitic nematode. If the wheat galls are kept in a dry condition, the nematode larvae within may remain viable for more than 25 years. From a single gall upto 90,000 nematodes were counted. Stunt, root-knot and lesion nematodes are reported as major pests of sorghum. Wheat gall nematode is the most economically important nematode parasite which has been disseminated through infested seed to many wheat-producing regions of the world.

Corn

Lesion nematodes and corn cyst nematodes are considered as economically important nematodes which cause severe damage to corn. The corn cyst nematode (*Heterodera zeae*) is widespread in India. White or yellow stage females or brown cysts of corn cyst nematode on roots of corn (*Zea mays*) can be seen on infested plants.

Coconut

Many different nematodes are found in diverse forms of association with the coconut palm. Some are found associated in different types of insect visitors of the palm (Govindankutty and Koshy 1979; Koshy and Banu 2002). The major nematode disease affecting the crop is red ring disease caused by *Bursaphelenchus cocophilus*. Red ring nematode in the roots, trunk and stem tissue of coconut palms causing lesions and the characteristic orange to red ring appears about 3 cm wide and 2.5 cm beneath the stem surface. The nematode invades through root tissue, stem and leaves. The red palm weevil is the main vector of nematodes from diseased to healthy trees. The only other nematode known to cause severe damage leading to malfunction in the coconut is *Radopholus similis* (Koshy *et al.*, 1975). The burrowing nematode, *R. similis*, occurs in most tropical and subtropical areas of the world and reported from coconut palms in Florida, Jamaica, Sri Lanka and India (Weischer, 1967; Koshy and Sosamma, 1987, 1996).

Arecanut

About 22 genera of parasitic nematodes are associated with the rhizosphere of the crop. The burrowing nematode (*Radopholus similes*) is the only nematode which causes significant damage to the crop. *R. similis* is a migratory root endoparasite causing lesions and rotting on root. The nematode produces elongate orange coloured lesions. A number of nematodes are reported from the rhizosphere of arecanut; but only *Radopholus similis* is an important parasite of the palm. The burrowing nematode(*R. similes*) was first reported from soil around roots of arecanut palm in Mysore, India by Kumar *et al.* (1971) and later by Koshy *et al.* (1975).

Banana

More than 134 species of nematodes, belonging to 54 genera, are reported in association with banana. *R. similis*, *Pratylenchus coffeae*, *Helicotylenchus multicinctus*, *H. dihystera*, *M. incognita*, *M. javanica*, *H. oryzicola* and *R. reniformis* are economically important (Sathiamoorthy, 2000). Among a number of nematodes reported in association with banana roots, the burrowing nematode *Radopholus similis* is the most important pest on banana. The disease of bananas caused by *R. similis* is known by different names *viz.*, *Radopholus* root rot, black head, black-head toppling disease and decline. From India, the first report of its occurrence on banana was made from Kerala (Nair *et al.*, 1966). Bioecology of the cyst nematode, *Heterodera oryzicola*, infesting banana was studied by Charles and Venkitesan (1993). Seven genera of plant parasitic nematodes *viz.*, *R. similis*, *Heterodera oryzicola*, *Helicotylenchus multicinctus*, *Pratylenchus* sp., *Meloidogyne incognita*, *Hoplolaimus indicus* and *Criconemoides* sp. were found in the rhizosphere of banana in different crop combinations like coconut+ banana, coconut+pepper+banana, coconut+ pepper+ banana+ vegetables and banana alone (Jiji *et al.*, 1999).

Citrus

Several nematode species are associated with citrus, and the most devastating one is the citrus nematode (*Tylenchulus semipenetrans*) which occurs in all the citrus producing regions of the world. This nematode causes the disease "slow decline" of citrus. Another nematode damaging citrus is the burrowing nematode (*Radopholus similis*), which causes a severe spreading decline disease of citrus. In addition, lesion, root-knot, sting, dagger, stubby root and other ecto-parasitic nematodes can damage citrus (Reddy and Singh, 1979). Occurrence of citrus nematode (*Tylenchulus semipenertans*) was first reported from Aligarh (Uttar Pradesh, India) by Siddiqi (1961). Out of more than 189 species, belonging to nematode genera, associated with citrus in world, approximately 122 species belonging to 57 genera are reported from India. Although more than ten species of phytonematodes are found pathogenic on citrus, the proof of pathogenicity, under Indian condition is available for *Tylenchulus semipenertans* (Siddiqi, 1961), *Pratylenchus coffeae* (Siddiqi, 1964), *Hoplolaimus indicus* (Gupta and Atwal, 1972) and *Meloidogyne javanica* (Mani, 1986). In the heavily infected feeder roots of citrus,adult females of nematode are found attached to the rootlets which often appear dark and thicker than the healthy roots due to soil particles which adhere to gelatinous egg masses produced by females on root surface. From soil and root washings around citrus trees second-stage larvae and males of this nematode are commonly encountered. Also, adults of vermiform migratory endoparasites of burrowing nematode can be recovered from the roots of citrus.

Papaya

Root knot nematode (*Meloidogyne incognita*), and reniform nematode, *Rotylenchulus reniformis*, are the major nematode species associated with papaya cultivation.

Apple

Pratylenchus, Meloidogyne, Xiphinema, Paratylenchus and *Longidorus* are the major nematode genera of economic importance as they cause deleterious effects on plant growth and productivity (Inagaki, 1978, Melton *et al.*, 1985).

Cotton

The two most important root diseases of cotton are root-knot caused by the root-knot nematode, and Fusarium wilt caused by the fungus *Fusarium oxysporum*. Infection by root-knot nematodes increases incidence and severity of Fusarium wilt. Other species which are pathogenic on cotton are reniform, lesion, sting, lance and dagger nematodes. The sting nematode is an aggressive pest of cotton but is restricted to soils with greater than 85 per cent sand content. Lance nematodes mostly feed in the cortical region of cotton roots causing cell damage and necrosis. The other species which are pathogenic on cotton are reniform nematode (*Rotylenchulus reniformis*), lesion nematode (*Pratylenchus brachyurus*) and sting nematode (*Belonolaimus longicaudatus*).

Legumes

The pea cyst nematode is an important parasite of peas and broad beans in many countries. The stem nematode is another important nematode on broad beans. Root-knot, cyst and reniform nematodes are the major nematode pests of chickpea and cowpea. Pea cyst nematode (*Heterodera goettingiana*) and pigeonpea cyst nematode on pigeonpea (*Cajanus cajan*) are important.

Nematodes damage peanuts in all production regions of the world. The annual loss caused by nematodes to peanuts is estimated 12 per cent. The nematodes which attack peanuts and cause damage are root-knot, lesion, sting, ring, stunt and potato-rot nematodes.

Galls on peanut pods, pegs and roots are caused by peanut root-knot nematode (*M. arenaria, M. javanica,* and *M. hapla.*). Lesion nematode (*P. brachyurus*) is the major lesion nematode parasitizing peanuts.

Soybean

The soybean cyst nematode (*Heterodera glycines*is) is the most serious pest of soybean throughout the world. Soybean production is not economically possible without effective control measures. Of the greater than 50 species of nematodes reported from soybeans, the soybean cyst and root-knot nematodes are the most important. Some other plant parasitic nematodes which attack soybeans are reniform, sting, lesion and lance nematodes.

Vegetables

Root-knot nematodes cause serious problems in vegetables. The disease complex caused by root-knot nematodes and bacterial or fungal wilt organisms is one of the most lethal known. Other plant-parasitic nematodes such as reniform, stubby root, sugar-beet cyst, false root-knot, sting and stunt nematodes are serious pests of vegetables. The magnitude of the losses range from 28 to 48 per cent in tomato, 26 to 50 per cent in brinjal, 18- 33 per cent in chillies, 6 to 90 per cent in okra and 38 to 47 per cent in bittergourd (Reddy 1989). Some of the plant parasitic nematodes act as a vector for carrying the plant diseases. Only the species of Dorylaimoid genera, *Xiphinema, Longidorus, Paralongidorus, Paratrichodorus* and *Trichodorus* are implicated as vectors. Presence of galls on the root system of vegetables is seen in root knot nematode. Potato cyst nematode (*Globodera pallida*) infects and damages tomato and egg plant. Stem nematode severely damages onion and garlic. Also, several species of stunt nematodes are often found associated with vegetables.

Potato

The most important nematode threat to potato production is potato cyst nematode, which causes severe damage to the crop. Two species of cyst nematodes infect potatoes - *Globodera rostochiensis* (the "golden" nematode) and *G. pallida*. The golden nematode is found in several countries. Yield loss up to 80 per cent is reported in some potato growing areas of the tropics where infestation levels due to golden nematode are high. Other major nematode parasites of potatoes are root-knot (*Meloidogyne*), false root-knot (*Nacobbus*), bulb and stem (*Ditylenchus dipsaci*), potato rot (*Ditylenchus destructor*) and lesion nematodes (*Pratylenchus*). The potato rot or tuber nematode and potato stem nematode are reported from temperate climates. Potato stem nematode is a parasite of foliage and attacks leaves, petioles and also injures tubers. The potato rot nematode mainly damages tubers and is a major pest of quarantine importance. The Columbia root-knot nematode (*M. chitwoodi*), decreases the quality of potato tubers by causing brown spots on the surface, rendering tubers unacceptable. Many other nematodes associated with potato include sting, dagger, reniform, burrowing and pin nematodes; most of them are of minor importance.

Sugar Beet

The sugar-beet cyst nematode (*Heterodera schachtii*) is the most devastating pest of all the sugar-beet growing areas of the world. The nematode causes less sugar production/area of land. It favours temperate regions and also tolerates broad range of climates.

Tea

M. brevicauda causes sever galling of roots in mature tea. The infected bushes grow slowly and produce small leaves which turn yellow. The roots show typical swelling and pitting.

Coffee

Quite a few nematodes attack coffee (*Coffea spp.*) throughout the world. The important ones are *Meloidogyne* spp., *Pratylenchus* spp., *Hemicriconemoides* spp., *Radopholus similis* and *Rotylenchulus reniformis*. Of these, only the species of *Pratylenchus* and *Hemicriconemoides* infect Arabica coffee (*C. arabica*) in India (Kumar, 1984; 1985). Robuta coffee (*C. canephora*) exhibits tolerance against nematode attack.

Pepper

Root-knot nematodes (*Meloidogyne* spp.) and the burrowing nematode (*Radopholus similes*) are the two important nematode species infesting rooted cuttings in the nursery (Venkitesan and Setty, 1978; 1979). The damage to feeder roots is caused by these nematodes and the fungus *Phytophthora capsici* either independently or together in combination. Studies under simulated conditions showed that *R. similis* and *P. capsici* alone or in association resulted in root rotting leading to slow decline disease (Ramana *et al.*, 1992). A new species *M. piperi* was recently described from Kerala, India

(Sahoo *et al.*, 2000). Another nematode of wide occurrence in black pepper tract is *Trophotylenchulus piperis* (Suderraju *et al.*, 1995).

Cardamom

The most important nematode problem is caused by the root knot nematode (*M. incognita*) though *P. coffeae* and *R. similis* are known to cause root rotting. *Rotylenchulus reniformis* is also reported from cardamom. The incidences of damping off and rhizome rot increase in the presence of *M. incognita* (Ali, 1986; Eapen, 1993).

Ginger

Plant parasitic nematodes belonging to 17 genera were reported from ginger. (Ramana and Eapen, 1998). The most important parasites are *Meloidogyne* spp., *R. similis* and *P. coffeae*. In Kerala, *M. incognita*, *R. similis* and *Pratylenchus* were the major nematode species found in the rhizosphere of ginger (Sheela *et al.*, 1995).

Turmeric

The root-knot nematode(*Meloidogyne* spp.) is widely prevalent in all the turmeric grown areas. The nematodes spread through the rhizomes. In case of severe infestation, stunting and yellowing of plants appear in patches. The infested plant roots show severe root galls and poor development of fingers.

Cassava

Meloidogyne incognita and *M. javanica* are the most widerspread among the nematodes infesting cassava (Mohandas, 1994). Other species of *Meloidogyne* reported on cassava are *M. arenaria* and *M. hapla*. Among *Pratylenchus* sp., *P.brachyurus* and *P. safaenis*- are reported. Other nematodes reported on cassava are *Rotylenchulus reniforms, Helicotylenchus erythrinae, H. dihystera* and *Scutellonema bradys* which are however of lesser importance.

Sweet Potato

Species of *Meloidogyne* and *Rotylenchulus reniformis* cause reduction in yield and quality of tubers the world over (Clark and Moyer, 1988). *Pratylenchus* spp. are reported very serious in Japan whereas *Ditylenchus destructor* and *D. dipsaci* are serious in China (Mohandas and Palaniswami,1990; Sharma *et al.*, 1997; Mohandas *et al.*,1998)

Yams

Three most important nematode pests of yams are Root-knot nematode, Yam nematode and Lesion nematode. Root-knot nematode infestation produces deformed tubers with uneven surface, whereas the other two nematodes produce typical 'dry rot' of tubers. Steiner (1937) reported *Rotylenchus blaberus* causing a disease in yams originating from Nigeria. *Scutellonema bradys* is one of the most important pests infesting yams. This nematode is recorded on yams from India, West Africa, the Caribbean and Brazil (Bridge, 1982). The nematode produces 'dry rot' on yams. In field, the damage is restricted to the outer thin layer of the tuber (Adesiyan, 1977). In the tubers, *S. bradys* may be found alone or associated with *Pratylenchus coffeae* (Castaynone-Sereno, *et al.*, 1988). However during storage, the nematode infection spreads and infection sites coalesce which encircle the tuber and results in heavy loss of yams (Adesiyan and Odihirin, 1975).

Coleus

Among the minor tuber crops, coleus (*Solenostemon rotundifolius*) is extensively grown in Kerala and Tamil Nadu and is highly susceptible to the root-knot nematode (Sathyarajan *et al.*, 1966). Nematode infestation is a major hurdle for coleus cultivation in Kerala.

Betelvine

About 38 species of plant parasitic nematodes are associated with betelvine in India. The most common and serious nematode is the root-knot nematode(*Meloidogyne incognita*), which prevails almost in all the states. This is closely followed by *Rotylenchulus reniformis* and *Hoplolaimus indicus*. *Helicotylenchus microcephalus*, *H. incisus*, *H. dihystera*, *M. javanica*, *Tylenchorhynchus indicus*, *Pratylenchus coffeae*, *P. zeae*, *Radopholus similis* etc. are more localized. According to Koshy and Bridge (1990) *M. incognita*, *R. reniformis* and *R. similis* are the principal nematode pests of betelvine in India.

Tobacco

The dominant nematodes are *Meloidogyne* species particularly *M. incognita*, *M. javanica* and *M. arenaria*. Others like *Globodera sp.*, *Pratylenchus sp.*, *Tylenchorhynchus* sp., *Ditylenchus* sp. and *Rotylenchulus reniformis* are restricted in distribution, causing losses in specific areas. A few species of *Xiphinema*, *Longidorus* and *Tichodours* are reported as vectors of virus diseases.

Ornamental Plants

The major nematode pests associated with ornamental plants are root-knot nematodes, lesion nematodes, reniform nematodes, spiral nematodes, needle nematodes, burrowing nematode, bud and leaf nematode, dagger nematode, apratylenchoides nematode etc.

Nematode Taxonomy

Cobb (1919) placed the nematodes under separate phylum Nemata/Nematoda, which consists of two classes *viz.*, Secernentea and adenophorea. Maggenti (1991) classified the nematodes.

Characteristics of the Phylum Nematoda

1. Nematode possesses elongate, unsegmented, cylindrical or worm like body tapering towards both the ends, unciliated and circular in cross section.
2. Body is bilaterally symmetrical.
3. They are aquatic, terrestrial and parasitic or free living.
4. The body is covered by tough and resistant cuticle secreted by epidermal (hypodermal) cells.
5. Terminal oral aperture surrounded with lips and papillae.
6. Digestive system consists of feeding apparatus, oesophagus, intestine and rectum.
7. Body consists of two tubes.
8. The central nervous system consists of, oesophageal nerve ring encircling isthumus and longitudinal nerves. Peripheral nervous system consists of sensory organs like cephalic papillae, amphids, cephalids etc.
9. Primitive excretory system, is devoid of protonephridial cilia or metanepharidial funnel.
10. Circulatory and respiratory systems are completely absent.
11. Females have a separate genital pore and the males have a common opening known as cloaca and well developed copulatory apparatus consisting of spicule and gubernaculum.

12. Females are oviparous or ovoviviparous or viviparous. The cleavage is terminal and growth is accompanied by moulting.

13. Life cycle is direct and there are four juvenile stages.

Table 16.2: Diagnostic Characters of Class Secernentea and Adenophorea

Secernentea (Phasmida)	Adenophorea (Aphasmida)
Amphidial opening is on the head near the lip region.	Amphids open behind the head *i.e.* post labial.
Laternal canals open into the excretory duct.	Lateral canals and excretory duct terminate in a cell.
Oesophagus is divided into-procorpus, median bulb, isthmus and basal bulb.	Oesophagus is cylindrical with-an enlarged glandular base.
Male tail with bursa (Caudal alae) genital papillae.	Male tail lacks bursa but possesses
Caudal glands are absent.	Caudal glands are present.
Phasmids are present.	Phasmids are absent.
The mesenterial tissues are less developed.	The mesenterial tissues are well developed.

The plant parasitic nematodes are included in the orders Tylenchida of class Secernentea and Dorylaimida of class Adenophorea.

Order: Tylenchida

Stoma armed with a protrusible spear on stomatostylet. Oesophagus consists of a procorpus, median bulb with a selerotized valvular apparatus, nerve ring encloses the narrow isthmus and with a basal bulb. It consists of two superfamilies *viz.*, Tylenchoidea with Tylenchina and Aphelenchina suborders and Criconematoidea.

Table 16.3: Differences Between Tylenchoidea and Criconematoidea

Characters	Tylenchoidea	Criconematoidea
Labial region	Lips are hexardiate, Labial frame work present.	Labial region is poorly developed, labial plate is present.
Stylet	Conus, shaft and knobs are variable in shape and size. in the base of metacarpus.	'Criconematoid' type stylet; long and anchor shape knob which lies
Oesophagus	Narrow procarpus, metacarpus with valve, isthmus followed by glandular basal bulb.	Pro and metacarpus amalgamated to a single unit, short isthumus The postcarpus reduced, appears as 'set-off'.
Deirids	Present (2 pairs)	Absent
Female gonad	Single or two ovary; post uterine sac (PUS) is present.	Single ovary with posterior vulva. PUS absent.
Male gonad	Single testis, caudal alae is present.	Single testis; caudal alae rare.
Phasmid	Eratically present in tail region.	Not known.

Table 16.4: Difference Between Tylenchina and Aphelenchina

Characters	Tylenchina	Aphelenchina
Lip	Varying in shape.	Set off.
Annules	Faint to strong annules. Faint annules.	
Stylet	Well developed; one dorsal and two subventral knobs.	Weakly developed no stylet knobs.
Oesophagus	Three parted.	Three parted with square shaped median bulb.
Gland bulb	Abutting, dorsal, ventral or dorso ventral overlapping on intestine.	Only dorsal overlapping.
Gland opening	Behind the stylet knob in procorpus.	Opens in the median bulb.
Female	One or two, vulval position vary.	Single ovary; vulva posterior.
Male	Bursa present.	Bursa rare.
Spicule	Weak to strong sclerotization is seen with gubernaculum.	Rose thorne shape spicule present

Interaction of Nematode with Microorganisms

Plant parasitic nematode infestation makes the host tissue more suitable for the establishment of secondary pathogens *viz.*, fungi, bacteria, virus etc. The nematodes alter the host in such a way that it encourages the colonization by the secondary pathogens. Even though the nematodes themselves are capable of causing considerable damage to the crops, their association with other organisms aggravate the disease. The nematodes cause mechanical wound which favours the entry of microorganisms. In some cases, the association of nematode and pathogen breaks the disease resistance in resistant cultivators of crop plants (Atkinson, 1892, Back *et al.*, 2003; Barbetti *et al.* 2006; Bertrand *et al.*, 2000; Ingham *et al.*, 1985).

Nematode – Fungus Interaction

Nematode – fungus interaction was first observed by Atkinson (1892) in cotton. It was observed that *Fusarium* wilt was more severe in the presence of *Meloidogyne* spp. Since then the nematode – fungus interactions received considerable attention on important crops like banana, cotton, cowpea, brinjal, tobacco and tomato. Some important examples are given in Table 16.5.

Nematode serves as a vector in nematode – virus complex. Numerous virus – nematode complexes are identified after the pioneer work by Hewit *et al.* (1958) who found that *Xiphinema* index was the vector of grapevine fan leaf virus. *Xiphinema* spp., *Longidorus* spp., and *Paralongidorus* spp. transmit the ring spot viruses which are called "NETU" derived from Nematode transmitted polyhedral shaped particles. *Trichodorus* spp. transmit the rattle viruses. All these nematodes have modified bottle shaped oesophagus with glands connected by short ducts directly to the lumen of the oesophagus. This actually results in the transmission of viruses.

Table 16.5

Crops	Name of the Disease	Nematode	Fungus	Role of Nematode
Cotton	Damping off	*Meloidogyne incognita acrita*	*Rhizoctonia solani*	Assists
		M. incognita acrita	*Pythium debaryanum*	Assists
	Vascular wilt	*M. incognita acrita*	*Fusarium oxysporum fasinfectum*	Assists
		Rotylenchulus reniformis	*F. oxysporum f. sp vasinfectum*	Assists
		Belonolaimus gracilis	*F. oxysporum f. sp vasinfectum*	Assists
		B. longicaudatus	*F.oxysporum f. sp vasinfectum*	Assists
	Black shank (vascular wilt)	*M. incognita acrita*	*Pytophthora parasitica var. nicotianae*	Assists
Tobacco	Damping off	*M. incognita acrita*	*P.debaryamum*	Assists
		M.incognita	*Alteraria tenuis*	Assists
	Vascular wilt	*M.incognita*	*F.oxysporum f.sp. nicotianae*	Assists
		M.incognita acrita	*P. parasitica var nicotianae*	Assists
Banana	Vascular wilt	*Radopholus similis*	*F.oxysporum f. sp. cubense*	Essential
Tomato	Cortical rot	*Globodera rostochiensis*	*R.solani*	Assists
	Vascular wilt	*Meloidogyne* spp.	*F.oxysporum f. sp. lycopersici*	Assists
Potato	Damping off	*Ditylenchus destructor*	*P. infestans*	Assists
	Cortical rot	*G. rostochiensis*	*R. solani* Assists	
		G. rostochiensis	*Verticillium dahliae*	Assists
Onion	Damping off	*D. dispsaci*	*Botrytis allii*	Assists
Brinjal	Vascular wilt	*P. penetrans*	*V.albo-atrum*	Assists
Pea	Vascular wilt	*Pratylenchus* spp.	*F. oxysporum f.sp.pisi*	Assists
		P. penetrans	*F. pisi*	Assists
		Hopololaimus spp.	*F.oxyporum.f.sp.pisi*	
Soyabean	Damping off	*M. javanica*	*R.solani*	Assists
	Vascular wilt	*Heterodera glycines*	*Fusarium sp.*	Assists
Cowpea	Vascular wilt	*M. javanica*	*F.oxysporum.f.sp. tracheiphylum*	Assists
Luceme	Vascular wilt	*M. hapla*	*F.oxysporum f.sp. vasinfectum*	Assists
Tulip, Narcissus	Cortical rot	*P. penetrans*	*Cylindrocarpon radicicola*	Assists
Camation	Vascular wilt	*Meloidogyne* spp.	*F.oxysporum f.sp. dianthi*	Assists
Wheat	Stem rot	*Auguina tritici*	*Dilophospora alopecuri*	Essential
Wheat rot		*H. avenae*	*R. solani*	Assists

Nematodes acquire and transmit the virus by feeding, which require as little as one day. The virus persists for longer period in the nematode body than *in vitro*. The grapevine fan leaf virus exists for as many as 60 days in *X. index*. Two mechanisms are observed in virus transmission (i) retention through close biological association between virus and vector as in Xiphinema; (ii) retention of virus mechanically as in Longidorus. Virus is retained in the inner surface of the guiding sheath of *Longidorus,* cuticle lining of the lumen of oesophagus in *Trichodorus* and *Paratrichodorus,* cuticle lining of stylet extension and eosophagus in Xiphinema. The virus particles are released into plant cell with the help of oesophagus.

Certain examples of the viral – nematode interactions in which nematode acts as vectors are given in Table 16.6.

Table 16.6

Viruses	*Nematode*
NEPO – Viruses	
Arabis mosaic	*Xiphinema diversicaudatum*
	X. paraelongatum
Arabis mosaic, Grapevine fan leaf	*X. index*
Arabis mosaic, Grapevine	*X. index*
Yellow mosaic	
Strawberry latent ring spot	*X. diversicaudatum*
Tobacco ring spot,	*X. americanum*
Tobacco ring spot, Peach Yellow	*X. americanum*
Bud mosaic	
Cowpea mosaic	*X. basiri*
Arabis mosaic, Raspberry ring spot,	*Paralongidorus maximus*
Strawberry latent ring spot	
Raspberry latent ring spot – Scottish strain	*Longidorus elongaatus*
Raspberry latent ring spot – English strain	*L. macrosoma*
Tomato black ring, Beet ring spot	*L. elongates*
Tomato black ring, Lettuce ring spot	*L. attenuatus*
NETU – Viruses	
Tobacco	*Paratrichodorus pachydermus,*
	P. allitus, P. nanus,
	P. porosus, P. teres
	Trichodorus christei
	T. primitivus, T. cylindricus
	T. hooperi
	T. minor, T. pachydermus
Pea early browning	*P. anemones, P. Pachydermus*
	P. teres, T. viruliferus

Nematode – Bacterium Interactions

Nematode–bacterium interactions are comparatively a fewer than the nematode-fungus interactions. Some examples of nematode–bacterial associations are presented in the Table 16.7.

Table 16.7

Crops	Name of the Disease	Nematode	Bacterium	Role of Nematode
Wheat	Tundu	A. tritici	C.tritici	Essential
Tobacco	Vascular wilt	M. incognita	Pseudomonas solanacearum	Assists
Tomato	Vascular wilt	M. hapla, M. incognita	P.solanacearum	Assists
	Vascular wilt	Helicotylenchus nannus	P. solanacearum	Assists
	Canker	M. incognita	C. michiganensis var michiganensis	Assists
Potato	Vascular wilt	Meloidogyne spp.	P.solanacearum	Assists
Lucerne	Crown buds (vascular wilt)	D. dipsaci	C.insidiosum	Essential and Assists
Rasberry	Crown gall	M. halpa	Agrobacterium tumefaciens	Assists
Strawberry	Cauliflower disease	Aphelenchodies ritzemabosi	C. faciens	Essential
Peach	Crown gall	M. javanica	A.syringae	Assists
Peach, plum canker		Criconemella xenoplex	P.syringae	Assists
Begonia	Leaf spot	A. fragariae	Xanthomonas begoniae	Assists
Carnation	Root (Vascular wilt)	Meloidogyne spp. h. dihystera	P.caryophylli	Assists
Rose	Hairy root	P.vulnus	A.rhizogenes	Assists
Gladiolus	Scab	M.javanica	P.marginata	Assists

Nematode Management

The nematode control aims to improve growth, quality and yield by keeping the nematode population below economical threshold level. The control measures to be adopted should be profitable and cost effective. The nematode control methods are:

1. Regulatory (Legal) control
2. Physical control
3. Cultural control
4. Biological control
5. Chemical control

Regulatory Control

Regulatory control of pests and diseases is the legal enforcement of measures to prevent them from spreading or having spread from multiplying sufficiently to become intolerably troublesome.

The principle involved in enacting quarantine is exclusion of nematode from entering into an area which is not infested to avoid spread of nematode.

Quarantine principles are traditionally employed to restrict movement of infected plant materials and contaminated soil into a state or country. Diseased and contaminated plant materials may be treated to kill the nematodes or their entry may be avoided.

In USA, the soyabean cyst nematode(*Heterodera glycines*) is subjected to Federal quarantine during 1954. State Quarantine was enforced against the potato cyst nematode(*Globodera rostochiensis*) during 1941 in New York state. The burrowing nematode of citrus(*Radopholus similes*) was brought under State Quarantine Act in Florida during 1953.

Plant Quarantine in India

The Destructive Insects and Pests Act, 1914 (DIP) was passed by the Government of India which restricts introduction of exotic pests and diseases into the country from abroad. The Agricultural Pests and Disease Acts of the various states prevent inter-state spread of pests within the country. The rules permit the Plant Protection Adviser to the Government of India or any authorized officer to undertake inspection and treatments.

Strict regulations are made against *G. rostochiensis*, the potato cyst nematode and *Rhadinaphelenchus cocophilus*, the red ring nematode of coconut. Domestic quarantine regulations are also imposed to restrict the movement of potato both for seed and table purposes to prevent spread of potato cyst nematode from Tamil Nadu to other states in India.

Physical Control

It is very easy to kill the nematodes in laboratory by exposing the nematodes to heat, irradiation, osmotic pressure, etc. It is extremely difficult to adopt these methods in field conditions. These physical treatments may be hazardous to plant or the men working with the treatments and the radiation treatments may have residual effects. Physical methods like the bare root treatment at 45° C for 25 min or 46.7° C for 10 min can disinfect citrus seedlings from citrus nematode (Bindra *et al.*, 1967). This treatment leaves no adverse effect on seedlings.

Plant Resistance

Use of nematode-resistant crop varieties is often viewed as the foundation of a successful integrated nematode management program. Commercially available nematode-resistant varieties are only for tomato, pepper, pea and sweet potato. In a resistant variety, nematodes fail to develop and reproduce normally within root tissues, allowing plants to grow and produce fruits even though nematode infection of roots occurs. Some crop yield loss can still occur, even though the plants are damaged less and are significantly more tolerant of root-knot infection than that of a susceptible variety.

In tomato, a single dominant gene (subsequently referred to as the Mi gene) is widely used in plant breeding efforts and varietal development which confer resistance to all of the economically important species of root-knot nematode, including *Meloidogyne incognita, arenaria,* and *javanica. Nema red, Nematex, Hisar Lalit,* and Atkinson are tomato varieties resistant to *M.incognita.* The potato variety *Kufri Swarna* is resistant to *G.rostochiensis.*

In bell pepper, two newly developed root-knot nematode resistant varieties (Carolina Belle and Carolina Wonder) were released by the USDA. Both varieties are open pollinated, and homozygous for the root-knot nematode resistant gene. Preliminary research demonstrated that these varieties conferred a high degree of resistance to the root-knot nematode. Expression of resistance is heat sensitive. Further research is necessary to characterize the usefulness of these varieties under the high soil temperature conditions. Like tomato, use of these varieties have to be restricted to spring plantings when cooler soil temperatures prevail.

In addition to problems of heat instability, the continuous planting of resistant varieties result in virulent races of *Meloidogyne* capable of overcoming resistance. Therefore the duration and utility of resistance will be time-limited. In previous studies with resistant tomatoes, resistance breaking nematode races develop within 1 to 3 years. Since new races of the nematode can develop so rapidly, a system of integrated control requires rotation of resistant and non-resistant varieties.

Use of resistant root stock offers cost efficient method for controlling citrus nematode. Varietal screening studies against citrus nematode revealed that trifoliate orange (*Poncirus trifoliate*) varieties, Carrizo and Trover were highly resistant. Rangpur lime, used as root stock for Blood Red, Campbell Valencia and Hamlim scions, proved resistant to citrus nematode. Hybrid citrus root stocks developed by crossing Rangpur lime (*C. limon*) with *P. trifoliata viz.*, CRH-3, CRH-5 and CRH-41 were reported highly resistant to nematodes (Reddy *et al.*, 1987). *C. senensis* x *P. trifoliata* var. Savage citrange was rated as moderately resistant to citrus nematode.

Ramana and Mohandas (1986, 1987) screened 101 cultivars and 140 intercultivar hybrids of black pepper and 74 accessions of wild *Piper* sp. against *M. incognita*. Out of these, only one cultivar CLT-P-812 was resistant. This accession was later released as 'Pournami' for cultivation in root knot infested areas (Ravindran *et al.*, 1992).

Cultural Control

Crop Rotation

Use of non- host cover crops, in the rotation sequence, is an effective strategy for nematode control. *Indigofera hirsuta* and *Aeschynomene americana* are two leguminous cover crops suitable for managing soil populations of sting or root-knot nematode. Sorghum is a good host for sting nematode but not for root-knot. Cover crop rotations with pasture land grasses (pangola digit grass, bahia grass and bermuda grass) significantly reduce root-knot nematodes. If the crop rotation period is shortened, nematode problems may intensify accordingly. Intercropping with *Crotalaria juncea* increased the yield and promoted better growth of nematode infested banana (Naganathan *et al.*, 1988; Subramaniyan and Selvaraj, 1990). Crop rotation with sweet potato variety Sree Bhadra reduced *Meloidogyne population* in vegetables significantly (Sheela *et al.*, 2000).). Asparagus, corn, onions and garlic are good rotation crops for reducing root-knot nematode populations. Crotalaria, velvet bean and rye are resistant to root-knot nematodes (Wang *et al.*, 2004).

Antagonistic Crops

Some plants produce allelochemicals which function as nematode-antagonistic compounds, like polythienyls, glucosinolates, cyanogenic glycosides, alkaloids, lipids, terpenoids, steroids, triterpenoids, and phenolics (Halbrendt, 1986; Patel *et al.*, 2004). Compounds from plants like castor bean, chrysanthemum, velvet bean, sesame, jack bean, crotalaria etc. are exuded during green manure decomposition. Sunnhemp and sorghum-sudan are popular nematode-suppressive cover crops which produce the allelochemicals known as monocrotaline and dhurrin (Chitwood, 2002).

Marigold (*Tagetes* species) was studied for its ability to suppress nematodes with antagonistic phytochemical exudates, namely the polythienyls. It was demonstrated that rhizobacteria living in association with marigold roots were suppressive to root lesion and other nematodes. African marigold

(*Tagetes erecta*) and French marigold (*Tagetes patula*) have several nematode-suppressive varieties (Dover *et al.*, 2003).

Brassicas like rapeseed and mustard have a nematode-supressive effect that benefits the following crops in rotation. This "mustard effect" is attributed to glucosinolate compounds contained in brassica residues. Toxicity is attributed to enzymatically induced breakdown products of glucosinolates, a large class of compounds known as isothiocyanates and nitriles which suppress nematodes by interfering with their reproductive cycle (Brown and Morra, 1997).

Soil Amendments

Many types of soil amendments were tried to suppress populations of plant parasitic nematode. Animal manure, poultry litter and disk-incorporated cover crop residues are examples of soil amendments to improve soil quality as well as a means for enhancing the bio control potential of soil. Some amendments which contain chitin and inorganic fertilizers which release ammoniacal nitrogen into soil suppress nematode populations and enhance the selective growth of microbial antagonists of nematodes. Neem is useful for the suppression of nematodes in tomato (Ajith and Sheela, 1996; Kumar and Khanna, 2006).

Fallowing

Clean fallow during the off-season is probably the most important and effective cultural control measure for nematodes. When food sources are not readily available, soil population of nematodes declines. Due to the wide host range of many nematode species, weeds and crop volunteers must be controlled during the fallow period to prevent nematode reproduction and further increase in population.

Flooding

Flooding suppresses nematode population. Alternating 2 to 3 week cycles of flooding and drying is more effective than long, continuous flooding cycles. Extended periods of off-season field flooding promote the decline of soil population of nematodes.

Soil Solarization

Soil solarization is a nonchemical technique in which transparent polyethylene sheets are laid over moist soil for 6 to 12 weeks to heat up non cropped soils to temperatures lethal to nematodes and other soil-borne pathogens. Soil temperature is magnified due to trapping of incoming solar radiation under the polyethylene panels. To be effective, soils must be wetted and maintained at high soil moisture content to increase the susceptibility (thermal sensitivity) of soil borne pests. Wet mulched soils increase soil temperatures due to elimination of heat loss by evaporation and upward heat convection, in addition to the greenhouse effect by prohibiting dissipation of radiation from soil. The most successful use of soil solarization appears to occur in heavier (loamy to clay soils) rather than sandy soils. Soils with poor water holding capacity and rapid drainage can significantly inhibit heat transfer to deeper soil horizons. Loss of pest control is directly correlated with soil depth. The depth to which lethal temperature can be achieved (18 to 24 cm) is also dependent on the intensity and duration of sunlight and ambient temperature.

Effective use of solarization for nematode control has required an integrated systems approach, coupling solarization with other chemicals or nonchemical approaches. The combined use of soil solarization with nematicide could improve nematode control and crop yield.

Solarization at 44°C at 10 cm soil depth resulted 100 per cent control of *M. incognita* in tomato nursery (Herrera *et al.*, 1999). Reddy *et al.* (2001) reported that soil solarization with clear transparent sheet for six weeks during hot summer showed an increase in soil temperature (8°C) resulting in significant reduction in population density of *M. incognito* (85.80 per cent) and *Pythium aphanidermatum* (85.40 per cent) in tomato. Soil solarization using 100 guage LLDPE clear polythene sheet for 15 days was very effective in reducing the nematode population (35 per cent) in tomato nursery.

Other Cultural Practices

Other cultural measures which reduce nematode problems include rapid destruction of the infested crop root system, following harvest. Since nematodes can be carried in irrigation water that has drained from an infested field, growers should avoid use of ditch or pond waters for irrigation or spray mixtures. In most cases, a combination of these management practices will substantially reduce nematode population.

Botanicals

The extracts of leaves, stem and buds of *Datura stramonium*, *Ipomea carnea*, *Tagetes patula* and *Lawsonia alba* at 4 mg/ml dilution can cause 75-100 per cent larval mortality of *T. semipenetans* (Kumari *et al.*, 1986). Leaf extracts of *Glycosmis pentaphylla* at 1:10 dilution was highly effective against *R. similis* in banana. (Sreeja and Charles, 1998). In banana, fresh leaf extracts of *Azadirachta indica*, *Crotolaria juncea* and *Vitex negundo* (80 per cent concentration) gave 73.4, 64.8 and 76.4 per cent mortality, respectively of *Pratylenchus coffeae* in banana, 20 hours after exposure. (Sundararaju and Cannayane, 2002). There was no larvae penetration from egg masses treated with argemone and lantana leaf extracts (Patel *et al.*, 2004).

Chemical Control

Kuhn 1881 first tested CS2 to control sugar beet nematode in Germany. In USA Bessey (1911) treated CS_2 for the control of root-knot nematodes. But the method proved impractical. Later formaldehyde, cyanide and quick lime were observed to have nematicidal properties. Mathews (1919) observed the effect of chloropicrin (tear gas) against plant parasitic nematodes in England. In the 1940s, the discovery that D-D (a mixture of 1,3-dichloropropene and1,2-dichloropropane) which controlled soil populations of phytoparasitic nematodes and led to substantial increases in crop yield which provided a great impetus to the development of other nematicides. In 1944,scientists from California and Florida states of USA reported efficacy of ethylene dibromide (EDB). In the same year,Dow Chemical company, USA introduced the chemical as a soil fumigant for the management of nematodes. In the 1960s, a new generation of nematicides like carbamates and organophosphates were introduced The important nematicides used as fumigants developed are D-D, Ethylene dibromide (EDB), Methyl bromide, Chloropicrin, metam sodium, dazomet, methyl Isothiocyanate (MITC) sodium tetrathiocarbonate, and sodium tetrathiocarbonate. The nematicides coming under carbamates are aldicarb, aldoxycarb, carbofuran and oxamyl. Ethoprop, fenamiphos, cadusafos, fosthiazate, terbufosphorate,thionazin, fosthietan, and isazofos belong to organophosphates.

Non-Fumigant Nematicides

Non-fumigant nematicides must be incorporated with soil or carried by water into soil to be effective. These compounds must be uniformly applied to soil, targeting the future rooting zone of the plant, where they will contact nematodes. Placement within the top 6 to 12 cm of soil should provide a zone of protection for seed germination, transplant establishment and protect initial growth of plant roots from seeds or transplants.

Pre Plant Soil Fumigation for Nematode Control

Fumigant Nematicides

Use of broadspectrum fumigants effectively reduces nematode populations and increases vegetable crop yields, compared to non fumigant nematicides. Since these products must diffuse through soil as gases to be effective, the most effective fumigation occurs when the soil is well drained.

All the fumigants are phytotoxic and hence must be applied at least 3 weeks before planting of crops. When applications are made during periods of low soil temperature, these products remain in the soil for an extended period, thus delaying planting or possibly causing phytotoxicity to a newly planted crop. Rainfall or irrigation which saturates the soil after treatment tends to retain phytotoxic residues for longer periods, particularly in deeper soil layers.

The Soil fumigant Methyl Bromide is identified as one of the chemicals depleting the ozone layer. First brought under the Montreal Protocol by the Copenhagen Amendment of 1992, its phase out is now considered a very important step. No single substance is found that can substitute for the wide range of methyl bromide use.

Nematicides of Biochemical Origin

DiTera

The nematode-parasitic fungus *Myrothecium verrucaria* produces a mixture of compounds registered in 1996 as a biologically based nematicide named DiTera. DiTera is active against many plant-parasitic nematodes.

ClandoSan

ClandoSan is a granular product made from processed crab and crawfish exoskeletons. The material contains large amounts of chitin and urea and registered in the United States in 1998 as a nematicide. Its nematicidal activity results from the stimulation of populations of nematode-antagonistic microorganisms, particularly those which produce chitinase, a major component of nematode egg shells.

Sincocin

Sincocin is the trade name of the mixture registered in 1997 as "Plant Extract 620" with the U.S. E.P.A. It consists of a blend of extracts from the prickly pear *Opuntia lindheimeri*, the oak *Quercus falcata*, the sumac *Rhus aromatica*, and the mangrove *Rhizophorangle*. Sincocin provides control of the citrus nematode on orange roots, the reniform nematode on sunflower and the sugarbeet cyst nematode.

Biological Control

Biological control of plant parasitic nematode has become a significant and intensive area of research because of environmental concerns and non availability of effective nematicides. Use of potential biocontrol agents like bacteria and fungi for the sustainable management of plant parasitic nematodes is gaining importance. Use of *Pseudomonas fluorescens*, *Trichoderma viride* and VAM fungus are effective for the management of root knot, cyst, reniform, citrus, rice root, white tip and lesion nematodes. Further research on this line to optimize the dosage and delivery system is now intensified. Native isolates of PGPR, *P. fluorescens* and *Bacillus* spp. were identified against root knot and lesion nematodes in banana and betelvine (Channabasappa *et al.*, 1995).

Bacteria

A large number of rhizobacteria (*Agrobacterium, Alcaligens, Bacillus, Clostridium, Desulfovibrio, Serratia*) reduce nematode populations. *Pseudomonas fluorescens, P. stutzeri, Bacillus subtilis thurigiensis,*

B. lichiniformis and *P. mindocina* effectively reduce nematode populations, when used as soil amendment or seed treatment. The cell free culture filtrate of *P. fluorescens* was toxic to *H. avenae*.

Depending on the mode of action,bacterial antagonists are of two types. First group includes the bacteria which release metabolites inhibitory to nematodes (*Bacillus, Clostridium, Pseudomonas, Azotobacter etc*). The metabolite may be toxic, antibiotic or inhibitory to nematodes. The second group includes parasitic bacteria such as *Pasteuria* which also produces exudates in colonized roots. It may be lectins or siderphores, acting against nematodes.

Paseturia penetrans parasitizes about 205 nematode species. It reduces root penetration by the nematode juveniles, inhibits egg formation and nematode population in soil and hence is the most potential bacterial biocontrol agent. *M.incognita* in pepper was suppressed by *Pasteuria penetrans* (Sosamma and Koshy, 1997), *Bacillus* spp. (Sheela *et al.*, 1993) and *Pseudomonas fluorescens* (Eapen *et al.*, 1997).

Fungi
Nematode destroying fungi play an important role in regulating nematode population dynamics. Nematodes belonging to genera like *Meloidogyne, Tylenchulus* and *Rotylenchulus,* at sedentary stages of their life cycle, are vulnerable to be attacked by fungi either within the host plant roots or when females and egg masses are exposed on the root surface.

Paecilomyces lilacinus is the most promising and practicable biocontrol agent for the management of root knot and cyst nematodes. The fungus grows abundantly in sterilized paddy seeds, oilcakes and mashed potato. *Paecilomyces lilacinus* increased growth of brinjal plants and root galling. *P. lilacinus* (Ramana, 1994) and *Pochonia chlamydosporia* (*Verticillium chlamydosporium*) (Sreeja *et al.*, 1996) were found suppressing *M. incognita* in pepper. Culture filtrates of *Aspergillus sp.* reduce root galling in tomato. *Glomus mosseae* was effective in reducing burrowing nematode population in banana (Sosamma *et al.*, 1998). *P. lilacinus* reduces *R. similis* population (Parvatha Reddy and Khan, 1998). In banana, application of *P. lilacinus* (multiplied in neem cake) @ 15 or 20g/plant significantly reduced the root gall index, egg masses, eggs, females and soil population of *M. incognita* (Kofoid and White, 1919; Chitwood 1949) and improved plant growth.

Mycorrhizae show antagonistic influence on the population of plant parasitic nematodes which may be due to increased amino acid contents (especially arginine) or the presence of additional amount of phosphorus in root. VAM can markedly alter the plant response to nematodes. Nematode reproduction is also suppressed on mycorrhizal plants. *G. mosseae* and *G. intrardices* in combination with neem cake were evaluated for their comparative efficacy to colonise crossandra root and suppress *M. incognita* infection.

Predaceous fungi, *Arthrobotrys* spp., *Dactylaria* spp., and *Dactylella* spp. can reproduce on citrus nematode and control its population build up on citrus. *Paecilomyces lilacinus*, endoparasitic fungus, is an excellent biocontrol agent. This fungus can effectively manage citrus nematode population. *P. lilacinus* in combination with carbofuran effeively reduced citrus nematode population in acid lime (Reddy *et al.*, 1990). Some native isolates of *Trichoderma hazianum* and other *Trichoderma* spp. are potential antagonists of root-knot nematodes. The effectiveness of this fungus was clearly proved in laboratory, greenhouse and nurseries (Eapen *et al.*, 2000).

Actinomycetes
The avermectins are a new class of macro cyclic lactones isolated from the soil actinomycetes, *Streptomyces avermitilis.* These compounds possess insecticidal, acaricidal and nematicide properties.

Avermectins have direct action on neurotransmission. They are antagonists of gama amino butryic acid. Avermectins B1 and B2a show high toxicity against *M. incognita*.

Integrated Nematode Management (INM)

Integrated nematode management (INM) procedures are implemented based on the principles of prevention, population reduction and tolerance. INM seeks to stabilize population of target nematodes at acceptable levels resulting in favourable long term socio-economic and environmental consequences. Integration of bioagents and botanicals, integration of mycorrhizae and botanicals, integration of bioagents and chemicals, integration of mycorrhizae and chemicals and integration of bioagents are the possibilities.

Sikora (1992) reviewed a range of control measures, including crop rotation, partial soil sterilization, soil amendments and nematicides which could be combined to increase the activity of naturally occurring biological control agents. Such measures could also be used to improve the performance of agents added to soil. Partial soil sterilization by methods like solarization reduces nematode infestations and also reduces competition from the residual soil microflora, enabling the biological control agent to establish more readily. Soil amendments may also reduce nematode infestations and increase number of facultative parasites. Pre-colonized substrates are the most effective in establishing nematophagous fungi in soil. Soils which are naturally suppressive to some nematodes may be used to shorten rotations of susceptible crops and improve the performance of nematicides and resistant cultivars (Kerry, 1990).

Integration of neem cake @ 400 g/plant, carbofuran @ 20g/plant and bioagents *G. fasciculatum* @ 50g culture (containing an average of 20 J_2 larvae of *M. incognita* race-1 infested with *P. penetrans* each of which had an average of 15 spores attached to the cuticle) was the most effective in reducing *R. similis* population both in root and soil by more than 50 per cent, in banana. The combination treatment improved the growth of pseudostem, plant height, number of leaves and leaf area (Channabarappa *et al.*, 1995).

Neem cake followed by carbofuran treated plots showed minimum gall index in ginger (Mohanty *et al.*, 1992). Sheela *et al.* (1995) reported that application of neem cake @ 2.50 t ha^{-1} at the time of planting and carbofuran 1 kg a.i. ha^{-1} forty five days after planting were effective in reducing *M. incognita* population in soil and root and in increasing the yield of ginger.

Eapen and Venugopal (1995) reported that integration of soil solarization and application of *P. lilacinus* and *Trichoderma* spp. significantly suppressed the *M. incognita* population by 58.30 to 86.90 per cent, resulting in improvement in growth and quality. *G. fasciculatum* was very effective in spices like ginger, turmeric, cardamom and black pepper (Siva Prasad and Sheela, 1998).

Integration of parring and hot water treatment of banana suckers along with application of carbofuran 3 G and neem cake at the time of planting was effective against the nematodes *Radopholus similis*, *Helicotylenchus multicinctus* and *M.incognita*. It reduced the soil and root population of nematodes besides increasing growth, development and yield of banana (Shreenivasa *et al.*, 2005).

Application of *P. fluorescens* and carbofuran in combination enhanced the yield and improved the vigour of tomato plants by suppressing the root-knot nematode (Khan, 2000) and cyst nematode, *H. cajani* in soil and root system (Sujatha *et al.*, 2000).

Some of the commercial products in this field, available in the market are given in Table 16.8.

Table 16.8

Botanical Nematicides	Producers or Distributors
Beneficial Nematodes	
(*Steinernema* sp.)	Nitron Industries, Johnny's Seed, BioLogic, Hydro-Gardens
Biocontrol Bacteria	
Deny, Blue Circle (*Burkholderia cepacia*)	Stine Microbial Products
Activate (*Bacillus chitinosporus*)	Rincon Vitova
Biocontrol Fungi	
DiT'era (*Myrothecium verrucaria*)	Valent USA, Peaceful Valley, Prophyta
MeloCon, BioAct (*Paecilomyces lilacinus*)	
Chitin	
ClandoSan	Igene Biotech, ARBICO, Peaceful Valley
Shrimp Shell meal	
Botanical Nematicide	
Nemastop (Organic extracts w/Fatty acids)	Soils Technology Corp
Dragonfire (sesame oil)	Poulenger USA
Ontrol (sesame meal)	Poulenger USA
Nemagard (ground up sesame plant)	Natural Organic Products
Neem cake	Monsoon, Peaceful Valley
Armorex (sesame oil, garlic, rosemary eugenol, white pepper)	Soils Technology Corp

Entomopathogenic Nematodes

Entomopathogenic nematodes (EPNs) are soil-inhabiting, lethal insect parasitoids belonging to the phylum Nematoda, commonly called roundworms. Although many other parasitic nematodes cause diseases in plants, livestock, and humans, entomopathogenic nematodes only infect insects. They live inside the body of their host, and hence designated as *endoparasitic*. They infect many different types of soil insects, including the larval forms of butterflies, moths, beetles and flies, as well as adult crickets and grasshoppers. Entomopathogenic nematodes are found in all inhabited continents and a range of ecologically diverse habitats, from cultivated fields to deserts. The most commonly studied genera are those useful in the biological control of insect pests (Gaugler, 2006). During the last decade, there is increased interest in the augmentative biological control of insects using EPN belonging to the families, Steinernematidae and Heterorhabditidae. These nematodes are mutually associated with insect pathogenic bacteria in the genus, *Photorhabdus/Xenorhabdus*. It is a nematode bacterium complex that works together as biological control unit to kill an insect host. These nematode-bacterium associations may meet many criteria for augmentative control of insect through inundative release including broad host range, ability to kill host rapidity, long duration storage of infective juveniles etc. There is no known negative effect on the environment and is amenable for genetic selection. These nematodes are exempted from registration in many countries. The intense interest in EPN as biological control agent of insect pest has resulted in a large number of research efforts and subsequent publications.

A major difference between the steinernematids and heterorhabditids is that all but one species in the former group are amphimictic, whereas species in the later group are hermaphrodites in the first generation. Thus steinernematids require a male and female infective juvenile to invade an insect host to produce progeny, whereas heterohabitids need only one infective juvenile to penetrate into a host as the resulting hermaphrodite adult is self-fertile.

The bacteria mutually associated with EPN are *Xenorhabdus* and *Photorhabdus* which are motile, gram negative, facultative, non spore forming anaerobicrods in the family Enterobacteriaceae. In the genus *Xenorhabdus*, five species are associated with *Steinernema* whereas in the genus *Photorhabdus*, three species are associated with *Heterorhabditis*, with one species *P. lumininescens*, divided into five sub species. The subspecies of *P. luminescens* are *luminescens, laumondii, akhursti, kayaii and thraciaensis*. One species, *P. asymbiotica* is also isolated from human clinical cases and is not associated with nematodes.

Effect of two native isolates of *Rhabditis* was assessed in terms of dipteran and coleopteran pests of bittergourd *viz.*, fruit fly, *Bacterocera cucurbitae* Coq. and Epilachna beetle, *Epilachna septima* Fab. The native isolates from the rhizosphere of vegetables using wax moth, *Galleria melonella* as trap were used for assessing the mortality of maggots of *B. cucurbitae* and grubs of *E. septima* of bittergourd. Active third stage juvenile (dauer juveniles) stored for seven days in sterile tap water were used for inoculation, following standard methods. Three levels of native isolates of *Rhabditis* were inoculated (50,100 and 200 DI/maggot or grub). The mortality of *B. cucurbitae* ranged 6-47 per cent at an inoculum range of 50, 100 and 200 DI/insect larvae. At the highest dose 87 per cent mortality was recorded at 48 h after treatment and it became 100 per cent at 72h after treatment. The mortality of grubs of epilachna beetle due to infestation by isolates of *Rhabditis* ranged 67-87 per cent at 48 h after treatment. At 72 h after inoculation, all the isolates showed 100 per cent mortality. The nematode multiplication was seen in insect cadaver 72 h after treatment.

Effect of two isolates of entomopathogenic nematodes, *Heterorhabditis* sp. was assessed in terms of larval mortality of leaf eating caterpillars of vegetables *viz.*, *Sylepta derogata* L. (Pyralidae:Lepidoptera), *Diaphania indica* Saund.(Pyralidae: Lepidoptera) and *Spilosoma oblique* W. (Arctiidae: Lepidoptera). All the three pests showed high mortality after 48 h of inoculation (40-100 per cent). Mortality of the above three larvae were comparatively higher in *Heterorhabditis* isolate at 24 and 48 h after inoculation while the effect of *Heterorhabditis* isolate was superior in all the above pests at 72h after inoculation. The mortality at 24, 48 and 72 h varied depending upon the isolate and insect pest, revealing the potential of nematodes as biological agents against lepidopteran pests. The efficiency of *Rhabitis* in managing pests of bittergourd was also reported (Sheela *et al.*, 2002). *S. carpocapsae* is suitable as a biological control agent for both *H. undalis* and *C. binotalis*. Steinernematids and Heterorhabditids are observed to infect over 200 species of insects from several orders. *S.carpocapsae* infected *S. liture* in both in lab and field conditions. *S. abbasi, S. carpocapsae, S. glaseri* and *H. indica* were infective to *A.ipsilon*. *H.indica* was effective against *H.armigera* in field condition (Gulsar Banu and Sheela, 2005).

Nematology vs Biotechnology

Biotechnology offers sustainable solutions to the problem of controlling plant parasitic nematodes. Several approaches are possible for developing transgenic plants with improved resistance; these include strategies against invasion and migration and against nematode feeding and development. Additive effect for durable resistance against nematodes can be provided by more than one independent basis for transgenic resistance. The approaches involve a cysteine proteinase inhibitor (a cystatin), a potato tuber serine/aspartic proteinase inhibitor and a repellent peptide. Cysteine proteinases are used by a wide range of plant parasitic nematodes to digest dietary protein. The cystatin prevents this digestion and slows nematode growth. Transgenic expression of both proteinase inhibitors provide effective control of both cyst and root-knot nematodes and cystatin shows to be effective against *Radopholus*. Cysteine proteinases are not present in mammals and those which are consumed lack toxicity or allergenicity for humans. They occur in common foods and are rapidly digested. Another approach is the use of a repellent. This is also not lethal to nematodes or other organisms. Nematodes do not invade roots applied with repellents because they fail to detect the host's presence. This approach is effective against a wide range of nematode species. A novel method, using RNA interference (RNAi) for functional analysis of plant parasitic nematode genes was first established in the University of Leeds. RNA interference (RNAi) was recently demonstrated in plant parasitic nematodes. It is a potentially powerful investigative tool for the genome-wide identification of gene function that should improve our understanding of plant parasitic nematodes. RNAi should help identify gene and, hence, protein targets for nematode control strategies. Prospects for novel resistance depend on the plant generating an effective form of double-stranded RNA in the absence of an endogenous target gene without detriment to itself. These RNA molecules must then become available to the nematode and be capable of ingestion via its feeding tube. If these requirements can be met, crop resistance could be achieved by a plant delivering a dsRNA that targets a nematode gene and induces a lethal or highly damaging RNAi effect on the parasite. The approach relies on the production of double-stranded RNA molecules by banana cells. When they are ingested by the nematode, they specifically interfere with the expression of the essential nematode gene they target. The advantage of the RNAi approach is that no novel protein production is required to achieve resistance to nematodes. This offers a considerable biosafety advantage, given that RNA molecules represent no food risk and there is little likelihood of non-target effects. The challenge is to provide an effective level of resistance to all banana nematodes by this approach. Genetic transformation of plantain using these approaches is in progress at International Institute of Tropical Agriculture.

Resistance against the aphid *Macrosiphum euphorbiae* previously was observed in tomato and attributed to a novel gene, designated *Meu-1*, tightly linked to the nematode resistance gene, *Mi*. Recent cloning of *Mi* made it possible to conclude that *Mi* and *Meu-1* are the same gene and that *Mi* mediates resistance against both aphids and nematodes, organisms belonging to different phyla. *Mi* is the first example of a plant resistance gene active against two such distantly related organisms. Furthermore, it is the first isolate-specific insect resistance gene to be cloned and belongs to the nucleotide-binding, leucine-rich repeat family of resistance genes.

Newer Molecules for Nematode Management

The avermectins, streptomycete-derived macrocyclic lactones originally isolated as antiparasitic agents, demonstrated high potencies in laboratory evaluations against the plant-parasitic nematode, *Meloidogyne*. Two new azaphilone metabolites, pseudohalonectrin A and B, were isolated from the culture of the aquatic fungus *Pseudohalonectria adversaria* YMF1.01019, originally separated from submerged wood in Yunnan Province, China. Pseudohalonectrin A and B were assessed for their nematicidal activity against the pine wood nematode *Bursaphelenchus xylophilus* and their structures were defined after spectral analysis. This is the first report of secondary metabolites from any member of the genus *Pseudohalonectria*. The search for new producers of biologically active compounds is underway among fungi growing under extreme conditions (Dong *et al.*, 2005).

Conclusions

Nematodes cause substantial crop loss in different crops world wide. The damages caused by nematodes are often overlooked due to the hidden nature of attack. Often the symptoms of nematode attack are confused with nutritional and water related disorders. Research conducted so far could bring forth effective management strategies against major nematode pests. Since many botanicals are effective under *in vitro* and *in vivo* conditions, the active principles have to be identified, synthesied and commericialised for eco-friendly nematode management. Research on biocontrol should be intensified to release commercial formulation of the promising biopesticides. More emphasis has to be on nematodes as a community in different cropping systems and in different agro climatic regions. Biotechnolgical tools should be utilized for unravelling the fundamental mechanisms behind nematode-plant interactions. Along with the study of plant parasitic nematodes, investigations on other groups of nematodes should be given priority and their role in soil health assessed.

References

Ali, S.S., 1986a. Occurrence of root–knot nematodes in cardamom plantations of Karnataka. *Indian Journal of Nematology* 16: 269–270

Ali, S.S., 1987. Effect of three systemic nematicides against root–knot nematodes in cardamom nursery. *Nematologia mediterranea*, 15: 155–158

Barbetti, M.J., Riley ,I.T. You, M.P Hua li, and Sivasithamparam., 2006. The association of necrotrophic fungal pathogens and plant parasitic nematodes with the loss of productivity of annual medic–based pastures in Australia and options for their management *Australasian Plant Pathology* 35: 691

Bertrand, B., Nunes, C and Sarah, J.L., 2000. Disease complex in coffee involving *Meloidogyne arabicida* and *Fusarium oxysporum. Plant Pathology* 49: 383–388

Back, M.A., Haydock, P.P.J, and Jenkinson, P., 2003. Disease complexes involving plant parasitic nematodes and soil borne pathogens *Plant Pathology* 51: 683–697

Channabasappa, B.S, Krishnappa, K and Reddy, B.M.R., 1995. Utilization of eco–friendly biological agents and bio components in the integrated management of *Radopholous similis* on banana. Paper presented in the National Symposium on Nematode Management with Eco–friendly Approaches and Bio components, New Delhi, Abstr. Pp. 42

Charles, J.S.K. and Venkitesan, T.S., 1993. Pathogenicity of *Heterodera oryzicola* (Nemata: Tylenchina) towards banana (*Musa AAB cv. Nendran*). *Fundamentals of Applied Nematology* 1: 359 385

Charles, J.S.K, Venkitesan, T.S. and Thomas, Y., 1985. Comparative efficiency of antagonistic intercrops with carbofuran in control of burrowing nematodes in banana cultivation. *Indian Journal of Nematology* 15: 241–242

Devarajan, K and Rajendran, G., 2001. Effect of the fungus *Paecilomyces lilacinus* (Thom.) Samson on the burrowing nematode. *Radopholus similis* (Cobb) Throne in banana. *Pest Management in Horticultural Ecosystems* 7: 171–173

Devarajan, K and Rajendran, G., 2002. Effect of fungal egg parasite, *Paecilomyces lilacinus* (Thom.) Samson n *Meloidogyne incognita* in banana. *Indian Journal of Nematology* 32: 78–101

Eapen, S.J., 1993. Seasonal variations of root–knot nematode population in a cardamom plantation. *Indian Journal* of Nematology 23: 63–8

Eapen, S.J., Beena, B. and Ramana, K.V., 2005. Tropical soil microflora of spice–based cropping systems as potential antagonists of root–knot nematodes. *Journal of Invertebrate Pathology* (3): 218–225

Eapen, S.J., Ramana, K.V. and Sarma, Y.R., 1997. Evaluation of *Pseudomonas fluorescens* isolates for control of *Meloidogyne incognita* in black pepper (*Piper nigrum* L). In: Edison, S. Raman, K.V. Sasikumar, B. Nirmal Babu, K. and Eapen, S.J. (Eds). Biotechnology of spices, medicinal and sromatic plants. *Proceeding of the National Seminar on Biotechnology of Spices and Aromatic plants, Indian Society for Spices,* Calicut, India. Pp. 129–133

Eapen, S.J. and Venugopal, M.N., 1995. Field evaluation of *Trichoderma* spp. and *Paecilomyces lilacinus* for control of root–knot nematodes and fungal diseases in cardamom nurseries. *Indian Journal of Nematology* 25: 15–16

Gopal Swarup, G., Dasguptha, D.R. and Gill, J.S., 1995. Nematode Pest Management– An Appraisal of Eco–friendly Approaches. Nematological Society of India. Pp. 300

Halbrendt, J.M., 1986. Allelopathy in the Management of Plant–Parasitic Nematodes. *Journal of Nematology* 28: 8–14

Jonathan, E.I and Rajendran, G., 2000. Biocontrol potential of the parasitic fungus *Paecilomyces lilacinus* against the root–knot nematode, *M. incognita* in banana. *Journal of Biological Control* 14: 67–69

Jinyan Dog1, Yongping Zhou1, Ru Li1, Wei Zhou1, Lei Li1, Yanhui Zhu1, Rong Huang and Keqin Zhang., 2006. New nematicidal azaphilones from the aquatic fungus. *Pseudohalonectria adversaria* YMF1.01019FEMS *Microbiology Letters* 264: 65–69

Krishnappa, K., Reddy, B., M.R, Shreenivasa, K.R, Ravichandra, N.G., Karuna, K. and Kantharaju, V., 2005. Integrated Management of nematode complex on banana. *Indian Journal of Nematology* 35: 37–40

Koshy, P.K. and Sosamma, V.K., 1979. Control of the burrowing nematode *Radopholous similis* on coconut seedlings with DBCP. *Indian J. Nematology,* 12: 200–203. Kuriyan, K.J. and Sheela, M.S., 191. Integrated control of root–knot nematode *Meloidogyne incognita* in brinjal. *Indian J. Nematol.* 11: 129

Koshy, P.K. and Sosamma, V.K., 1987. Pathogenicity of *Radopholus simils* on coconut (*Cocos nucifera* L.) seedlings under green house and field conditions. *Indian Journal of Nematology* 17: 108–118

Koshy, P.K. and Sosamma, V.K., 1996. Effect of *Radopholus similis* on growth, flowering and yield of coconut. *Journal of Plantation Crops* 24: 157–165

Kumar, S. and Khanna, A.S., 2006. Role of *Trichoderma harzianum* and neem cake separately and in combination against root–knot nematode on tomato. *Indian Journal of Nematology*: 36: 2

Luc, M., Sikora, R.K. and Bridge, J., 1990. *Plant Parasitic Nematode in Subtropical and Tropical Agriculture.* CAB International Institute of Parasitology, UK. Pp. 462

Luc, M., *Sikora,* R. A. and Bridge, J., 2005. Plant Parasitic Nematodes in Subtropical and Tropical Agriculture. 2nd Edn. CABI Publishing, UK. Pp. 896

Maggenti, A. R., 1991. Nemata: higher classification. (In): W. R. Nickle (Ed.). *Manual of Agricultural Nematology.* Marcel Dekker, Inc. New York. Pp. 147–187

Mohandas, C. and Palaniswami, M.S., 1990. Resistance in sweet potato (*Ipomoea batatas* L.) to *Meloidgyyne incognita* (Kofoid and White) Chitwood in India. Journal of Root Crops 148–149

Mohandas, C., Kumar, P.S.A., Sreeja, P. and Nagaeswari, S., 1998. Pre and post infectional resistance in sweet potato germplasm to the root–knot nematode, *Meloidgyyne incognita* (Kofoid and White, 1919). Chitwood (Abstract. Third International Symposium of Afro–Asian Society of Nematologists, 16–19 April, 1998 Coimbatore)

Naganathan, T.G., Arumagam, R., Kulasekaran, M. and Vadivelu, S., 1988. Effect of antagonistic crops as intercrops on the control of banana nematodes. *South Indian Horticulture* 36: 268–269

Nisha, M.S. and Sheela, M.S., 2002. Effect of green leaf mulching for the management of root–knot nematode in Kacholam. *Indian Journal of Nematology* 2: 211

Patel, A.D, Patel, D.J and Patel, N.B., 2004. Effect of aqueous leaf extracts of botanicals on egg hatching and larval penetration of *Meloidogyne incognita* in banana. *Indian Journal of Nematology* 34: 33–36

Ramana, K.V., 1994. Efficacy of *Paecilomyces lilacinus* (Thom.) Samson in suppressing nematode infestations in black pepper (*Piper nigrum L.*). *Journal of Spices and Aromatic Crops* 3: 130–34

Rao, P.S., Krishna. M.R., Srinivas, C., Meenakumari, K. and Rao, A.M., 1994. Short duration, disease-resistant turmerics for northern Telangana. *Indian Horticulture* 39: 55–56

Ramana, K.V., 1994. Efficacy of *Paecilomycs lilacinus* (Thom.) Samson in suppressing nematode infestation in black pepper (*Piper nigrum L.*) *Journal of Spices and Aromatic Crops* 3: 130–134

Russell E. Ingham, J. A. Trofymow, Elaine R. Ingham and David C. Coleman., 1985. Interactions of Bacteria, Fungi, and their Nematode Grazers: Effects on Nutrient Cycling and Plant Growth. *Ecological Monographs* 55: 119–140

Ravi, K, Nanjegowda, Dand Reddy, P.P., 2000. Integrated management of the burrowing nematode,*Radopholous similis* (wbb,1893) Thorne, 1949 on banana. *Pest Management in Horticultural Ecosystems* 6: 124–129

Reddy, P., Nagesh, M., Rao, M.S. and Devappa, V., 1998. Integrated management of the burrowing nematode *Radopholous similis*, using endomycorhiza, *Glomus mosseae* and oil cakes. *Pest Management in Horticultural Ecosystem* 4: 25–29

Siddiqi, M.R., 1986. *Tylenchida: Parasites of Plants and Insects*. Commonwealth Institute of Parasitology, UK. p. 645

Sheela, M.S., Jiji, T. and Nisha, M.S., 2002. Evaluation of different control strategies for the management of nematodes associated with vegetables (brinjal). Proceedings on International Conference on Vegetables. P. 268–269

Sheela, M.S., Bai, H., Jiji, T. and Kuriyan, K.J., 1995. Nematodes associated with ginger rhzosphere and their management in Kerala. *Pest Management in Horticultural Ecosystems* 1: 43–48

Sheela, M.S. Jiji, T. Nisha, M.S.and Joseph Rajkumar, 2005. A new record of Meloidogyne graminicola on *rice*, *Oryza sativa* in Kerala. *Indian Journal of Nematology* 35: 2

Sheela, M.S. and Venkitesan, T.S., 1994. Effect of Pesticides on the nematode pathogenic bacteria, *Bacillus macerans in vitro*. *Journal of Biological Control* 8: 102–104

Sosamma, V.K. and Koshy, P.K., 1997. Biological control *of Meloidogyne incognita* on black pepper by *Pasteuria penetrans* and *Paecilomyces lilacinus Journal of Plantation Crops* 25: 72–76

Sreeja, P. and Charles, J.S.K., 1998. Screening of botanicals against the burrowing nematode, *Radopholous similis* (wbb.1983) Throne. Pest Management in Horticultural Ecosystems 4: 36–39

Sundararaju, P. and Cannayane, I., 2002. Antinemic activity of plant extracts against *Pratylenchus coffeae* infecting banana. *Indian Journal of Nematology* 32: 121–124

Sundararaju, P and Sudha, S., 1998. Effect of neem oil cake and nematicide for the control of burrowing nematode, *Radopholous similis* in the arecanut based on cropping system. *Proceedings of the Third Internal Symp.Afro–Asian Society of Nematologists*, Coimbatore 16: 4–7

Stirling, G.R. and Nikulin, A., 1998. Crop rotation, organic amendments and nematicides for control of root–knot nematodes (*Meloidgyne incognita*) on ginger. *Australiasian Plant Pathololgy* 27: 234–243

Thomas, G.V., Sundararaju, P., Ali, S.S. and Ghai, S.K., 1989. Individual and interactive effects of VA mycorrhizal fungi and root–knot nematodes, *Meloidogyne incognita* on cardamom. *Tropical Agriculture* 66: 21–24

Vadhera, I., Tiwari, S.P. and Dave, G.S., 1998b. Integrated management of root–knot nematode, *Meloidogyne incognita* in ginger. *Indian Phytopathol*ogy 51: 161–163

Venkitesan, T.S. and Jacob, A., 1985. Integrated control of root–knot nematode, infecting pepper vines in Kerala. *Indian Journal of Nematology* 15: 261–262

Venkitesan, T.S. and Setty, K.G.H., 1978. Reaction of 27 black pepper cultivars and wild forms to the burrowing nematode, *Radopholous similis* (Cobb) thorne. *J ournal of Plantation Crops* 6: 81–83.

Appendix 1

For Further Reading

Nematode Taxonomy

1. Goodey, J.B. 1963. Soil and fresh water nematodes. John Wiley and Sons, IWC. New York, U.S; Pp. 544

2. Jairajpuri, M.S. and Khan, W.U. 1982. Predatory nematodes (Mononchida) with special reference to India. Associated Publication, New Delhi. Pp.131

3. Zuckermen, G.M., Lod, R.A. and Mai, W.F. 1980. Plant parasitic nematodes. Vol.1. Academic Press.

4. Mohammed Rafirq, Siddiquie. 1993. Tylenchida parasites of plants and animals. CAB International

5. Nickele, W.R. 1991. *Manual of Agricultural Nematology*. Marshel Dekker INC. New York. p. 1034

6. Mayer, E. 1969. *Principles of Systematic Zoology*. Macro-Hill Book Company, New York. Pp.428

7. Mayer, E. 1982. *The Growth of Biological Thoughts: Diversity-Evolution and inheritance*. The Belknepe press of Harvard University, Cambridge, Mass. U.S.A. Pp.974

8. Philip, J.Y.N., and Schunran Steakonon, J.H. 1941. Manual of Agricultural Helminthology. E.J. Bril., Holland.

9. Platcher, U.G. 1981. Potential use of protein pattern and DNA nucleotypes sequences in nematode taxonomy In: *Plant Parasitic Nematodes*. Vol.III (Ed.) Zuckermen and Rhode. R.A. Academic press, Newyork. Pp.580

10. Sokkal, R.R. and Sneath. P.H.A. 1963. *Principles of Nematode Taxonomy: International Code of Zological Nomenclature*. W.H. Kreemen, London

11. Sasser, J.N. and Carter, C.C. 1985. *An Advance Treatise on Meloidogyne Vol. I Biology and Control*. North Carolina State University, Releigh. Pp. 422

12. Stone, A.R., Platt, H.M. and Khalil, L.F. 1983. *Concepts in Nematode Systematics*. Academic Press, New York, Pp. 388

13. Geratt, R.P. 1978, 1979. On growth and form of nematode. Nematologica 24: 171-151. Nematologica 24: 344-360; Nematologica 25:1-25.

14. Hussay, R.S. 1979. Biochemical systemetics of nematodes. A review; Helminthological Abstract Series B, *Plant Nematology* 48; 141-148

Nematode Ecology

1. Andrewartha, H.G. 1971. Introduction to the study of animal population.

2. Andrewarthe, H.G. and Birch., L.C. 1954. The distribution and abundance of animals.

3. Clegg, M.T. and Epperson. 1985. Advances in genetics.

4. Chitwood, B.G. and Chitwood. M.R. 1950. An introduction to Nematology.

5. Norton, D.C. 1978. Ecology of plant parasitic nematodes.

6. Pinaka, E.R. 1975. Evolutionary Ecology.

7. Rollinson, D. and Anderson, R.M. 1985. Ecology and genetics of host-parasitic interactions.

8. Ricklef, R.E. 1990. Ecology – 3rd Edition.

9. Wallace, H.R. 1963. The biology of plant parasitic nematodes.

10. Zuckermann, B.M., Mai, W.F. and Rohde, R.A. 1971. Plant parasitic nematodes.

11. Freekman and Edward, P, Caswell. 1985. The Ecology of nematodes in Agro ecosystems. Ann. Rev. Phyto. Path. 23. P.275.

Entomophilic Nematodes

1. Nickle, R.W. 1984. *Plant and Insect Nematodes*. Marcel Dekker, Inc. New York. Pp.925

2. Poinor, J.G. 1979. *Nematodes for Biological Control of Insects*. CRC Press

Nematode Management

1. Mrinal, K. Dasgupta. 1996. *Phytonematology*. Naya Prakash, Calcutta. Pp.846

2. Nickle, R.W. 1984. *Plant and Insect Nematodes*. Marcel Dekker, Inc. New York. Pp.925

3. Luc, M., Richard A. Sikora and Bridge. 1990. Plant Parasitic Nematodes in Sub tropical and Tropical Agriculture. CAB International Institute of Parasitology, UK Pp.629

4. Evans, K., Trudgill, D.L. and Webster. J.M. 1991. Plant Parasitic Nematodes in Temperate Agriculture. CAB International, Walling Fork, UK. Pp.647

5. Bridge, J. Nematode management in sustainable and subsistence agriculture. *Annual Review of Phytopathology* 34: 201-226

6. Mcsorley, R. and Duncan, L.W. 1995. Economic thresholds and nematode management. *Advances in Plant Pathology* 11: 147

7. Larry W. Duncan. 1991. Current options for nematode management. *Annual Review of Phytopathology* 29: 469

Nematological Techniques

1. Bird, A.F. 1971. The Structure of Nematodes.

2. Bird, A.F. and Bird, J. 1991. The Structure of Nematodes.

3. Nickle, W.R. 1991. Manual of Agricultural Nematology.

4. Zuckerman, B.M., Mai, W.F. and Rohde, R.A. (eds.) 1971. Plant parasitic nematodes.

5. Southey, J.F. 1959. Plant Nematology.

Nematode Association and Interaction with Microorganisms

1. Powell, N.T. 1971. Interaction of plant parasitic nematodes with other disease causing agents. In: B.M. Zuckerman, W.F. Mai. and R.A. Rohde (Eds.). *Plant Parasitic Nematodes* Vol.II. Academic Press. Pp.347

2. Wajidkhan, M. 1993. *Nematode Interactions*. Chapman and Hall. Pp. 377

3. Brown, D.J.F., Robertson, W.M. and Trudgill, D.L. 1995. Transmission of viruses by plant nematodes. *Annual Review of Phytopathology* 33:223-250

4. Powell, N.T. 1971. Interactions between nematodes and fungi in disease complex. *Annual Review of Phytopathology* 9: 253-274

5. Mai, W.F. and Abawi, G.S. 1987. Interactions among root-knot nematodes and Fusarium wilt fungi on host plants. *Annual Review of Phytopathology* 25: 317

Biological Control of Plant Parasitic Nematodes

1. Sikora, R.A. 1992. Management of the antagonistic potential in agricultural ecosystems for the biological control of plant-parasitic nematodes. *Annual Review of Phytopathology* 30: 245-270

2. Jatala, 1986. Biological control of plant parasitic nematodes. *Annual Review of Phytopathology* 24: 453-489

3. Singh, S.B. and Hussaini,S. 1998. *Biological suppression of plant diseases, phytoparasitic nematodes and weeds*. Project Directorate of Biological Control, Bangalore.

Host-Parasite Relationships of Nematodes

1. Zuckerman, Mai and Rohde. 1971. *Plant-parasitic Nematodes* Vol.I. Academic Press. New York, USA.

2. Zuckerman, Mai and Rohde. 1980. *Plant-parasitic Nematodes* Vol.II. Academic Press, New York, USA.

3. Wallace, H.R. 1963. *Biology of Plant Parasitic Nematodes*. Arnold, London.

4. Webster, J.M. 1972. *Economic Nematology*, Academic Press, New York.

5. Nickle, W.R. 1984. *Plant and Insect Nematodes*. Marcel Decker Inc. New York, USA. Pp. 925

6. Trivedi, P.C. 2003. *Advances in Nematology*. Scientific Publishers. Pp. 317

Appendix 2

Nematological Societies and Federations

- ☆ Afro-Asian Society of Nematologists (AASN)
- ☆ Australian Association of Nematologists (AAN)
- ☆ Brazilian Nematological Society (Sociedade Brasileira de Nematologia) (SBN)
- ☆ Chinese Society of Plant Nematologists (CSPN)
- ☆ Egyptian Society of Agricultural Nematology (ESAN)
- ☆ European Society of Nematologists (http://www.esn-online.org/)
- ☆ International Federation of Nematology Societies (IFNS)
- ☆ Italian Society of Nematologists (Societa Italiana di Nematologia) (SIN) Japanese Nematological Society (JNS)
- ☆ Nematological Society of India (NSI)
- ☆ Nematological Society of Southern Africa (Nematologiese Vereniging van Suidelike Afrika) (NSSA)
- ☆ Organization of Nematologists of Tropical America (ONTA)
- ☆ Pakistan Society of Nematologists (PSN)
- ☆ Russian Society of Nematologists (RSN)
- ☆ Society of Nematologists (SON)

Nematology Journals and Other Publications (Publishing Society or Organization)

- ☆ International Journal of Nematology (ASSN)
- ☆ Nematologia Brasileira (SBN)
- ☆ Journal of Nematology (SON)
- ☆ http://en.wikipedia.org/wiki/Nematology
- ☆ Nematology Newsletter (SON)
- ☆ The Egyptian Journal of Agronematology (ESAN)
- ☆ Egyptian Society of Agricultural Nematology Newsletter (ESAN)
- ☆ Nematology News (ESN)
- ☆ Nematologia Mediterranea (Istituto di Nematologia Agraria of the C.N.R.)
- ☆ Japanese Journal of Nematology (JSN)
- ☆ Indian Journal of Nematology (NSI)
- ☆ African Plant Protection (NSSA)
- ☆ Nematropica (ONTA)
- ☆ Organization of Nematologists of Tropical America Newsletter (ONTA)
- ☆ Pakistan Journal of Nematology (PSN)
- ☆ Pakistan Society of Nematologists Newsletter (PSN)
- ☆ Russian Journal of Nematology (RSN)
- ☆ Nematologica (Brill Academic Publishers)

2012, Nutri-Horticulture
Editor: Professor K.V. Peter
Published by: DAYA PUBLISHING HOUSE, NEW DELHI

Pages 309–321

Chapter 17

Marketing Management

K.R. Ashok* and K. Mani

Department of Agricultural Economics,
Tamil Nadu Agricultural University, Coimbatore – 641 003, T.N., INDIA
*E-mail: ashok10tnau@yahoo.com

Marketing is a set of business activities which facilitate movement of goods and services from producer to consumer. It is an ongoing process of discovering and translating consumer needs into products and services, creating demands for them. Customer's needs and wants are very important aspects of today's marketing. Customer focus is the very essence of marketing.

The American Marketing Association (AMA) defined, "Marketing is an organizational function and a set of processes for creating, communicating, and delivering value to customers and for managing customer relationships in ways that benefit the organization and its stakeholders (AMA 2004)"The AMA's new definition of marketing is: "Marketing is the activity involving set of institutions and processes for creating, communicating, delivering and exchanging offerings that have value for customers, clients, partners and society at large(AMA 2008)"

Unit I

Marketing Management

Marketing Management refers to all the activities which the marketing managers, executives and personnels have to undertake to carry out the marketing function of the firm. It involves (i) analyzing the market opportunities by undertaking consumer needs and changes taking place in the marketing environment, (ii) planning the marketing activities, and (iii) implementing marketing plans and settings control mechanisms to ensure smooth and successful accomplishment of the organizations goals.

Marketing Environment

A firm's performance is affected by a number of variables, of which some are controllable and some are uncontrollable. Controllable variables are internal variables and uncontrollable variables are external variables. The 4P's of marketing, *i.e.*, the marketing mix, are controlled by the marketing manager or internal variables. There are variables, however, which are uncontrollable and managers have to take them as given. This external environment influences company's strategies in two levels *i.e.* external macro environment and external micro environment. The macro environment involves political and legal, economic and natural, social and cultural and technological. The micro environment consists of supply chain, customer and competitor. These factors are uncontrollable by the organization.

A good marketer scans these various environments for threats and/or opportunities. Developments in internal and external environment can have an adverse effect on an organization if no action is taken. Consider the effect of electronic commerce on music industry and effect of e- ticketing on travel agents.

Marketing Mix

The *marketing mix* is generally accepted as the use and specification of the 'four Ps' describing the strategic position of a product in the marketplace. The 'marketing mix' is a set of controllable, tactical marketing tools which work together to achieve company's objectives. Elements of the marketing mix are often referred to as 'the four Ps':

☆ *Product* - A tangible object or an intangible service that is mass produced or manufactured on a large scale with a specific volume of units. Intangible products are often service based like the tourism industry, the hotel industry, etc.

☆ *Price* – The price is the amount a customer pays for the product. It is determined by a number of factors including market share, competition, material costs, product identity and the customer's perceived value of the product.

☆ *Place* – Place represents the location where a product can be purchased. It is often referred to as the distribution channel. It can include any physical store as well as virtual stores on the Internet.

☆ *Promotion* – Promotion represents all of the communications which a marketer may use in the marketplace. Promotion has four distinct elements - advertising, public relations, word of mouth and point of sale. **Advertising** covers any communication that is paid for, which includes cinema commercials, radio and internet, print media and billboards. In **Public relations,** communication is not directly paid for and includes press releases, sponsorship deals, exhibitions, conferences, seminars or trade fairs and events. **Word of mouth** is any apparently informal communication about the product by ordinary individuals, satisfied customers or people specifically engaged to create word of mouth momentum. Sales staff often plays an important role in word of mouth and Public Relations.

Broadly defined, optimizing the marketing mix is the primary responsibility of marketing. By offering the product with the right combination of the four Ps marketers can improve their results and marketing effectiveness.

Extended Marketing Mix

There are attempts to develop an 'extended marketing mix' to better accommodate specific aspects of marketing. For example, in the 1970s, **Nickels and Jolson** suggested the inclusion of **packaging**. In the 1980s **Kotler** proposed **public opinion** and **political power. Booms and Bitner** included three additional 'Ps' to accommodate trends towards a service or knowledge based economy like people, process and physical evidence.

Market Segmentation

Undifferentiated marketing is using one marketing strategy to go after an entire market, *i.e.*, mass marketing. This does not usually work too well since it is unlikely that every individual will have the same needs. It means market is not monolithic, *i.e.*, not everyone has the same needs. Contrast to mass marketing, the new trend is micro marketing at four levels namely segments, niches, local areas and individuals.

A market segment consists of a group of consumers who share a similar set of needs and wants. The shampoo preferred by a worker is probably quite different from the shampoo preferred by a young lady. A niche is more narrowly defined customer group and a niche market is usually identified by dividing a segment into sub segments. In local marketing, programs are tailored to the needs of local customer groups and individual stores. The ultimate segmentation leads to individual marketing.

Some approaches to segment the consumer markets are:

1. *Geographic segmentation* – by country, by region within the country, by state, by rural or urban. For example a firm selling bicycles might sell one bicycle for rural areas (mountain bike) and another kind of bike for urban areas. Most firms recognize that they have to use segmentation when marketing their products to foreign countries.

2. *Demographic segmentation* – demographics refers to characteristics of people such as sex, age, ethnicity, income, family size, occupation, education, marital status, social class, and stage of the family life cycle. For example **Sex**: shampoos, hairsprays, deodorants, **Age**: cereals, vitamins, and games for young people and older people. **Income**: luxury cars for high-income individuals; restaurants which cater to high-income and low-income people.

3. *Lifestyle segmentation*, which is sometimes called **psychographics**. Lifestyle is thus a person's way of living. Knowing the psychographics of one's customers enables a firm to create the correct advertising message.

Market Targeting

After identifying market segments, a company should decide on market targeting. Market targeting refers to the evaluation of each segment to determine to which segment the firm should enter. The potential segment is evaluated in terms of size of segments, potential for growth, profitability, scale economies and risk. After evaluating different segments a firm can consider market targeting such as selective specialization, single segment concentration, product specialization and market specialization.

Market Positioning

Market positioning means how a firm wants its brand to be perceived in relation to other brands in the market. This involves creating a distinct image for the brand relative to other brands. For example, if a firm wants to introduce a health drink, it can position it as a good diet or low fat or good taste, low price etc

Marketing Information System

A Marketing Information System(MIS) is a set of procedures to collect, analyze and distribute accurate, prompt and appropriate information to different levels of marketing decision makers. Philip Kotler defines MIS as "a system that consists of people, equipment and procedures to gather, sort, analyze, evaluate and distribute needed, timely and accurate information to marketing decision makers. A good market information system has the following characteristics.

☆ It is a planned system developed to facilitate smooth and continuous flow of information.

☆ It provides relevant information, collected from internal and external sources.

☆ It provides right information at the right time to the right person.

Marketing Organization and Control

The modern marketing department has evolved to the years from a simple sales department to an organizational structure where marketing personnel work mainly on cross disciplinary teams.

Modern marketing departments can be organized in a number of ways. Some companies are organized by functional specialization, while others focus on geography and regionalization. Still others emphasize product and brand management or market segment management. Some companies established a matrix organization consisting of both product and market managers. Finally, some companies have strong corporate marketing, others have limited corporate marketing, and still others place marketing only in the divisions.

A fact of modern marketing organizations is marked by a strong cooperation and customer focus among the company's departments like Marketing, Research and Development, Engineering, Purchasing the Line Manufacturing, Operations and Finance and accounting

Companies must practice social responsibility through their legal, ethical, and social words and actions. Social marketing is done by a non-profit or government organization to directly address essential problem or cause.

The marketing department has to monitor and control marketing activities continuously. Efficiency control focuses on finding ways to increase the efficiency of the sales force, average rising, sales promotion, and distribution. Strategic control entails a periodic reassessment of the company and its strategic approach to the marketplace using the tools of the marketing effectiveness and marketing excellence reviews, as well as the marketing audit.

Unit II

Marketing Potential

Market potential is the maximum amount of sales available to a firm during a given period, under a given level of marketing effort and market environment. Or market potential is the limit approached by the market demand as industry marketing expenditure approach infinity for a given market environment.

Kotler and Keller define potential market as the set of consumers who profess a sufficient level of interest in a market offer. However consumer interest is not enough to define a market unless they also have sufficient income and access to the product. The available market is the set of consumers who have interest, income and access, to a particular offer. The target market is part of the available market in which a company decides to concentrate its marketing efforts while penetrated market is the set of consumers who are already buying the company's product.

Forecasting

Forecasting is the science and art of anticipating what buyers are likely to do under a given set of conditions. Forecasting is important since it provides a basis for scheduling production, determining personnel needs, determining plant and equipment needs, establishing sales quotas for salespeople, making pricing decisions, scheduling purchases of raw materials, etc. A company should be making short-run forecasts (3 to 12 months) as well as long-run forecasts (1 to 5 years and even beyond). There are many methods used in sales forecasting. Some are judgmental, others involve surveys and statistical techniques. Judgmental Methods used in sales forecasting are:

Expert Opinion

In this method, a panel of experts from various fields in marketing such as dealers, distributors, marketing consultants and trade associations make forecasts. The advantage of this method is that it is quick and inexpensive. However, the experts need to be familiar with the business under study.

Composite of Sales Force Opinion

In this method, sales force provides forecasts for their territories. The belief is that salespeople talk to their customers and should have some ideas what the future will be liked. Such forecasts sometimes need adjustments as the sales force tends to overestimate or under estimate the demand based on their outlook.

Quantitative Methods Used in Sales Forecasting

Survey

In this method, a survey is conducted with a representative sample of customers on the likely future demand.

Extrapolation

In this method past data are used to project the future trend. The implicit assumption is that what happened in the past will continue to happen in the future. The technique that is usually used to draw a trend line is regression.

Time Series Analysis

This statistical technique assumes that time is the major predictor variable, *i.e.*, that sales are a function of time. Time series data are usually decomposed into Trend, Cyclical variation, Seasonal variation, and Random variation. By breaking down the time series, it makes it possible to forecast sales.

Classification of Products

Products are generally classified into three groups based on durability and tangibility of goods. These are non-durable goods, durable goods and services. Non-durable goods are tangible goods, which are normally consumed in one or a few uses like bread or soap. Durable goods are tangible goods that last for use for a relatively longer time and for number of times. For example, bike or a refrigerator. Services are intangible, inseparable from the provider and variable products. Examples include service of doctors and lawyers. Consumer goods are classified into:

Convenience Goods

These are goods which consumers usually purchase frequently with minimum efforts. Eg. soaps, soft drinks, pen etc. Convenience goods are further classified into staples (milk, bread), impulse items (candy, gum, soda, magazine), and emergency goods (umbrella during rainy day, ambulance service).

Shopping Goods

Customers tend to shop around, *i.e.*, make price and/or quality comparisons. Examples of shopping goods: computer, suit, printer, sofa, etc.,

Specialty Goods

Customers are willing to make an extended search to find these goods. In some cases, a customer might be willing to travel 30 or 50 km to find them. Some examples: wedding dress, car, camera etc.

Product Life Cycle

Product life cycle refers to the distinct stages through which product sales pass through. Understanding these stages is important to device different marketing strategies for each of these stages. Most product life cycle is divided into four stages: introduction, growth, maturity and decline.

Introduction

At this stage, the sales grow at slow pace since the product is new in the market and profits may not be there until the end of this stage.

Growth

Sales grow rapidly and profits tend to peak during this stage.

Maturity

Market shares become fairly stable, the product's sales growth slows down because product has achieved acceptance by most potential buyers. The maturity stage is usually the longest stage in terms of time.

Decline

Sales show a downward trend and profits erode.

New Products Development

New product development has increased in recent years with globalization of economies. Companies which fail to develop new products put themselves at risk as their existing products are vulnerable to changing consumer preferences, technological advancements and domestic and foreign competition. Stages of New Product Development include:

Idea Generation

New product development starts with idea generation. Ideas for new products can come from the company's own research and development department, customers, competition, employees, sales people, and top management. Conducting focus groups with customers will help in identifying their problems with products and get ideas for new products. Top management conducts brainstorming sessions to come up with ideas for new products.

Screening

In this stage, ideas which are not technologically feasible or economically viable are eliminated. Some ideas are eliminated because they do not fit the company's mission or objectives.

Concept Development and Testing

A concept test is used to test the idea of the product. Consumers are shown a drawing of the proposed product with a description which includes the price and advantages/disadvantages of the proposed product. Consumers are asked whether they would buy the product. Some firms use virtual reality for testing during the concept testing stage.

Marketing Strategy Development

Market strategy development includes three stages. First stage describes size of the market, structure, market share and product positioning. Second stage analyzes the price, distribution strategy, and marketing budget for the new product. The third stage outlines sales goals, profit targets and marketing mix strategy.

Business Analysis

Techniques taught in finance such as the Net Present Value (NPV) method and breakeven analyses are used to determine whether the idea has the potential of making profit.

Product Development

So far the product existed as a description or drawing or as a prototype. Product development is the process of translating the product idea into a technically and commercially feasible product.

Market Testing

Market testing is done to get information on buyers, dealers, marketing effectiveness, and market potential. There are different methods of market testing for consumer goods such as simulated test marketing, sales-wave research, controlled test marketing, and test markets.

Commercialization

Commercialization is the introduction of the product in the market and involves the largest cost. In commercialization the market entry timing is the most critical.

Product Line

Kotler and Keller identify a product hierarchy which stretches from basic needs to a particular item that satisfies those needs. Product family, product class, product line and product type are part of that hierarchy.

Product line refers to a group of products which perform a similar function, are sold to same customer groups, marketed through same outlets are channels, or fall within given price ranges. A product line may consist of different brands or single brand.

Product Mix

A product mix is the set of all products, a particular seller offers for sale. A product mix consists of various product lines. For example the product mix of company A has two product lines, communication products and computer products.

Various dimensions of product mix is further analyzed through product-mix width, product-line length, the depth of product mix and consistency of product mix.

Product-mix width

Product-line length	Tooth paste	Soaps	Detergents	Shaving creams
	Colgate	Hamam	Surf	Gillette
	Pepsodent	Pears	Ariel	
		Lux	Sunlight	
		Dove		

Width of the product –mix refers to the number of product lines handled by a company. Above example shows a product mix width of four lines. The length of product mix refers to total number of items in the mix. In the example, it is 10. The average product line length is 2.5 [*i.e.* 10/4]. The depth of a product mix refers to how many variants are offered in each product in the line.*e.g.*Colgate tooth paste variants for adults and children. Consistency of product mix refers to how closely the various product lines are related in end use and in production requirements. These four dimensions of product mix permit a company to expand its business in four ways.

Branding

A brand consists of a brand name and a brand mark. The brand name can be vocalized and the brand mark can be recognized. Trademark means the brand name or mark which has legal protection and has been registered. Some names which started out as trademarks and are now generic include: aspirin, corn flakes, nylon, kerosene, gold card and escalator. The public may use some registered brand names in a generic sense but they are still protected trademarks. For example xerox is a brand name and generic name is photocopy.

Packaging and Labeling

Packaging is defined as all activities involved in designing and producing a container for the product. Package is the buyer's first encounter with the product and is capable of turning the buyer on or off. Packaging as a marketing tool has gained wide popularity due to consumer affluence, as a means to build brand equity, increase in selling on self service basis and innovation opportunity in packaging. Packaging also helps the firm and consumer in number of ways.

1. To identify the brand
2. Display the information about the product
3. Facilitate in transportation and protection of the product
4. Helps in storing the product at home and in consumption.

Label may be a simple tag attached to the product or a detailed design which is part of the package. Labels perform several functions. The label identifies the product or brand. Label might give information about the product like who made it, when it was made, what it contains and how it is to be used. Label will also promote the product through attractive graphics.

Unit III

Pricing Policy

Selecting the right price for a product or service is both an art and a science. The McKinsey and Co showed that the effect of pricing on profitability was even stronger than that of increasing sales volume or reducing costs. According to Kotler, price is the one element of the marketing mix that produces revenue, the other elements produce costs. Prices are perhaps the easiest element of a marketing program to adjust.

Pricing Objectives

Pricing methods are based on an organization's overall mission and purpose. Economists usually claim that the objective of firms is to maximize profits. Not every firm has maximization of profits as their objective. Some companies try to maximize their current profit sacrificing long-run performance. Some companies pricing objective will be on maximizing market share. Such companies believe higher sales volume will result in lower unit costs and higher long run profit. Yet another pricing objective is called market-skimming pricing in which products are priced very high when it is launched and prices drop over time. Some firms might want to keep their price low because they feel that the product or service is important for society (*e.g.*, education, medicines). Non-profit organizations do not have the maximization of profits as their objective. Some organizations have a particular rate of return as their objective, say for example achieve 15 per cent return before taxes. Some common pricing objectives include maximization of profits, sales volume objectives (market share), social and ethical considerations (some may feel that their product/service is important for society like in education, medicines), etc.,

Factors Affecting Prices

Demand

Demand is the relationship between the price of a good and the quantity purchased on the assumption that prices of other commodities and the consumer's income is held constant. The demand function is obtained by maximizing the consumer's utility subject to the constraint that the customer's budget.

Elasticity

Elasticity is the responsiveness of one variable to the changes in another variable. Elasticity of demand is the responsiveness in demand to the changes in the determinants of demand like own price of the commodity, prices of its substitutes, income etc.,

Own-price elasticity of demand or simply price elasticity of demand is the measure of the responsiveness of demand for a good to a change in price of that good. Represented by the ratio between percentage change in quantity demanded and percentage change in price:

$$\text{Price elasticity of demand} = \frac{\text{Per cent change in quantity demanded of good X}}{\text{Per cent change in price of good Y}}$$

If the percent change in the quantity demanded is greater than the percent change in the price of a good, demand is said to be price elastic or more responsive to price changes. (Example: A 1 per cent change in price induces a change in quantity demanded by more than 1 per cent). If the per cent change in the quantity demanded is less than the per cent change in the price of a good, demand is said to be price inelastic, or less responsive to price changes. (Example: A 1 per cent change in price induces a change in quantity demanded by less than 1 per cent). If the response is exactly equal to 1 per cent, the demand is said to be unitary, where a 1 per cent decrease in price results in a 1 per cent increase in demand. The higher the price elasticity, the more sensitive consumer demand is to price changes.

Cross-Price Elasticity of demand is a measure of responsiveness of *demand for one good* to a change in the *price of another good*. Represented by the ratio of the per cent change in the quantity demanded of good X to a per cent change in the price of some other good Y:

$$\text{Cross - price elasticity of demand} = \frac{\text{Per cent change in quantity demanded of good X}}{\text{Per cent change in the price of some other good Y}}$$

If the per cent change in the quantity demanded of good X is greater than the per cent change in the price of good Y, the demand for good X is said to be cross-price elastic with respect to good Y, or responsive to changes in the price of good Y. (Example: A 1 per cent change in cross price induces a change in quantity demanded by more than 1 per cent.) If the per cent change in the quantity demanded of good X is less than the per cent change in the price of good Y, the demand for good X is said to be cross-price inelastic with respect to good Y, or not responsive to changes in the price of good Y. (Example: A 1 per cent change in cross price induces a change in the quantity demanded by less than

1 per cent.) Cross-price elasticities can be complements or substitutes. If the cross-price elasticity of demand is positive, the goods X and Y are substitutes. If the cross-price elasticity of demand is negative, the goods X and Y are complements.

Income Elasticity of Demand is a measure of the responsiveness of demand to changes in income. It shows how the quantity purchased changes (how sensitive it is) in response to a change in the consumer's income. Represented by the ratio between percentage change in quantity demanded and percentage change in income:

$$\text{Income elasticity of demand} = \frac{\text{Per cent change in quantity demanded of good X}}{\text{Per cent change in consumer income}}$$

If the percent change in the quantity demanded is greater than the per cent change in consumer income, the demand is said to be income elastic, or responsive to changes in consumer income. (Example: One per cent change in income induces a change in quantity demanded by more than 1 per cent.) If the percent change in the quantity demanded is less than the per cent change in consumer income, the demand is said to be income inelastic, or not responsive to changes in consumer income. (Example: One per cent change in income induces a change in quantity demanded by less than 1 per cent) The higher the income elasticity, the more sensitive consumer demand is to income changes.

Assumption

If the income elasticity of demand are positive, the good is a normal good, and if the income elasticity of demand is negative, the good must be an inferior good. Negative income elasticity is common with staple foods in developing countries, generally considered inferior goods. As income increases, consumers substitute traditional staple foods (such as rice, wheat, minor millets) for higher value foods such as meats, fruits, and processed products.

Costs and Levels of Production

Prices of commodities are also affected by cost of production. Cost of production is mainly influenced by labor costs and input costs. Apart from these direct costs changes in external environment like taxes and subsidies also affect the cost of production. The levels of production affect the cost through economies of scale in production.

Pricing Methods and Strategies

Cost-based Pricing

Most retailers use *markup pricing*. Suppose a store uses a 50 per cent markup on cost on commodity A and the commodity costs the retailer Rs. 80, the selling price of commodity will be Rs. 120 [80 + 40 = 120].

Rate of Return Pricing or Target Return Pricing

In this method the firm fixes a price that would yield its desired rate of return or the target return. Price = Desired Total Revenue/Quantity. Suppose a firm manufactures 800 garments at a cost of Rs. 10000. The firm desires a 20 per cent return on investment (ROI). The price the firm should charge is: 12000/800 = Rs. 15/garment.

Buyer-based Pricing

This method is mainly adopted for pricing the services of lawyers, doctors, consultants etc. For example lawyers may charge a portion of benefit his client is getting through the legal suit. Doctors sometimes charge as per the ability of the client to pay in pricing their service.

Competitor-based Pricing or Going–Rate Pricing

In this method the firm's pricing method is based on competitor's price and charge a price which is same as that of the competitor or more or less than the competitor. In oligopolistic industries all firms charge the same price. Smaller firms follow the leader irrespective of the changes in their demand or costs.

Demand-based Pricing

In this method, the firms consider their demand at different periods when pricing products. When demand is heavy price will be more like tickets in airline companies (**Peak-User Pricing**). Eg: pricing during peak periods and off-peak periods.

Segmented Pricing

Segmented pricing is done based on the elasticity of demand of the customer. The demand for business travel is inelastic and can be charged high, whereas a vacation traveler will have elastic demand.

Smart Pricing

Smart pricing is dynamic pricing and means that a firm will not use a fixed price but will adjust prices, even on a daily basis, as market conditions, consumer demand, and product valuations change. Smart prices take into account differences in the costs of serving different segments and also the different valuations of one's product by different segments.

Perceived Value Pricing

In this method companies base their pricing on the customer's perceived value of their product rather than on demand for or cost of their product. In this method company try to convince the customer that their product gives them more total value in terms of product performance, warranty, customer support, reputation, trustworthiness etc.,

Auction –Type Pricing

This method of pricing is popular with the advent of marketing through the internet. Seller put a price in the internet and bidders raise the offer price until the maximum price is reached. Highest bidder gets the commodity.

Shrouding

The term shrouding was coined by Prof. David Laibson of Harvard University. Firms attract the consumer to the product or service by promoting low prices by hiding additional fees in the fine print. Some hotels offer low prices but charge additional fees for parking the car, using the pool, using the Internet, room service, etc.

Unit IV

A marketing channel is an organized network of agencies and institutions which, in combination, perform all the activities required to link producers with users to accomplish the marketing task.

The Distribution Channel

Frequently there may be a chain of intermediaries; each passing the product down the chain to the next organization, before it finally reaches the consumer or end-user. This process is known as the 'distribution chain' or the 'channel.' Each of the elements in these chains will have their own specific needs, which the producer must take into account, along with those of the all-important end-user.

The *Distribution Channel Management* program builds a coherent framework uniting marketing and sales efforts in a collaborative learning environment. A well-designed distribution channel strategy takes into account the linkages between both the salespeople's activities with channel partners and the marketing managers' efforts to better reach and serve end-users.

Types of Distribution Channels

A number of alternative 'channels' of distribution are available through the channel members like agents, distributors, dealers, telemarketing and internet. Depending on the number of intermediaries channels are classified as zero-level channel (direct selling), one-level channel (a channel with one intermediary) and two level channel and so on. A channel may have one or more of the following channel members. Direct selling

- ☆ Agent, who typically sells direct on behalf of the producer
- ☆ Distributor (also called wholesaler), who sells to retailers
- ☆ Retailer who sells to end customers

Functions of Channel Members

Each member in the channel should get an opportunity to be profitable. The main elements in the functions of channel members include establishing an equitable price list and schedule of discounts, payment terms and guarantees to the channel members, distribution of territorial rights, and clear specification of mutual services and responsibilities.

Channel Management Decisions

The channel management decision involves a form of trade-off: the cost of using intermediaries to achieve wider distribution. The firm's job extends to managing all the processes involved in the chain, until the product or service arrives with the end-user. This may involve a number of decisions:

- ☆ Channel membership
- ☆ Channel motivation
- ☆ Monitoring and managing channels

In addition to marketing considerations, channel design has important financial and risk impacts that need to be considered as part of management decision.

Unit V

Promotion Mix

Promotion is all about companies communicating with customers. To generate sales and profits, the benefits of products have to be communicated to customers. In marketing, this is commonly known as "promotion". In a business the total marketing communications programme is called the "promotional mix". There are four main aspects of a **promotional mix.**

Advertising

Any paid form of non-personal communication of ideas or products in the media like television, newspapers, magazines, billboard posters, radio, cinema etc is called advertising. It is used to develop attitudes, create awareness, and transmit information in order to gain a response from the target market. The two basic aspects of advertising are the message (what you want your communication to say) and the medium (how you get your message across)

Personal Selling

Oral communication with potential buyers of a product with the intention of making a sale. Can be face-to-face or via telephone. Personal Selling is an effective way to manage personal customer relationships. The sales person acts on behalf of the organization. They tend to be well trained in the approaches and techniques of personal selling. However sales people are very expensive and should only be used where there is a genuine return on investment. For example salesmen are often used to sell cars or home improvements where the margin is high.

Sales Promotion

Providing incentives to customers or to the distribution channel to stimulate demand for a product. Examples include competitions, free accessories (such as free blades with a new razor), introductory offers and so on. Each sales promotion should be carefully costed and compared with the next best alternative.

Publicity

The communication of a product, brand or business by placing information about it in the media without paying for the time or media space directly, otherwise known as "public relations" or PR.

The other elements of the promotions mix are Trade Fairs, Exhibitions and Sponsorship. Different aspects of the promotions mix are integrated to deliver a unique campaign

Direct Marketing

Direct marketing is the sale of goods and services to the consumer without intermediaries. Direct marketers use number of channels to reach the consumer like telemarketing, direct mail, websites and mobile services. Direct marketers create long term relationship with customers by sending birthday cards giving premiums etc.

Customer Relationship Management

The objective of customer relationship management (CRM) is to maximize customer loyalty by carefully managing detailed information about the individual customer. This information will enable firms to customize their services to individual customers. The basic requirement of CRM is identification of the potential customers, grouping of customers based on their needs and their value to the firm, interaction with individual customers, and customization of products.

References

American Marketing Association, 2008. http;//www.marketingpower.com.

Kotler, Philip and Keller, Keven Lane, 2008. 13th edn. Prentice Hall of India, New Delhi.

Index